T0137092

Advances in Intelligent Systems and Computing

Volume 942

The series "Advances in Intelligent Systems and Computing" contains publications on theory, applications, and design methods of Intelligent Systems and Intelligent Computing. Virtually all disciplines such as engineering, natural sciences, computer and information science, ICT, economics, business, e-commerce, environment, healthcare, life science are covered. The list of topics spans all the areas of modern intelligent systems and computing such as: computational intelligence, soft computing including neural networks, fuzzy systems, evolutionary computing and the fusion of these paradigms, social intelligence, ambient intelligence, computational neuroscience, artificial life, virtual worlds and society, cognitive science and systems, Perception and Vision, DNA and immune based systems, self-organizing and adaptive systems, e-Learning and teaching, human-centered and human-centric computing, recommender systems, intelligent control, robotics and mechatronics including human-machine teaming, knowledge-based paradigms, learning paradigms, machine ethics, intelligent data analysis, knowledge management, intelligent agents, intelligent decision making and support, intelligent network security, trust management, interactive entertainment, Web intelligence and multimedia.

The publications within "Advances in Intelligent Systems and Computing" are primarily proceedings of important conferences, symposia and congresses. They cover significant recent developments in the field, both of a foundational and applicable character. An important characteristic feature of the series is the short publication time and world-wide distribution. This permits a rapid and broad dissemination of research results.

** Indexing: The books of this series are submitted to ISI Proceedings, EI-Compendex, DBLP, SCOPUS, Google Scholar and Springerlink **

More information about this series at http://www.springer.com/series/11156

Ana Maria Madureira · Ajith Abraham ·
Niketa Gandhi · Catarina Silva ·
Mário Antunes
Editors

Proceedings of the Tenth International Conference on Soft Computing and Pattern Recognition (SoCPaR 2018)

 Springer

Editors
Ana Maria Madureira
School of Engineering
Instituto Superior de Engenharia (ISEP/IPP)
Porto, Portugal

Ajith Abraham
Machine Intelligence Research Labs (MIR)
Auburn, WA, USA

Niketa Gandhi
Machine Intelligence Research Labs (MIR Labs)
Auburn, WA, USA

Catarina Silva
Politécnico de Leiria
Leiria, Portugal

Mário Antunes
Politécnico de Leiria
Leiria, Portugal

ISSN 2194-5357   ISSN 2194-5365   (electronic)
Advances in Intelligent Systems and Computing
ISBN 978-3-030-17064-6   ISBN 978-3-030-17065-3   (eBook)
https://doi.org/10.1007/978-3-030-17065-3

Library of Congress Control Number: 2019936516

This Springer imprint is published by the registered company Springer Nature Switzerland AG
The registered company address is: Gewerbestrasse 11, 6330 Cham, Switzerland

Preface

Welcome to Porto, Portugal, and to the tenth International Conference on Soft Computing and Pattern Recognition (SoCPaR 2018) and the fourteenth International Conference on Information Assurance and Security (IAS 2018) held at Instituto Superior de Engenharia do Porto (ISEP) during December 13–15, 2018.

SoCPaR 2018 is organized to bring together worldwide leading researchers and practitioners interested in advancing the state of the art in soft computing and pattern recognition, for exchanging knowledge that encompasses a broad range of disciplines among various distinct communities. The themes for this conference are thus focused on "Innovating and Inspiring Soft Computing and Intelligent Pattern Recognition." The conference is expected to provide an opportunity for the researchers to meet and discuss the latest solutions, scientific results, and methods in solving intriguing problems in the fields of soft computing and pattern recognition. SoCPaR 2018 received submissions from 16 countries, and each paper was reviewed by at least five reviewers in a standard peer-review process. Based on the recommendation by five independent referees, finally 22 papers were accepted for the conference (acceptance rate of 47%).

Information assurance and security have become an important research issue in the networked and distributed information sharing environments. Finding effective ways to protect information systems, networks and sensitive data within the critical information infrastructure is challenging even with the most advanced technology and trained professionals. IAS 2018 aims to bring together researchers, practitioners, developers, and policy makers involved in multiple disciplines of information security and assurance to exchange ideas and to learn the latest development in this important field. The conference provided an opportunity for the researchers to meet and discuss the latest solutions, scientific results, and methods in solving intriguing problems in the fields of IAS. IAS 2018 received submissions from 12 countries, and each paper was reviewed by at least five reviewers in a standard peer-review process. Based on the recommendation by five independent referees, finally 16 papers were accepted for the conference (acceptance rate of 45%).

Conference proceedings are published by Springer Verlag, Advances in Intelligent Systems and Computing Series. Many people have collaborated and worked hard to produce this year successful SoCPaR-IAS 2018 conference. First and foremost, we would like to thank all the authors for submitting their papers to the conference, for their presentations and discussions during the conference. Our thanks to program committee members and reviewers, who carried out the most difficult work by carefully evaluating the submitted papers. We are grateful to our three plenary speakers:

* *Petia Georgieva, University of Aveiro, Portugal*
* *J. A. Tenreiro Machado, Polytechnic of Porto, Portugal*
* *Henrique M. Dinis Santos, University of Minho, Portugal*

Our special thanks to the Springer Publication team for the wonderful support for the publication of these proceedings. Enjoy reading!

Ana Maria Madureira
Ajith Abraham
General Chairs

Catarina Silva
Mário Antunes
Oscar Castillo
Simone Ludwig
Program Chairs

Organization

General Chairs

Ana Maria Madureira   Instituto Superior de Engenharia do Porto, Portugal

Ajith Abraham   Machine Intelligence Research Labs (MIR Labs), USA

Program Chairs

Catarina Silva   Politécnico de Leiria, Portugal

Mário Antunes   Politécnico de Leiria, Portugal

Oscar Castillo   Tijuana Institute Technology, Mexico

Simone Ludwig   North Dakota State University, USA

Advisory Board Members

Albert Zomaya   University of Sydney, Australia

Andre Ponce de Leon F. de Carvalho   University of Sao Paulo at Sao Carlos, Brazil

Bruno Apolloni   University of Milan, Italy

Imre J. Rudas   Óbuda University, Hungary

Janusz Kacprzyk   Polish Academy of Sciences, Poland

Marina Gavrilova   University of Calgary, Canada

Patrick Siarry   Université Paris-Est Créteil, France

Ronald Yager   Iona College, USA

Salah Al-Sharhan   Gulf University for Science and Technology, Kuwait

Sebastian Ventura   University of Cordoba, Spain
Vincenzo Piuri   Universita' degli Studi di Milano, Italy
Francisco Herrera   University of Granada, Spain
Sankar Kumar Pal   ISI, Kolkota, India

Publication Chairs

Niketa Gandhi   Machine Intelligence Research Labs (MIR Labs),
                              USA
Azah Kamilah Muda   Universiti Teknikal Malaysia Melaka, Malaysia

Web Services

Kun Ma   Jinan University, China

Publicity Committee

Bruno Cunha   Instituto Superior de Engenharia do Porto,
                              Portugal
Diogo Braga   Instituto Superior de Engenharia do Porto,
                              Portugal
Santoso Wibowo   CQUniversity Melbourne, Australia
Nesrine Baklouti   University of Sfax, Tunisia
Isabel Jesus   Institute of Engineering of Porto, Portugal
Marjana Prifti Skenduli   University of New York, Tirana

Local Organizing Committee

Ana Maria Madureira   Instituto Superior de Engenharia do Porto,
                              Portugal
Diogo Braga   Instituto Superior de Engenharia do Porto,
                              Portugal
Duarte Coelho   Instituto Superior de Engenharia do Porto,
                              Portugal
André Santos   Instituto Superior de Engenharia do Porto,
                              Portugal

Judite Ferreira	Instituto Superior de Engenharia do Porto, Portugal
Luis Coelho	Instituto Superior de Engenharia do Porto, Portugal
Isabel Sampaio	Instituto Superior de Engenharia do Porto, Portugal
Ricardo Almeida	Instituto Superior de Engenharia do Porto, Portugal
Marina Sousa	Instituto Superior de Engenharia do Porto, Portugal

Technical Program Committee

Ahmad Samer Wazan	Paul Sabatier University, France
A. Galicia	Pablo de Olavide University, Spain
Akira Asano	Kansai University, Japan
Alessio Merlo	DIBRIS, University of Genoa, Italy
Alberto Cano	University of Córdoba, Spain
Alberto Fernández	University of Granada, Spain
Alicia Troncoso	Universidad Pablo de Olavide, Spain
Amparo Fuster-Sabater	Institute of Physical and Information Technologies (CSIC), Spain
Ana Madureira	Instituto Superior de Engenharia do Porto, Portugal
Angel Arroyo	UBU, Spain
Arash Habibi Lashkari	University of New Brunswick (UNB), Canada
Andries Engelbrecht	University of Pretoria, South Africa
Antonio Bahamonde	Universidad de Oviedo, Gijón, Asturias, Spain
Arun Kumar Sangaiah	Vellore Institute of Technology, India
Aswani Kumar Cherukuri	Vellore Institute of Technology, India
Atta Rahman	University of Dammam, Dammam, Saudi Arabia
Azah Muda	UTeM, Malaysia
Candelaria Hernández-Goya	Universidad de La Laguna, Spain
Carlos Pereira	ISEC, Portugal
Chian C. Ho	National Yunlin University of Science and Technology, Taiwan
Clay Palmeira	Francois Rabelais University of Tours, France
Constantino Malagón	Nebrija University, Spain
Christian Veenhuis	HELLA Aglaia Mobile Vision GmbH, Germany
Corrado Mencar	University of Bari "A. Moro," Italy
Daniela Zaharie	West University of Timisoara, Romania
Donato Impedovo	Dipartimento di Informatica, UNIBA, Italy

Dilip Pratihar   Federal University of Rio Grande do Norte,
                                    Brazil
Eiji Uchino   Yamaguchi University, Japan
Elizabeth Goldbarg   Federal University of Rio Grande do Norte,
                                    Brazil
Fernando Tricas   Universidad de Zaragoza, Spain
Ficco Massimo   Second University of Naples (SUN), Italy
Francisco Valera   Universidad Carlos III de Madrid, Spain
Francisco Chicano   Universidad de Málaga, Spain
Gabriel López   University of Murcia, Spain
Gustavo Isaza   University of Caldas, Colombia
Gregorio Sainz-Palmero   Universidad de Valladolid, Spain
Ilkka Havukkala   IPONZ, New Zealand
Intan Ermahani A. Jalil   Universiti Teknikal Malaysia Melaka (UTeM),
                                    Malaysia
Isaac Chairez   Instituto Politécnico Nacional, Mexico
Isabel S. Jesus   Institute of Engineering of Porto, Portugal
Joan Borrell   Universitat Autònoma de Barcelona, Spain
João Paulo Magalhaes   ESTGF, Porto Polytechnic Institute, Portugal
Jose Luis Imana   Universidad Politécnica de Madrid, Spain
Jose M. Molina   Universidad Carlos III de Madrid, Spain
Jose Vicent   Universidad de Alicante, Spain
Jose-Luis Ferrer-Gomila   University of the Balearic Islands, Spain
Juan Jesús Barbarán   University of Granada, Spain
Juan Pedro Hecht   University of Buenos Aires - FCE/FCEyN/FI,
                                    Argentina
Jung-San Lee   Feng Chia University, Taiwan
Jerry Chun-Wei Lin   Western Norway University of Applied Sciences
                                    (HVL), Bergen, Norway
Jerzy Grzymala-Busse   University of Kansas, USA
Joana Costa   CISUC, IPLeiria, Portugal
José Everardo Bessa Maia   State University of Ceará, Brazil
José F. Torres   Pablo de Olavide University, Spain
José Raúl Romero   University of Cordoba, Spain
Jose Santos   University of A Coruña, Spain
Jose Tenreiro Machado   ISEP, Portugal
Joseph Alexander Brown   Innopolis University, Canada
Juan A. Nepomuceno   University of Seville, Spain
Kazumi Nakamatsu   University of Hyogo, Japan
Kin Keung Lai   International Business School, Shaanxi Normal
                                    University, Xian, China
Leandro Maciel Almeida   Federal University of Pernambuco, Brazil
Lin Wang   Jinan University, China
Leocadio G. Casado   University of Almeria, Spain
Manuel Grana   University of the Basque Country, Spain

Miguel Frade   Politécnico de Leiria, Portugal
Mohd Faizal Abdollah   University Technical Malaysia Melaka, Malaysia
M. C. Nicoletti   FACCAMP and UFSCar, Brazil
M. J. Ramírez   Universitat Politècnica de València, Spain
Manuel Grana   University of the Basque Country, Spain
Maria Ganzha   Warsaw University of Technology, Poland
Mario Giovanni C. A. Cimino   University of Pisa, Italy
Martin Lukac   Nazarbayev University, Kazakhstan
Michal Wozniak   Wroclaw University of Technology, Poland
Mohammad Shojafar   Sapienza University of Rome, Italy
Niketa Gandhi   Machine Intelligence Research Labs (MIR Labs),
                                  USA

Oscar Gabriel Reyes Pupo   UCO, Spain
Patrick Siarry   Université Paris-Est Créteil, France
Pedro Gago   Politécnico de Leiria, Portugal
Pedro Gonzalez   Universidad de Jaén, Spain
Prabukumar Manoharan   Vellore Institute of Technology, India
Ramon Rizo   Universidad de Alicante, Spain
Rolf Oppliger   eSECURITY Technologies, Switzerland
Romain Laborde   IRIT/SIERA, France
Rosaura Palma-Orozco   CINVESTAV - IPN, Mexico
Radu-Emil Precup   Politehnica University of Timisoara, Romania
Ricardo Tanscheit   PUC-Rio, Brazil
Salvador Alcaraz   Miguel Hernandez University, Spain
Salvatore Venticinque   Seconda Università di Napoli, Italy
Sorin Stratulat   Université de Lorraine, Metz, France
Sye Loong Keoh   University of Glasgow, Singapore
Sabri Pllana   Linnaeus University, Sweden
Simone Ludwig   North Dakota State University, USA
Stefano Cagnoni   University of Parma, Italy
Sylvain Piechowiak   LAMIH, University of Valenciennes, France
Thomas Hanne   University of Applied Sciences Northwestern
                                  Switzerland, Switzerland
Umberto Villano   University of Sannio, Italy
Varun Ojha   Swiss Federal Institute of Technology,
                                  Switzerland
Victor Manuel Rivas Santos   University of Jaen, Spain
Wenjian Luo   University of Science and Technology of China,
                                  China

Contents

Shaping the Music Perception of an Automatic Music Composition: An Empirical Approach for Modelling Music Expressiveness

Michele Della Ventura[(✉)]

Department of Technology, Music Academy "Studio Musica",
Via Andrea Gritti, 25, 31100 Treviso, Italy
michele.dellaventura@tin.it

Abstract. Expressiveness is an important aspect of a music composition. It becomes fundamental in an automatic music composition process, a domain where the Artificial Intelligent Systems have shown great potential and interest. The research presented in this paper describes an empirical approach to give expressiveness to a tonal melody generated by computers, considering both the symbolic music text and the relationships among the sounds of the musical text. The method adapts the musical expressive character to the musical text on the base of the "harmonic function" carried by every single musical chord. The article is intended to demonstrate the effectiveness of the method by applying it to some (tonal) musical pieces written in the 18th and 19th century. Future improvements of the method are discussed briefly at the end of the paper.

Keywords: Automatic music composition · Functional harmony ·
Music expressiveness

1 Introduction

Everyday many people listening to (live or recorded) music. But while listening to music, sounds have the ability to foster communication that goes beyond the use of language, promoting the expression of emotions [1, 2]. The ways of musical imagination are endless and it is in the process of interaction between imagination and sound event that the expression takes shape losing its apparent "arbitrariness" and giving meaning to the things heard. The expressivity is an important part of the music and without it the music would not attract people [3].

Artificial Intelligence (AI) is now present in many areas, but among these, the field of music is the area that has always attracted researchers and composers. As early as 1950, composer Iannis Xenakis designed a system (GENDY: GENeral DYnamic) to compose a piece of music using a computer [4].

In the past 20 years, some interesting papers have been published on the different aspects of automatic music composition research. Consequently, the AI research area has developed on other aspects of music such as "musical expressivity". Approaches to this problem were based on:

- probabilistic method [5],
- statistical analysis [6, 7],

© Springer Nature Switzerland AG 2020
A. M. Madureira et al. (Eds.): SoCPaR 2018, AISC 942, pp. 1–10, 2020.
https://doi.org/10.1007/978-3-030-17065-3_1

- mathematical models [8],
- analysis by synthesis [9, 10, 11].

They are usually empirical methods, which allow obtaining results expressed by numbers, therefore easy to analyze. All these approaches have an algorithm created by a person who conceived a mathematical model able to seize the musical expressiveness elements of a performance.

Another interesting approach is the one based on the theory according to which people acquire the affective aspects implicit in music, through a process of observation and imitation [12], that means based on inductive learning of the rules [9, 10, 11, 13]: instead of manually creating a model for the recognition of the elements related to musical expressiveness, the computer must automatically discover these elements through certain learning rules.

This research focuses on the problem of how to manage the absence of indications in the domain of music expressiveness. It has been created an algorithm able to investigate the musical expressiveness of a musical piece, by reading the score on its symbolic level. Unlike the aforementioned studies that are based on the analysis of an execution, the algorithm tries to define the musical expressiveness on the basis of a musical grammar which is reflected in the functional harmony. The algorithm created for such purpose has the task of reading a certain melody on its symbolic level (this is why scores in MIDI format were used, without any indication on dynamics); to identify the harmonic structures (through a melody segmentation process) and the corresponding harmonic functions (see paragraph 3); finally, to render in graphic format a diagram related to the musical dynamics to apply to the melody.

The effectiveness of the method was tested by analysing tonal piano pieces, the results of which were compared with the scores reviewed by important musicians. These results allow applying this method to the algorithms for the automatic generation of a tonal melody so as to render it pleasant and interesting.

This paper is structured as follows. Section 2 describes the theory of Functional Harmony. Section 3 describes the approach to consider the music score to model the musical expressiveness. We discuss the methods and initial results in Sect. 4. Finally, Sect. 5 presents some conclusions and indicates some future improvements of the method.

2 Functional Harmony

Functional Harmony Theory tends to go beyond the sound event as it occurs, to interpret what lies behind to what appears in a certain instant, to seize the meaning or "role" that it covers in relation to other events which precede it or follow it [14, 15, 16], i.e. the "function" that it performs within the context where it is inserted. With respect to the chord, functional theory tries to identify its harmonic function and the relationship that it establishes with the preceding and with the following one [14]. When it comes to an individual, isolated chord, we may define its structure (that is its disposition and the type of intervals[1] starting from the bottom) (see Fig. 1) and identify its inversion state (see Fig. 2) [15]; but we may not fully understand its meaning, which essentially derives from the musical *context* where it is found.

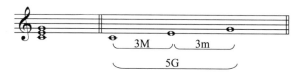

Fig. 1. Chord structure.

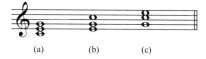

Fig. 2. State of a chord: (a) fundamental state, (b) first inversion, (c) second inversion.

Such context is *temporally oriented* [14, 16], in fact, except for the first and the last sound, all the sounds are between other sounds which represent, respectively, their *past* and their *future*. Reference is made to sounds, before mentioning chords, in order to underline that the latter are none other than *temporary aggregates of notes on a linear motion*, or rather a contrapuntal motion, and that to sign/write down/note a chord (see Fig. 3) actually corresponds to taking *a snapshot in a directed flow* [14, 15].

Fig. 3. Contrapuntal motion and functional motion.

---

[1] The interval between the various sounds is the distance separating a sound from another. The classification of an interval consists in the denomination (generic indication) and in the qualification (specific indication). The denomination corresponds to the number of degrees that the interval includes, calculated from the low one to the high one; it may be of a 2nd, a 3rd, 4th, 5th, and so on.; the qualification is deduced from the number of tones and semi-tones that the interval contains; it may be: perfect (G), major (M), minor (m), augmented (A), diminished (d), more than augmented (A+), more than diminished (d+), exceeding (E), deficient (df).

According to functional theory, there are three **harmonic functions**, the tonic (T), the subdominant (S) and the dominant (D). Only two of these act dynamically "pushing forward" the musical discourse: the **dominant** and the **subdominant** [14]. These two functions express respectively the *conduction of tension* towards the tonic T and the *preparation* of such conduction. The tonic, being a "static" function, is a mere rest point, the end of the movement (Fig. 4).

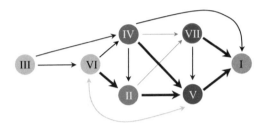

Fig. 4. Resolution tendency of the chords based on the harmonic function.

The three harmonic functions of I, IV and V degree are called main because they are linked by a relation based on the interval of the perfect 5th that separates the keynotes of the three corresponding chords; the chords relating to the rest of degrees on the scale are considered "representatives" of the I, IV and V degree (with which there is an affinity of the third - two sounds in common - because the 3rd is actually the interval that regulates the distance between the respective keynotes) and secondary harmonic functions rest with it (see Fig. 4) [14].

The identification of these harmonic functions allows the executant to have an image of the character of the piece so as to better define the field of dynamics.

This latter concept is represented by a sort of musical crescendo, understood not only in terms of intensity (i.e. an increase of the sound volume), but also as a change of timbre (i.e. a change of sound register, from low-pitch to high-pitch and the other way around).

3   Music Score Analysis

The identification of the harmonic functions, indispensable to describe the dynamics of a score, is performed by an algorithm realized for the occasion.

Initially the algorithm has to read a score from a MIDI file (Musical Instruments Digital Interface). The MIDI protocol is a symbolic language of event-based messages used to represent music. It permits to divide into levels the various voices of a melody, considering the pitch and the duration of each sound of the melody in numerical form. A monodic score may therefore be represented as a sequence S_m of N notes n_i indexed on the basis of the appearance order i:

$$S_m = (n_i)_{i \in [0, N-1]}$$

A polyphonic score may be considered as the overlapping of two or more monodic sequences $S_{m1}, S_{m2, \dots}$ which may be represented by a matrix $P_{x,y}$

$$P(k) = \begin{pmatrix} p_{1,1}(k) & p_{1,2}(k) & \cdots & p_{1,y}(k) \\ p_{2,1}(k) & p_{2,2}(k) & \cdots & p_{2,y}(k) \\ p_{3,1}(k) & p_{3,2}(k) & \cdots & p_{3,y}(k) \\ p_{4,1}(k) & p_{4,2}(k) & \cdots & p_{4,y}(k) \\ \cdots & \cdots & \cdots & \cdots \\ p_{x,1}(k) & p_{x,2}(k) & \cdots & p_{x,y}(k) \end{pmatrix}$$

where x represents the number of voices (or levels) and y the number of rhythmic movements existing in the musical piece [17]. Figure 5 shows a polyphonic musical segment (4 voices or levels) and the corresponding matrix.

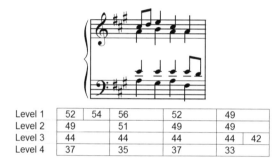

Level 1	52	54	56		52	49	
Level 2	49		51		49	49	
Level 3	44		44		44	44	42
Level 4	37		35		37	33	

Fig. 5. Score representation matrix.

Every sound is identified by a number: the MIDI protocol takes into consideration the piano keyboard and it assigns to the lowest note the value 1 and to every subsequent sound an increasing value (Fig. 6).

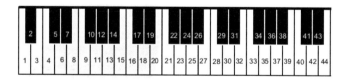

Fig. 6. MIDI piano keyboard representation.

Every sound has its own duration which is defined considering the shortest duration existing in the piece and calculating the other durations proportionally [18]. In the example shown in Fig. 4 the shortest duration is represented by the eighth (·) which assumes the value 1 and therefore the quarter (·) assumes the value 2.

After reading the musical score, the algorithm proceeds with the recognition of the tonal system and therefore of the chords. At this point, for every single rhythmic movement, the sounds that make up a chord are identified (keeping in mind that the sounds must be distant from each other by a third interval) and the sounds which are extraneous to the harmony must be eliminated (Fig. 7) [19].

Fig. 7. Simplification of the score by eliminating the notes which are extraneous to a harmony.

Finally, the algorithm identifies for each chord the harmonic function (T, S, D) associated to it, so as to have for each movement a functional indication to be used for the definition of the dynamics.

4   Obtained Results

Based on the above considerations, an algorithm was designed and developed. It is a prototype of a desktop computer application aimed at monitoring and supporting the analysis of a musical text. The algorithm reads the initial musical notes to identify the tonal system; after that, it identifies the musical chords and defines the harmonic function for each one of them. Input parameters are not necessary for the elaboration.

The method may be applied only to tonal musical pieces. The initial tests were carried out on a set of musical segments of various lengths, specifically selected, in order to verify the "validity" of the analysis. Then, each analysis was compared with scores already reviewed by important musicians, in order to verify the validity of the method.

The results of the analysis are indicated in an isometric graphic which compares the harmonic function of every single chord with the harmonic function of rest, expressed by T, which represents the chord of the reference tonal system and with respect to which the comparisons and classifications may be made: otherwise, every single chord may not provide any information. In an isometric graphic, if two musical chords have the same functional harmony, the diagram will contain a column having only the color of the corresponding information bracket. If the two functional harmonies are different, the color of the column changes, passing from the color of the preceding functional harmony to the color of the current functional harmony. In this regard, Table 1 shows the color associated with the three main functional harmonies (T, S and D), based on what is shown in Fig. 4. Every harmonic function gathers the various degrees together: (a) I, III and VI, (b) II and IV, (c) V and VII. At the same time, each degree was identified by a numeric value, which depends on the number of degrees contained in it: function T contains the degrees I, III and VI (therefore the values 1, 2 and 3), function S contains the degrees IV and II (values 4 and 5) and, finally, function D contains the degrees V and VII (values 6 and 7).

Table 1. Information value of every single chord.

T			S		D	
1	2	3	4	5	6	7
I	III	VI	II	IV	V	VII

In Fig. 8 it is possible to see an example of analysis made up by the algorithm, with the corresponding graphic analysis. In the upper part of the isometric graphic colours are used to represent the harmonic functions of the single chords, while the lower part of the diagram shows only the colour of the functional chord of the Tonic chord (to which the comparison is made).

Following the gradualness of the colour fad-out it is possible to create:

- a "crescendo", in case the transition is from green (T) to orange (S) or to red (D); this effect is represented by the graphic symbol < the length of which varies based on the fade-out duration;
- a "diminuendo", in case the transition goes from red (D) or orange (S) to green (T); this effect is represented by the graphic symbol > the length of which varies based on the fade-out duration;
- an "accento" or "appoggiato", in case a transition occurs in succession from T–D–T (green-red-green): the "accento" or "appoggiato" would be on the harmonic function D (see Fig. 7, bar 3, second movement); this effect is represented by the graphic symbol "-".

On the basis of this information the algorithm then proposes a dynamic to apply to the piece (see Fig. 8).

Fig. 8. Brahms "Valzer" n.3 op.39: functional analysis.

5 Conclusions and Discussion

This article presented an empirical method for shaping the music perception of an automatic tonal music composition generated by computer. The model defined the expressive musical terms analysing the harmonic structure of the musical piece. The proposed features in terms of dynamics, derived from the functional harmony, gave a good performance compared with a human performance. Furthermore, the proposed method observed both the musical grammar related to the succession of harmonies and the musical phrase syntax. The computer played a growing role in the entire process, both in extracting the harmonic function from the symbolic music text, and in analysing and modelling the expressiveness of the musical piece.

Intelligent computational methods find many applications in music education, supporting teachers and students. These methods are able to support a learning process helping students (with disability) advance their understanding of a complex phenomenon such as the expressive interpretation of a musical piece.

In other words, the method presented in this paper represents a tool to stimulate the recovery of not-fully-acquired abilities or as simple tool of consultation and support to the explanation of the teacher.

A future study could concern the possibility of uniting the Harmonic Functional Theory and the Information Theory to quantify the information content of each detected harmonic function. On the basis of these values it would be possible to modify the musical expressivity analysing the events that precede and follow a certain musical chord (or harmonic function).

References

1. Scruton, R.: The Aestheics of Music, pp. 80–170. Clarendon Press, Oxford (1997)
2. Bigand, E., Vieillard, S., Madurell, F., Marozeau, J., Dacquet, A.: Multidimensional scaling of emotional responses to music: the effect of musical expertise and excerpts' duration. Cogn. Emot. **19**(8), 1113–1139 (2005)
3. Blood, A.J., Zatorre, R.J., Bermudez, P., Evans, A.C.: Emotional responses to pleasant and unpleasant music correlate with activity in paralimbic brain regions. Nat. Neurosci. **2**(4), 382–387 (1999)
4. Serra, M.: Stochastic composition and stochastic timbre: GENDY3 by Iannis Xenakis. Perspect. New Music **31**(1), 236–257 (1993)
5. Moorer, J.A.: iMusic and computer composition. In: Schwanauer, S.M., Levitt, D.A. (eds.) Machine Models of Music, pp. 167–186. The MIT Press, Cambridge (1993)
6. Amatriain, X., Bonada, J., Loscos, A., Arcos, J., Verfaille, V.: Content-based transformation. J. New Music Res. **32**(1), 95–114 (2003)
7. Bresin, R., Battel, G.U.: Articulation strategies in expressive piano performance analysis of legato, staccato, and repeated notes in performances of the andante movement of Mozart's sonata in g major (k 545). J. New Music Res. **29**(3), 211–224 (2000)
8. Todd, N.: The dynamics of dynamics: a model of musical expression. J. Acoust. Soc. Am. **91**, 3540–3550 (1992)
9. Friberg, A.: A quantitative rule system for musical performance, Ph.D. thesis, KTH, Sweden (1995)
10. Grachten, M., Widmer, G.: Linear basis models for prediction and analysis of musical expression. J. New Music Res. **41**(4), 311–322 (2012)
11. Rodà, A., Canazza, S., De Poli, G.: Clustering affective qualities of classical music: beyond the valence arousal plane. IEEE Trans. Affect. Comput. **5**(4), 364–376 (2014)
12. Dowling, W.J., Harwood, D.L.: Music Cognition. Academic, San Diego (1986)
13. Lindgren, T., Bostrom, H.: Classification with intersecting rules. In: Proceedings of 13th International Conference on Algorithmic Learning Theory. Springer (2002)
14. de la Motte, D.: Manuale di armonia. Bärenreiter (1976)

15. Coltro, B.: Lezioni di armonia complementare. Zanibon (1979)
16. Schonber, A.: Theory of Harmony. University of California Press, Berkeley (1983)
17. Della Ventura, M.: Toward an analysis of polyphonic music in the textual symbolic segmentation. In: Proceedings of the 2nd International Conference on Computer, Digital Communications and Computing (ICDCC 2013), Brasov, Romania (2013)
18. Della Ventura, M.: Rhythm analysis of the "Sonorous Continuum" and conjoint evaluation of the musical entropy. In: Proceedings of the 13th International Conference on Acoustics & Music: Theory & Applications (AMTA 2012), Iasi (Romania) (2012)
19. Della Ventura, M.: Automatic tonal music composition using functional harmony. In: Social Computing, Behavioral - Cultural Modeling and Prediction. Springer (2015)

Diverse Ranking Approach in MCDM Based on Trapezoidal Intuitionistic Fuzzy Numbers

Zamali Tarmudi[1]([⊠]) and Norzanah Abd Rahman[2]

[1] Faculty of Computer and Mathematical Sciences,
Universiti Teknologi MARA (Johor Branch),
KM12, Jalan Muar, 85000 Segamat, Johor, Malaysia
`zamalihj@sabah.uitm.edu.my`
[2] Faculty of Computer and Mathematical Sciences,
Universiti Teknologi MARA (Sabah Branch),
Locked Bag 71, 88997 Kota Kinabalu, Sabah, Malaysia

Abstract. Intuitionistic fuzzy set (IFS) is a generalization of the fuzzy set that is characterized by the membership and non-membership function. It is proven that IFS improves the drawbacks in fuzzy set since it is designed to deal with the uncertainty aspects. In spite of this advantage, the selection of the ranking approach is still one of the fundamental issues in IFS operations. Thus, this paper intends to compare three ranking approaches of the trapezoidal intuitionistic fuzzy numbers (TrIFN). The ranking approaches involved are; expected value-based approach, centroid-based approach, and score function-based approach. To achieve the objective, one numerical example in prioritizing the alternatives using intuitionistic fuzzy multi-criteria decision making (IF-MCDM) are provided to illustrate the comparison of these ranking approaches. Based on the comparison, it was found that the alternatives MCDM problems can be ranked easily in efficient and accurate manner.

Keywords: Intuitionistic fuzzy set · Trapezoidal intuitionistic fuzzy numbers · Multi-criteria decision-making · Ranking approach

1 Introduction

Intuitionistic fuzzy set (IFS) is one of the fuzzy set generalizations introduced by Atanassov in 1986. This IFS claims to be more precise in dealing with the uncertainty aspects since it considers both membership and non-membership function. Ever since the introduction of Intuitionistic fuzzy numbers (IFN), a number of other types of IFN have as well been introduced such as triangular IFN (TIFN), trapezoidal IFN (TrIFN), and interval-valued trapezoidal IFN. The development of the arithmetic operation of and the ranking approach of the IFN has become one of the fundamental issues in fuzzy environment. For instance, Mitchell [1] adopted a statistical viewpoint and interpreted each IFN as an ensemble of ordinary fuzzy numbers to define a fuzzy rank and a characteristic vagueness factor of each IFN. Nayagam, Venkateshwari and Sivaraman [2] proposed new method of IF scoring that involved a hesitation for both membership and non-membership function. This method was further developed and applied by

© Springer Nature Switzerland AG 2020
A. M. Madureira et al. (Eds.): SoCPaR 2018, AISC 942, pp. 11–21, 2020.
https://doi.org/10.1007/978-3-030-17065-3_2

Kakarontzas and Gerogiannis [3] to rank the web services. In 2010, Shen, Wang and Feng [4] introduced two ranking processes (i.e. based on probabilities and based on hesitation-probabilities) of IFN in MCDM. Biswas and De [5] continued by proposing a ranking technique based on probability function of the membership and non-membership function and applied it in solving linear bilevel programming.

Meanwhile, the studies on both TIFN and TrIFN's ranking approach had received worldwide attention. Li, Nan and Zhang [6] introduced new ranking approach based on ratio and ambiguity for TIFN. Their ranking approach was improved by Das and De [7, 8] in terms of using the sum of value and ambiguity index to satisfy linear property. By focusing on the -cut and -cut of the membership and non-membership function, Rezvani [9] also proposed similar approach. Likewise, Kumar and Kaur [10] claimed that their proposed ranking approach based on comparisons of fuzzy numbers and IF numbers could produce IF optimal solution. Ranking approach based on magnitude of membership and non-membership functions was introduced by Roseline and Amirtharaj [11]. As opposed to Nehi's ranking approach based on characteristic values of membership and non-membership function for trapezoidal (triangular) IFN [12] and Grzegorzewski's expected value for TrIFN [13], their method produced similar results to Nehi's approach. In the same year, they also proposed a new ranking method based on circumcenter of centroid of membership and non-membership function for TrIFN. This method was proven to provide the exact ordering and ranking of IFN in solving different fuzzy optimization problems.

In recent times, the ranking approach for TrIFN has been developed to produce better solution. For instance, Zeng, Li and Yu [14] stated their proposed method to rank the TrIFN using values and ambiguity was effective since it produced similar results with the existing method. Li and Yang [15] also proposed a ranking method based on difference index of value-index to the ambiguity-index for TrIFN to satisfy the linearity. They extended this method to solve the MADM problems. Their proposed method was not dependent on the form or shape of its membership and non-membership functions, natural appealing interpretation, and possessed some good linearity properties. Keikha and Nehi [16] improved the arithmetic operation and ranking based on centroid points of IFN since the outcomes were closer to the IFN reality. Prakash, Suresh, and Vengataasalam [17] also took advantage on centroid concept to develop ranking approach of TIFN and TrIFN that claimed to be time-consuming, suitable for both TrIFN and TIFN, and flexible in ranking index of their attitudinal analysis. Das and Guha [18] proposed a centroid point as a ranking method for TrIFN that showed some advantages such as reasonableness with human intuition and consistent with other approach and ranking results, as well as flexible algorithm when incorporated into decision making process. Most recently, Velu et al. [19] proposed new ranking approach for TrIFN using eights different scores function namely imprecise score, non-vague score, incomplete score, accuracy score, spread score, non-accuracy score, left area score and right area score. This proposed method improved the results in MADM problems, fuzzy information systems, and artificial intelligence. It was also easy to understand and operationally easy to use.

Based on the literature, it shows that there are several improvements in selecting the suitable ranking approach for IFN that can produce the best solutions. Thus, we intend to compare the ranking approaches of the trapezoidal intuitionistic fuzzy numbers

(TrIFN). The ranking approaches are; expected value-based approach, centroid-based approach, and score function-based approach. In order to achieve the objective, one numerical example in prioritizing the alternatives using intuitionistic fuzzy multi-criteria decision making (MCDM) is provided to illustrate the comparison of these ranking approaches. The rest of this paper is organized as follows; following the introduction in the first section, preliminaries of the IFS, IFN, and the ranking approaches will be presented. Then, the weighted values of ranking approaches (i.e. centroid and score function) will be proposed according to the Ye's weighted expected value [20]. Based on these weighted values, one illustrative example is provided in Sect. 4. Finally, conclusion on the comparison of these ranking approaches based on the alternative ranking of trapezoidal IF-MCDM is presented.

2 Preliminaries

In this section, some basic definition and concept are introduced. These include IFS, IFN, and the ranking approaches.

2.1 Intuitionistic Fuzzy Set

An IF is a generalization of the fuzzy set that is characterized by both the degree of membership and non-membership.

Definition 2.1.1 [21]: Let a set X be fixed. An IFS, A in X is defined as an object of the $A = \{\langle x, \mu_A(x), v_A(x) \rangle | x \in X\}$, where the functions: $\mu_A : X \rightarrow [0,1]$ and $v_A : X \rightarrow [0,1]$ is the degree of membership and the degree of non-membership of the element $x \in X$, respectively, and for every $x \in X$; $0 \leq \mu_A(x) + v_A(x) \leq 1, x \in X$.

2.2 Intuitionistic Fuzzy Numbers

Definition 2.2.1 [22]: A trapezoidal Intuitionistic Fuzzy Number (TrIFN), A with parameters $b_1 \leq a_1 \leq b_2 \leq a_2 \leq a_3 \leq b_3 \leq a_4 \leq b_4$ is represented as $A = (a_1, a_2, a_3, a_4; b_1, b_2, b_3, b_4)$ in a real set of X. The membership and non-membership are defined as follows:

$$\mu_A(x) = \begin{cases} 0 & \text{if} \quad x < a_1 \\ \frac{x-a_1}{a_2-a_1} & \text{if} \quad a_1 \leq x \leq a_2 \\ 1 & \text{if} \quad a_2 \leq x \leq a_3 \\ \frac{x-a_4}{a_3-a_4} & \text{if} \quad a_3 \leq x \leq a_4 \\ 0 & \text{if} \quad a_4 < x \end{cases} \quad v_A(x) = \begin{cases} 0 & \text{if} \quad x < b_1 \\ \frac{x-b_2}{b_1-b_2} & \text{if} \quad b_1 \leq x \leq b_2 \\ 1 & \text{if} \quad b_2 \leq x \leq b_3 \\ \frac{x-b_3}{b_4-b_3} & \text{if} \quad b_3 \leq x \leq b_4 \\ 0 & \text{if} \quad b_4 < x \end{cases} \quad (1)$$

When $a_2 = a_3$, it will change into triangular IFN.

2.3 Expected Value-Based Approach

Definition 2.3.1 [13]: Let $A = (a_1, a_2, a_3, a_4; b_1, b_2, b_3, b_4)$ be a TrIFN in X, $X \in \Re$. Thus, when $\frac{x-a_1}{a_2-a_1}, \frac{x-a_4}{a_3-a_4}, \frac{x-b_2}{b_1-b_2}, \frac{x-b_3}{b_4-b_3}, a_1, a_2, a_3, a_4, b_1, b_2, b_3, b_4 \in \Re$

The expected value can be calculated using the following formula:

$$EV(A) = \frac{1}{8}(a_1 + a_2 + a_3 + a_4 + b_1 + b_2 + b_3 + b_4) \tag{2}$$

Theorem 2.3.1: Let $\succ_L$ and $\succ_U$ denote the quasi-order with respect to the lower and upper horizon, respectively, based on the metric d_1 (i.e. d_p for $p = 1$). Then, the following order of two TrIFNs, $A, B \in X$ can be derived:

i. $A \succ_L B \Leftrightarrow EV(A) \geq EV(B)$
ii. $A \succ_U B \Leftrightarrow EV(A) \geq EV(B)$

2.4 Centroid-Based Approach

Definition 2.4.1 [17]: The centroid of a symmetric TrIFN, $A = (a_1, a_2, a_3, a_4; b_1, b_2, b_3, b_4)$ are defined as follows:

$$\tilde{x}_\mu(A) = \frac{1}{2}\left(\frac{a_3^2 + a_4^2 - a_1^2 - a_2^2 - a_1 a_2 + a_3 a_4}{a_4 + a_3 - a_2 - a_1}\right)$$

$$\tilde{x}_v(A) = \frac{1}{3}\left(\frac{2b_4^2 - 2b_1^2 + 2b_2^2 + 2b_3^2 + b_1 b_2 - b_3 b_4}{b_3 + b_4 - b_1 - b_2}\right)$$

$$\tilde{y}_\mu(A) = \frac{1}{3}\left(\frac{a_1 + 2a_2 - 2a_3 - a_4}{a_1 + a_2 - a_3 - a_4}\right)$$

$$\tilde{y}_v(A) = \frac{1}{3}\left(\frac{2b_1 + b_2 - b_3 - 2b_4}{b_1 + b_2 - b_3 - b_4}\right)$$

Definition 2.4.2: The ranking function of TrIFN, $A = (a_1, a_2, a_3, a_4; b_1, b_2, b_3, b_4)$ for membership and non-membership function are defined as follows:

$$R(A) = \sqrt{\frac{1}{2}\left([\tilde{x}_\mu(A) - \tilde{y}_\mu(A)]^2 + \left([\tilde{x}_v(A) - \tilde{y}_v(A)]^2\right)\right)} \tag{3}$$

Consider that there are two TrIFNs, $A, B \in X$, $X \in \Re$ can be defined as follows:

i. $R(A_\mu) > R(B_\mu)$ iff $A < B$
ii. $R(A_\mu) < R(B_\mu)$ iff $A \prec B$
iii. $R(A_\mu) = R(B_\mu)$ and $R(A_v) = R(B_v)$ iff $A \sim B$
iv. $R(A_\mu) = R(B_\mu)$ and $-R(A_v) > -R(B_v)$ iff $A < B$
v. $R(A_\mu) = R(B_\mu)$ and $-R(A_v) < -R(B_V)$ iff $A < B$

2.5   Score Function Based-Approach

Definition 2.5.1 [19]: Let $A = (a_1, a_2, a_3, a_4; b_1, b_2, b_3, b_4)$ be a TrIFN in X, $X \in \Re$ with $a_1, a_2, a_3, a_4, b_1, b_2, b_3, b_4 \in [0, 1]$. Then,

Imprecise Score $$J_1(A) = \frac{(a_1 + a_2) - (a_3 + a_4) + (b_1 + b_2) - (b_3 + b_4)}{8} \quad (4)$$

Nonvague Score $$J_2(A) = \frac{(a_1 + a_2) - (a_3 + a_4) - (b_1 + b_2) - (b_3 + b_4)}{8} \quad (5)$$

Incomplete Score $$J_3(A) = \frac{(b_1 + b_2 + b_3 + b_4) - (a_1 + a_2 + a_3 + a_4)}{8} \quad (6)$$

Accuracy Score $$J_4(A) = \frac{(a_1 + a_2 + a_3 + a_4) + (b_1 + b_2 + b_3 + b_4)}{8} \quad (7)$$

Spread Score $$J_5(A) = \frac{(a_1 + a_3) - (a_2 + a_4) - (b_2 + b_4) + (b_1 + b_3)}{8} \quad (8)$$

Nonaccuracy Score $$J_6(A) = \frac{(a_2 + a_4) - (a_1 + a_3) + (b_1 + b_3) - (b_2 + b_4)}{8} \quad (9)$$

Left Area $$J_7(A) = \frac{(b_3 - b_1) + (b_2 - b_4) - (a_3 - a_1) - (a_2 - a_4)}{8} \quad (10)$$

Right Area $$J_8(A) = \frac{(a_1 - a_3) + (a_4 - a_2) + (b_1 - b_3) + (b_4 - b_2)}{8} \quad (11)$$

The ranking of TrIFN can be summarized as follows:

i. $J_1(A) < J_1(B)$; $A < B$
ii. $J_1(A) = J_1(B)$ then if $J_2(A) < J_2(B)$; $A < B$
iii. $J_1(A) = J_1(B)$ and $J_2(A) = J_2(B)$, then if $J_3(A) < J_3(B) A < B$
iv. $J_1(A) = J_1(B)$, $J_2(A) = J_2(B)$, and $J_3(A) = J_3(B)$, then if $J_4(A) < J_4(B) A < B$
v. $J_1(A) = J_1(B)$, $J_2(A) = J_2(B)$, $J_3(A) = J_3(B)$, and $J_4(A) = J_4(B)$, then if $J_5(A) < J_5(B) A < B$
vi. $J_1(A) = J_1(B)$, $J_2(A) = J_2(B)$, $J_3(A) = J_3(B)$, $J_4(A) = J_4(B)$ and $J_5(A) = J_5(B)$, then if $J_6(A) < J_6(B) A < B$
vii. $J_1(A) = J_1(B)$, $J_2(A) = J_2(B)$, $J_3(A) = J_3(B)$, $J_4(A) = J_4(B)$, $J_5(A) = J_5(B)$ and $J_6(A) = J_6(B)$, then if $J_7(A) < J_7(B) A < B$

viii. $J_1(A) = J_1(B), J_2(A) = J_2(B), J_3(A) = J_3(B), J_4(A) = J_4(B),$
$J_5(A) = J_5(B), J_6(A) = J_6(B)$ and $J_7(A) = J_7(B),$ then if
$J_8(A) < J_8(B) A < B$

ix. $J_1(A) = J_1(B), J_2(A) = J_2(B), J_3(A) = J_3(B), J_4(A) = J_4(B),$
$J_5(A) = J_5(B), J_6(A) = J_6(B), J_7(A) = J_7(B)$ and $J_8(A) = J_8(B),$ then
$A = B$

3 Trapezoidal IF-MCDM Method Based on Ranking Approaches of TrIFN

Based on the weighted expected value proposed by Ye [20] in (12)

$$WEV(A_i) = \sum_{j=1}^{n} W_i EV(p_{ij}); \ W_j = \frac{EV(W_j)}{\sum_{j=1}^{n} EV(W_j)} \tag{12}$$

We propose a decision procedure for trapezoidal IF-MCDM with centroid-based approach and score function-based approach.

Let $A = (A_1, A_2, \ldots, A_m)$ be a set of alternatives and let $P = (P_1, P_2, \ldots, P_n)$ be a set of criteria. The preference value of an alternative on criterion $P_j(j = 1, 2, \ldots, n)$ is an TrIFN $P_{ij} = \langle (a_{ij1}, a_{ij2}, a_{ij3}, a_{ij4}; b_{ij1}, b_{ij2}, b_{ij3}, b_{ij4}) \rangle$, $a_1, a_2, a_3, a_4, b_1, b_2, b_3, b_4 \in \Re$, $j = 1, 2, \ldots, n$ and $i = 1, 2, \ldots, n$ which indicates the degree that the alternative satisfies or does not satisfy the criterion P_j given by the domain expert according to the linguistic values of TrIFN of linguistic variables. Then, the decision matrix $D = (P_{ij})_{n \times n}$ which is expressed by TrIFN can be produced. The weight of criterion $P_j(j = 1, 2, \ldots, n)$, is the TrIFN weight $W_j = (c_1, c_2, c_3, c_4; d_1, d_2, d_3, d_4)$. This can be obtained using (2) to (11) for each one of the alternatives, respectively. The following weighted value for each ranking approach to rank the alternative $A_1(i = 1, 2, \ldots, n)$ is proposed, as follows:

Weighted Centroid Value: $$WCV(A_i) = \sum_{j=1}^{n} W_i C(p_{ij}); \ W_j = \frac{C(W_j)}{\sum_{j=1}^{n} C(W_j)} \tag{13}$$

Weighted Score Function: $$WJ_iV(A_i) = \sum_{j=1}^{n} W_i J(p_{ij}); \ W_j = \frac{J(W_j)}{\sum_{j=1}^{n} J(W_j)} \tag{14}$$

Step-by-step decision process for Trapezoidal IF-MCDM using (12), (13), and (14) are demonstrated in the figure below:

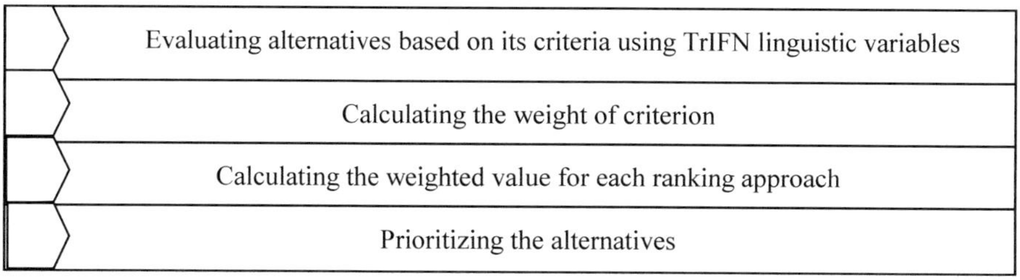

| Evaluating alternatives based on its criteria using TrIFN linguistic variables |
| Calculating the weight of criterion |
| Calculating the weighted value for each ranking approach |
| Prioritizing the alternatives |

Fig. 1. The step-by-step decision making process for Trapezoidal IF-MCDM

Table 1. The linguistic variables and TrIFN values

Linguistic variable	Linguistic values/TrIFN values
Very high influence (VH)	(1.00, 1.00, 1.00, 1.00; 1.00, 1.00, 1.00, 1.00)
High influence (H)	(0.70, 0.80, 0.90, 1.00; 0.70, 0.80, 0.90, 1.00)
Low influence (L)	(0.30, 0.40, 0.50, 0.60; 0.20, 0.40, 0.50, 0.70)
Very low influence (VL)	(0.00, 0.10, 0.20, 0.30; 0.00, 0.10, 0.20, 0.30)
No influence (NI)	(0.00, 0.00, 0.00, 0.00; 0.00, 0.00, 0.00, 0.00)

(Sources: Nikjoo and Saeedpoor, [22] and Vafarnikjoo et al. [23])

4 Illustrative Example on Trapezoidal IF-MCDM

In order to demonstrate the comparisons of the ranking approaches, one illustrative example of prioritizing the alternatives in MDCM problem is provided in this section. Suppose that the government wants to select the best flood mitigation measures $A = \{A_1, A_2, A_3\}$ to be implemented based on three criteria, $P = \{P_1, P_2, P_3\}$. Three experts, $k = \{k_1, k_2, k_3\}$ are involved in the decision making process. According to step-by-step decision making process in Fig. 1, the application and results are defined as follows:

Step 1: Evaluating alternatives based on respective criteria using linguistic variables of TrIFN

The criteria of each alternative were evaluated by experts using linguistic variables of TrIFN as in Table 1. Table 2 shows the raw data (linguistic variables), the initial decision matrix (TrIFN values), and the weights of criterion for each alternative.

Table 2. The raw data and the initial decision matrix for each alternative

A	k		P_1		P_2		P_3
A_1	k_1	H	(0.70, 0.80, 0.90, 1.00; 0.70, 0.80, 0.90, 1.00)	L	(0.30, 0.40, 0.50, 0.60; 0.20, 0.40, 0.50, 0.70)	V H	(1.00, 1.00, 1.0, 1.00; 1.00, 1.00, 1.00, 1.00)
	k_2	H	(0.70, 0.80, 0.90, 1.00; 0.70, 0.80, 0.90, 1.00)	L	(0.30, 0.40, 0.50, 0.60; 0.20, 0.40, 0.50, 0.70)	H	(0.70, 0.80, 0.90, 1.00; 0.70, 0.80, 0.90, 1.00)
	k_3	VH	(1.00, 1.00, 1.00, 1.00; 1.00, 1.00, 1.00, 1.00)	V L	(0.00, 0.10, 0.20, 0.30; 0.00, 0.10, 0.20, 0.30)	V H	(1.00, 1.00, 1.0, 1.00; 1.00, 1.00, 1.00, 1.00)
A_2	k_1	VH	(1.00, 1.00, 1.0, 1.00; 1.00, 1.00, 1.00, 1.00)	V H	(0.70, 0.80, 0.90, 1.00; 0.70, 0.80, 0.90, 1.00)	V L	(0.00, 0.10, 0.20, 0.30; 0.00, 0.10, 0.20, 0.30)
	k_2	H	(0.70, 0.80, 0.90, 1.00; 0.70, 0.80, 0.90, 1.00)	L	(0.30, 0.40, 0.50, 0.60; 0.20, 0.40, 0.50, 0.70)	H	(0.70, 0.80, 0.90, 1.00; 0.70, 0.80, 0.90, 1.00)
	k_3	V	(1.00, 1.00, 1.0, 1.00; 1.00, 1.00, 1.00, 1.00)	V L	(0.00, 0.10, 0.20, 0.30; 0.00, 0.10, 0.20, 0.30)	L	(0.30, 0.40, 0.50, 0.60; 0.20, 0.40, 0.50, 0.70)
A_3	k_1	L	(0.30, 0.40, 0.50, 0.60; 0.20, 0.40, 0.50, 0.70)	H	(0.70, 0.80, 0.90, 1.00; 0.70, 0.80, 0.90, 1.00)	H	(0.70, 0.80, 0.90, 1.00; 0.70, 0.80, 0.90, 1.00)
	k_2	H	(0.70, 0.80, 0.90, 1.00; 0.70, 0.80, 0.90, 1.00)	L	(0.30, 0.40, 0.50, 0.60; 0.20, 0.40, 0.50, 0.70)	H	(0.70, 0.80, 0.90, 1.00; 0.70, 0.80, 0.90, 1.00)
	k_3	VL	(0.00, 0.10, 0.20, 0.30; 0.00, 0.10, 0.20, 0.30)	H	(0.70, 0.80, 0.90, 1.00; 0.70, 0.80, 0.90, 1.00)	V H	(1.00, 1.00, 1.0, 1.00; 1.00, 1.00, 1.00, 1.00)
Weights	k_1	H	(0.70, 0.80, 0.90, 1.00; 0.70, 0.80, 0.90, 1.00)	L	(0.30, 0.40, 0.50, 0.60; 0.20, 0.40, 0.50, 0.70)	H	(0.70, 0.80, 0.90, 1.00; 0.70, 0.80, 0.90, 1.00)
	k_2	VH	(1.00, 1.00, 1.00, 1.00; 1.00, 1.00, 1.00, 1.00)	L	(0.30, 0.40, 0.50, 0.60; 0.20, 0.40, 0.50, 0.70)	L	(0.30, 0.40, 0.50, 0.60; 0.20, 0.40, 0.50, 0.70)
	k_3	H	(0.70, 0.80, 0.90, 1.00; 0.70, 0.80, 0.90, 1.00)	L	(0.30, 0.40, 0.50, 0.60; 0.20, 0.40, 0.50, 0.70)	H	(0.70, 0.80, 0.90, 1.00; 0.70, 0.80, 0.90, 1.00)

$$D = \begin{pmatrix} (0.8000, 0.8667, 0.9333, 1.0000; & (0.2000, 0.3000, 0.4000, 0.5000; & (0.9000, 0.9333, 0.9667, 1.0000; \\ 0.800, 0.8667, 0.9333, 1.0000) & 0.1333, 0.3000, 0.4000, 0.5667) & 0.9000, 0.9333, 0.9667, 1.0000) \\ (0.9000, 0.9333, 0.9667, 1.0000; & (0.3333, 0.4333, 0.5333, 0.6333; & (0.3333, 0.4333, 0.5333, 0.6333; \\ 0.9000, 0.9333, 0.9667, 1.0000) & 0.3000, 0.4333, 0.5333, 0.6667) & 0.3000, 0.4333, 0.5333, 0.6667) \\ (0.3333, 0.4333, 0.5333, 0.6333; & (0.5667, 0.6667, 0.7667, 0.8667; & (0.8000, 0.8667, 0.9333, 1.0000; \\ 0.3000, 0.4333, 0.5333, 0.6667) & 0.5333, 0.6667, 0.7667, 0.9000) & 0.8000, 0.8667, 0.9333, 1.0000) \end{pmatrix}$$

$$\text{Weights} = \begin{bmatrix} (0.8000, 0.8667, 0.9333, 1.0000; & (0.3000, 0.4000, 0.5000, 0.6000; & (0.5667, 0.6667, 0.7667, 0.8667; \\ 0.8000, 0.8667, 0.9333, 1.0000) & 0.2000, 0.4000, 0.5000, 0.7000) & 0.5333, 0.6667, 0.7667, 0.9000) \end{bmatrix}$$

Step 2: Calculating the weight of criterion

Table 3 presents the weights of criterion for each ranking approach obtained using (2) to (11), respectively.

Table 3. The weights of criterion using each ranking approach

Weights	Expected value	Centroid value	Score function							
P_1	0.4355	0.6005	0.2222	0.0000	0.0000	0.4355	0.2000	0.0000	0.0000	0.0000
P_2	0.3468	0.1441	0.4167	0.7500	1.0000	0.2177	0.4500	0.7500	1.0000	5.0000
P_3	0.2177	0.2554	0.3611	0.2500	0.0000	0.3468	0.3500	0.2500	0.0000	4.0000

Step 3: Calculating the weighted values for each ranking approach

In this step, the weighted values are computed using (12) to (14) as illustrated in Table 4.

Table 4. The weighted values of each ranking approach

A	WEV	WCV	WJ_1V	WJ_2V	WJ_3V	WJ_4V	WJ_5V	WJ_6V	WJ_7V	WJ_8V
A_1	1.0161	6.3427	−0.0463	0.0000	0.0000	1.0161	−0.0217	0.0000	0.0000	0.0000
A_2	0.3375	0.3946	−0.1389	0.0250	0.0000	0.3375	−0.0825	−0.0250	0.0000	0.0000
A_3	0.8091	2.6973	−0.0752	0.0021	0.0000	0.8091	−0.0379	−0.0021	0.0000	0.0000

Step 4: Prioritizing the alternatives

Finally, based on the weighted values, the ranking of the alternatives can be achieved (refer to Table 5).

Table 5. The ranking of the alternatives

Ranking approaches	WEV, WCV, WJ_1V, WJ_4V, WJ_5V and WJ_6V	WJ_2V	WJ_3V, WJ_7V, and WJ_8V
Ranking alternatives	$A_1 > A_3 > A_2$	$A_2 > A_3 > A_1$	$A_1 = A_2 = A_3$

5 Conclusion

In this paper, three ranking approaches of the trapezoidal intuitionistic fuzzy numbers (TrIFN) are studied. These include expected value-based approach, centroid-based approach, and score function-based approach. Based on Ye's weighted expected value, we have proposed the weighted value for centroid-based approach and score function-based approach, so that the alternatives in the MCDM problems can be ranked in efficient and accurate manner. The rank of the alternatives shows that the WEV, WCV, WJ_1V, WJ_4V, WJ_5V, and WJ_6V formed a similar ranking which is $A_1 > A_3 > A_2$. Otherwise, the ranking produced by WJ_2V is $A_2 > A_3 > A_1$, where the first and last rank varied, though the second rank held constant. Meanwhile, the ranking of WJ_3V, WJ_7V, and WJ_8V produced an equal ranking $(A_1 = A_3 = A_2)$ which is irrelevant to the real life.

Acknowledgements. This research was supported by grant of Fundamental Research Grant Scheme (FRGS) from Ministry of Education (formerly known as Ministry of Higher Education (MOHE) Malaysia and Universiti Teknologi MARA (UiTM), Malaysia, reference no.: 600-IRMI/FRGS 5/3(84/2016).

References

1. Mitchell, H.B.: Ranking intuitionistic fuzzy numbers. Int. J. Uncertain. Fuzziness Knowl.-Based Syst. **12**(3), 377–386 (2004)
2. Nayagam, V.L.G., Venkateswari, G., Sivaraman, G.: Modified ranking of intuitionistic fuzzy numbers. NFIS **17**(1), 5–22 (2011)
3. Kakarontzas, G., Gerogiannis, V.C.: An intuitionistic fuzzy approach for ranking web services under evaluation uncertainty. In: IEEE International Conference on Services Computing, pp. 742–745 (2015)
4. Shen, L., Wang, H., Feng, X.: Ranking methods of intuitionistic fuzzy numbers in multicriteria decision making. In: 3rd International Conference on Information Management Innovation Management and Industrial Engineering, pp. 143–146 (2010)
5. Biswas, A., De, A.K.: An efficient ranking technique for intuitionistic fuzzy numbers with its application in chance constrained bilevel programming. Adv. Fuzzy Syst. **2016**, 1–12 (2016)
6. Li, D.F., Nan, J.X., Zhang, M.J.: A ranking method of triangular intuitionistic fuzzy numbers and application to decision making. Int. J. Comput. Intell. Syst. **3**(5), 522–530 (2010)
7. De, P.K., Das, D.: Ranking of trapezoidal intuitionistic fuzzy numbers. In: 12th International Conference on Intelligent Systems Design and Applications (ISDA), pp. 184–188 (2012)
8. De, P.K., Das, D.: A study on ranking of trapezoidal intuitionistic fuzzy numbers. Int. J. Comput. Inf. Syst. Ind. Manag. Appl. **6**, 437–444 (2014)
9. Rezvani, S.: Ranking method of trapezoidal intuitionistic fuzzy numbers. Ann. Fuzzy Math. Inform. **5**(3), 515–523 (2012)
10. Kumar, A., Kaur, M.: A ranking approach for intuitionistic fuzzy numbers and its application. J. Appl. Res. Technol. **11**, 381–396 (2013)
11. Roseline, S.S., Amirtharaj, E.C.H.: A new ranking of intuitionistic fuzzy numbers. Indian J. Appl. Res. **3**, 1–2 (2013)
12. Nehi, H.M.: A new ranking method for intuitionistic fuzzy numbers. Int. J. Fuzzy Syst. **12**(1), 80–86 (2010)
13. Grzegrorzewski, P.: The hamming distance between intuitionistic fuzzy sets. In: Proceedings of the IFSA 2003 World Congress, Istanbul, pp. 35–38 (2013)
14. Zeng, X.-T., Li, D.-F., Yu, G.-F.: A value and ambiguity-based ranking method of trapezoidal intuitionistic fuzzy numbers and application to decision making. Sci. World J. **2014**, 1–8 (2014)
15. Li, D.-F., Yang, J.: A difference-index based ranking method of trapezoidal intuitionistic fuzzy numbers and application to multiattribute decision making. Math. Comput. Appl. **20**(1), 25–38 (2015)

16. Keikha, A., Nehi, H.M.: Operation and ranking methods for intuitionistic fuzzy numbers, a review and new method. Int. J. Intell. Syst. Appl. **1**, 35–48 (2016)
17. Prakash, A.A., Suresh, M., Vengataasalam, S.: A new ranking of intuitionistic fuzzy numbers using a centroid concept. Math. Sci. **10**, 177–184 (2016)
18. Das, S., Guha, D.: A centroid-based ranking method of trapezoidal intuitionistic fuzzy numbers and its application to MCDM problems. Fuzzy Inf. Eng. **8**, 41–74 (2016)
19. Velu, L.G.N., Selvaraj, J., Ponnialagan, D.: A new ranking principle for ordering trapezoidal intuitionistic fuzzy numbers. Complexity **2017**, 1–24 (2017)
20. Ye, J.: Expected value method for intuitionistic trapezoidal fuzzy multicriteria decision-making problems. Expert Syst. Appl. **38**, 11730–11734 (2011)
21. Atanassov, K.T.: Intuitionistic fuzzy set. Fuzzy Sets Syst. **20**, 87–96 (1996)
22. Nehi, H.M., Maleki, H.R.: Intuitionistic fuzzy numbers and it's applications in fuzzy optimization problem. In: Proceedings of the 9th WSEAS International Conference on Systems, Athen, Greece, pp. 1–5 (2015)
23. Nikjoo, A.V., Saeedpoor, M.: An intuitionistic fuzzy DEMATEL methodology for prioritising the components of SWOT matrix in the Iranian insurance industry. Int. J. Oper. Res. **20**(4), 439–452 (2014)
24. Vafadarnikjoo, A., Mobin, M., Allahi, S., Rastegari, A.: A hybrid approach of intuitionistic fuzzy set theory and DEMATEL method to prioritize selection criteria of bank. Int. J. Comput. Intell. Syst. **8**(4), 637–666 (2015)

Decision Tree and MCDA Under Fuzziness to Support E-Customer Satisfaction Survey

Houda Zaim[1(✉)], Mohammed Ramdani[1], and Adil Haddi[2]

[1] Informatics Department, LIM Laboratory,
Faculty of Sciences and Techniques of Mohammedia,
University Hassan II, Casablanca, Morocco
houdazaime@gmail.com, ramdani@fstm.ac.ma
[2] Informatics Department, LAMSAD Laboratory,
Superior School of Technology of Berrechid,
University Hassan I, Settat, Morocco
adil.haddi@gmail.com

Abstract. The proposed work extends existing approaches by analyzing customer click stream data and online reviews to implicitly identify satisfaction level when customer's rate is not available and find the website criteria score that positively influence e-customer satisfaction. Fuzzy mining customer navigation data is our task to set up inputs of the two proposed supervised evaluation approaches; a multi criteria analysis approach for the website assessment and a new decision tree algorithm to classify customers. A case study from the B2C Chinese website "TMALL" has been used for validating our proposal, and a comparison between the proposed approaches has shown promising results.

Keywords: Website quality determinants · Fuzzy membership degree · Similarity · Class prototype · Confusion matrix

1 Introduction

Authors of this paper survey e-customer satisfaction within the context of enhancing e-commerce website quality. This requires identifying the key strategic issues. First, Customer Profiling: Which forms of classifying customers are most appropriate? How should criteria of classification be determined? Second, E-commerce Website Quality: How are online products/services? Which decisions should be made by managers to satisfy online customers? And how do we analyze customer's pathways to evaluate website criteria? Answering these questions requires providing some prerequisites as following: 1. Specifying critical factors that have the most impact on customer satisfaction. 2. Presenting e-customers as alternatives to the critical factors. 3. Applying customer profiling strategy. In order to achieve the solution for these raised considerations, authors as mentioned in Fig. 1 have benefited from mining customer's navigation data to first, build customer's profiles using decision tree generation algorithm, and second, perform a multi criteria analysis approach which affects scores to different e-commerce website features using similarity and dissimilarity to ideal satisfaction class prototype.

© Springer Nature Switzerland AG 2020
A. M. Madureira et al. (Eds.): SoCPaR 2018, AISC 942, pp. 22–32, 2020.
https://doi.org/10.1007/978-3-030-17065-3_3

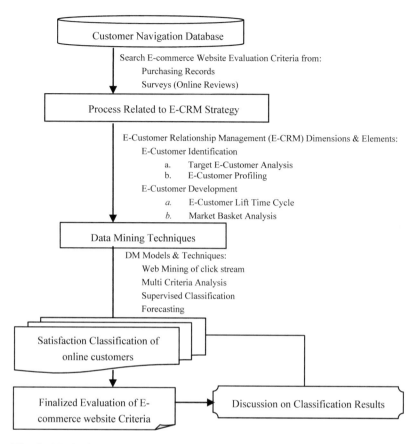

Fig. 1. Navigational data in accordance with CRM elements and DM techniques.

Issues that are addressed such a motivation of the proposed approach include Behavioral Data to Load Online Customer Profile and Approaches for Identifying E-commerce Evaluation Indexes and Estimating their Weights.

Vercamer et al. [1] combined commercial and internal company data to predict load profiles using random forests and stochastic boosting algorithms for classification. User profiles can also be created through multi-resolution clustering designed for smart metering data [2, 3]. Lu et al. [4] investigated the load profile clustering using an adaptive weighted fuzzy algorithm where Fuzzy C-Means is adopted to cluster big data. Ghuman et al. [5] segmented consumers using a cluster analysis based on consumer's social risk perception and their psychological, cultural, and socio demographic variables. Even though researches reflect various data in creating customer profile, it ignores data from customer's click and reviews. Moreover customer's profile-based feedback for evaluating website dimensions is ignored. Therefore our approach will first try to find which dimensions should prioritize to build customer profile and to maximize electronic commerce outcome. For Rai et al. [6], the e-commerce software made possible to easily manage inventory, add or remove products, calculate taxes, and everything else required to manage a website an fulfill orders. Sriram et al. [7] focused on security protocols and their achievement in providing effective communication. According to authors of [8–13], a set of antecedents that positively influence consumer's satisfaction (Efficiency, content and logistics of order processing, customer payment and billing) have arisen. Our work extends existing works taking into account new set of evaluation attributes; categorical, continuous and fuzzy ones, keeping in mind the areas where sequence of web events generated by each user were required to assess the online service quality.

For assigning the relative importance score to each e-commerce website attribute, The DANP-DEMATEL method combining Decision Making Trial and Evaluation Laboratory and Analytic Network Process was used to obtain influential weights based on expert opinions [15]. Rouyendegh et al. [14] presented a hybrid framework of AHP (Analytic Hierarchy Process) which was applied to establish a set of all judgments based on decision-makers in the comparison matrix to obtain a local weight of each criterion and sub criterion. As stated by Qin et al. [16], the decision maker can express his decision in terms of preference using LINMAP method. A paucity of these approaches quantified customer behavioral data in order to assign weights of evaluation criteria according to the e-customer satisfaction level.

After expressing the literature study and assumptions, the rest of paper explains in Sect. 2 the system architecture and pre-processing steps with algorithm pseudo code of the proposed supervised classification approaches. Section 3 discusses performance evaluation and comparison result analysis. Section 4 summarizes the overall work with future outcomes.

2   Proposed Approaches for E-Customer Satisfaction Survey

2.1   E-Customer Profiling

The algorithm selects learning data, calculates dissimilarity between any two values of the selected attribute for a unique class and calculates the average similarity of the attribute according to the sum of calculated dissimilarities. The global similarity is then computed for all classes (Fig. 2).

Generate_Decision_Tree (T,A_j, C_i)

Inputs: T: Set of continuous, categorical and fuzzy valued attributes: A_j, C_i: Target attribute

Output: Decision tree to classify e-customers where nodes are the website criteria to be evaluated, branches values are deduced based on e-customer behavior and satisfaction class of e-customer presents the leaf of the tree (Target attribute)

Start

D ← **Node_Selection (T,A_j)** //D is the root of the tree

    If D is continuous type **Then**

        x_u←**Calculate splitting threshold (T, A_j)**

    End if

$\{x_1,x_2,..., x_u, ...x_v\}$ ← Values of the attribute D

Return the constructed tree whose root is D and arcs are labeled by $x_1, x_2,... x_v$ and going to sub-trees $(T_1, A_j\text{-}\{D\}, C_i)$, $(T_2, A_j\text{-}\{D\}, C_i)$, .., $(T_v, A_j\text{-}\{D\}, C_i)$

End

Fig. 2. Pseudo code of decision tree induction algorithm

Node_Selection (T, A_j). Selects the root node which has the maximum similarity. If the splitting attribute A_j is continuous, the function **Calculate splitting threshold (T, A_j)** sorts attribute's values and for each value x_u of A_j, it calculates the similarity of two subsets that divided by x_u and considers the value x_u with the maximum similarity as the splitting threshold [17].

2.2 Proposed MCDA Method

MCDA problems are said to involve prioritization of a set of alternatives in situations that involve multiple conflicting criteria. Steps and functions of suggested weighing model are below (Fig. 3):

Inputs:

A_j:E-commerce website criteria to be evaluated, C_i: Satisfaction class of E-customer

$V_{q,j}^i$: Values of e-commerce website criteria for the class C_i based on E-customer behavior

E-commerce website criteria score $\alpha_j^i \leftarrow 0$

The prototype of the attribute A_j for the class C_i $C_j^i \leftarrow 0$

    For each class C_i

        For each criterion A_j

          If value criterion type=numeric **Then**

$$\alpha_j^i = {1}\Big/{1 + \sum_{q=1}^n \left| V_{q,j}^i - C_j^i \right|}$$

$$C_j^i = {\sum_{q=1} V_{q,j}^i}\Big/{n}$$

        End If

          If value criterion type=nominal **Then**

$$\alpha_j^i = \text{freq}(C_j^i) \in [0,1]$$

$$C_j^i = V_{q,j}^i \text{ where } f(A_j^i) = \max_{k=1,\ldots m}(f(V_{q,j}^i))$$

        End If

          If value criterion type=fuzzy **Then**

$$\alpha_j^i = \frac{\sum_{k=1, k \neq q}^5 (T_{q,j}^i - T_{k,j}^i)}{4}$$

$$C_j^i = V_{q,j}^i \text{ where } T_{q,j}^i = \max_{k=1,\ldots 5}(T_{k,j}^i)$$

$$T_{k,j}^i = \frac{(S_{k,j}^i + D_{k,j}^i)}{2}$$

$$S_{k,j}^i = \frac{1}{n}\sum_{q=1}^n s(V_{k,j}, V_{q,j}^i), D_{k,j}^i = \frac{1}{n}\sum_{q=1}^n d(V_{k,j}, V_{q,j}^r)$$

$(*(V_{k,j})_{k=1,\ldots 5}$: value of the fuzzy criterion ("very weak", "weak", "medium", "strong", and "very strong")*)

$$s(V_{k,j}, V_{q,j}^i) = \frac{M(V_{k,j} \cap V_{q,j}^i)}{M(V_{k,j} \cup V_{q,j}^i)}$$

$$d(V_{k,j}, V_{q,j}^r) = \frac{M(V_{k,j} - V_{q,j}^i) + M(V_{q,j}^i - V_{k,j})}{M(V_{k,j} \cup V_{q,j}^i)}$$

        End If

      End For

    End For

Fig. 3. Pseudo code of the proposed weighing model

Let $(C_i)_{i=1,\ldots 3}$ be the consumer's satisfaction level over the website: "very satisfied" (C_1), "moderately satisfied" (C_2), and "not satisfied" (C_3). $(A_j)_{j=1,\ldots p}$ denotes the set of p evaluation criteria. $V_{q,j}^i$ represents the possible value of A_j for the class C_i. We denote by $\left(C_j^i\right)_{j=1,\ldots p}$ the prototype of the attribute A_j for the class C_i. The weight α_j^i denotes the importance of attribute A_j to determine the class C_i. The prototype of numeric criterion is the average of its values given by all the n members of class C_i. For nominal criterion, the prototype is the value that has a high frequency among all possible values of this criterion.

We could measure the similarity and dissimilarity between two fuzzy sets by: $s(A,B) = \dfrac{M(A \cap B)}{M(A \cup B)}$, $d(A,B) = \dfrac{M(A-B)+M(B-A)}{M(A \cup B)}$. The prototype C_j^i of fuzzy criterion is one that has the maximum typicality: $C_j^i = V_{q,j}$ where $T_{q,j}^i = \max_{k=1,\ldots 5}(T_{k,j}^i)$. More details are in [18].

3 Experimental Validation Overview

3.1 Criteria for Evaluation

According to the state of art and our previous work [19], input data includes a set of five website features which most influence the customer satisfaction: Usability V_U, Content Adequacy V_{CA} and Reliability V_R as fuzzy features, order Fulfillment as a categorical variable and Delivery Time which is numeric one.

Figure 4 presents an algorithm used to generate triangular membership function on numerical value of fuzzy data. The membership function spans the interval [0, 1] for five linguistic terms: VW (Very Weak), W (Weak), M (Medium), S (Strong) and VS (Very Strong):

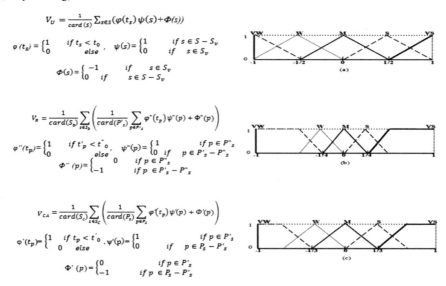

Fig. 4. MF for linguistic labels of: (a) usability (b) reliability (c) content adequacy.

t_0: The timeout threshold for website access without discovering product(s).

t_0': the time period before the current time without adding any product to the basket.

t_p: is the length of product consultation time.

t_p': is a time spent before purchase that must not be greater than t_0''.

For each user, the algorithm finds the time difference between the website access and abandonment. If this difference exceeds t_0, it assigns a "session visit" S_v to the user.

If t_p exceeds t_0', consultation sessions S_c is assigned to the user who consults product(s) without any placement to online basket. The algorithm affects the session basket S_b to the user who adds items to his online shopping cart, but exits without completing the purchase. Session of purchase S_P has an end when user pays consulted product(s).

3.2 Experimental Settings

The aim is to uncover characteristic description of satisfaction classes of clients comparing the proposed decision tree induction algorithm with the proposed multi criteria classification method. We choose reviews and click stream data from the top one B2C platform in China "TMALL". Clients are divided in three decision classes C1: Very Satisfied, C2: Moderately Satisfied and C3: Not Satisfied. Classes are deduced from online reviews rates. Corresponding values of evaluation criteria are calculated based on navigational data processing as explained previously. Data in Table 2 is just a small portion of the processed original data (Table 1) used as training data to adjust the weights. It contains 25432915908 records of user- interactions. Data is randomly partitioned into three parts: a training set (say, 60%), a validation set (e.g. 20%), and test data (e.g. 20%).

Table 1. Original tmall data sampling

Used Id	Item Id	User action	V-time	Review
u41	i161	Click	30/09/2014 15:19:00	★★★★★
	i534	Alipay	03/09/2014 15:46:01	
u57	i135	Cart	26/09/2014 20:32:01	★★★
u1641	i109	Alipay	02/09/2014 11:50:52	★★

Table 2. Portion of training, validation and test data

Data	User-Id	Class	Usability					Content adequacy					Reliability					Order fulfillment	Delivery time
			VS	S	M	W	VW	VS	S	M	W	VW	VS	S	M	W	VW		
Training	u41	C1	1	0	0	0	0	0	1	0	0	0	1	0	0	0	0	Available-to-promise	3
	u13	C1	1	0	0	0	0	0	0.5	0.5	0	0	1	0	0	0	0	Available-to-promise	5
	u48	C2	0	0	1	0	0	1	0	0	0	0	1	0	0	0	0	Available-to-promise	4
	u21	C2	0	0	1	0	0	0	1	0	0	0	0	0	0	0.6	0.4	Available-to-promise	3
	u24	C3	0	0	0	1	0	0	0	1	0	0	0	0	1	0	0	Not available-to-promise	12
	u422	C3	0	0	0	0	1	0	0	0	0.5	0.5	1	0	0	0	0	Not available-to-promise	10
	u644	C3	0	0	0	0	1	0	0	1	0	0	0	0	0	1	0	Available-to-promise	6
	u32	C1	1	1	0	0	0	0	1	0	0	0	0	1	0	0	0	Available-to-promise	6
	u39	C2	0	1	0	0	0	0	0	1	0	0	0	1	0	0	0	Available-to-promise	3
Validation	u328	C3	1	0	0	0	0	0	0	0.5	0.5	0	0	0	0	1	0	Available-to-promise	12
	u54	C1	0.5	0.5	0	0	0	1	0	0	0	0	1	0	0	0	0	Available-to-promise	3
	u14	C1	1	0	0	0	0	0	1	0	0	0	0	1	0	0	0	Available-to-promise	4
	u26	C2	0	0	1	0	0	0	1	0	0	0	0	1	0	0	0	Available-to-promise	6
Test	u4	C1	1	0	0	0	0	0	0	1	0	0	1	0	0	0	0	Available-to-promise	3
	u1136	C3	0	0	1	0	0	0	0	0	0.5	0.5	0	0	0	1	0	Not available-to-promise	14
	u31	C2	0	1	0	0	0	0	0	0	0.5	0.5	1	0	0	0	0	Not available-to-promise	10
	u40	C2	0	0	1	0	0	0	0	0.7	0.3	0	1	0	0	0	0	Available-to-promise	8

3.3   Primary Results and Discussion

Results are implemented in JAVA using Eclipse neon.3 where customer interactions data are in CSV Files. The output of the first proposed algorithm of building customer's profiles resembles to an orientation diagram where each leaf is a decision. A path of each leaf can be converted into a production rule IF-THEN. A part of simulation results is summarized in Fig. 5.

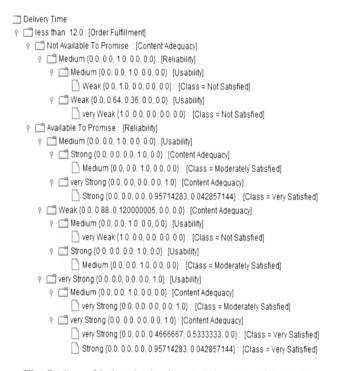

Fig. 5. Part of induced rules from training data of Table 2

Tests show that the algorithm selects appropriate threshold for stopping growth respecting the three types of features and gives good classification rates. The multi criteria classification method have made possible to calculate importance of each criterion to determine different e-customer's satisfaction levels as mentioned in Table 3.

Table 3. Prototypes and criteria scores obtained from training data of Table 2

Categories	E-commerce website criteria				
	Usability	Content adequacy	Reliability	Order fulfillment	Delivery time
Very satisfied	VS 0.975	M 0.763	S 0.727	Available-to-promise	5
Moderately satisfied	M 0.760	M 0.760	S 0.727	Available-to-promise	6
Not satisfied	W 0.785	VW 0.785	W 0.564	Not available-to-promise	9

The principle is to deduce a system of weights of various e-commerce website criteria within each satisfaction class. Results show that usability of the website is fundamental. The content factor was rated second, reinforcing that an e-commerce website has to manage and guarantee data preservation and update content. Moreover, there is a preference for e-commerce websites that have security resources against the payment process. The validation is performed both for the decision tree induction algorithm for classification and multi criteria classification method. Table 4 allows visualization of the performance of these algorithms for comparison in way of which classification model is getting right.

Table 4. Classification methods confusion matrix

Classification model	Predicted			
Proposed decision tree algorithm for classification		C1	C2	C3
	C1	1	1	0
	C2	0	1	0
	C3	0	1	0
Proposed multi criteria classification	C1	2	0	0
	C2	0	1	0
	C3	0	0	1

Results of Table 4 serves as comparative study where the total number of incorrectly classified items present accuracy measure for the methods under comparison. The number of misclassifications reached by decision tree is greater than those achieved by the proposed multi criteria classification model. If we use the minimum error as classification model selection, we would select the proposed multi criteria classification model (Table 5).

Table 5. Misclassification performance comparison

Model	Validation	Test
Proposed decision tree algorithm for classification	50%	—
Proposed multi criteria classification	0%	25%

Based on the above, an estimate of generalization error is 25%, what this means is that if we classify future customers for which only the attributes will be know, not the class labels, we are likely to make incorrect classification about 25%.

4 Conclusion

The proposed approach quantifies the store navigational click stream and online reviews resulting from online customer experience and proposing an adequate decision tree algorithm and a MCDA model under fuzziness which allow creating customer satisfaction groups. We hope this research shows promising results for machine learning techniques adoption and encourages managers to consider whether customer pathways might provide guidance in regards to which e-commerce website's criteria they should prioritize. As perspectives, we will first choose more navigational and behavioral data to be included in user profiles generation process. Second, we assign customers to satisfaction groups based on online reviews filtered by stars, without considering text mining of these reviews. We hope that future work might address these limitations.

References

1. Vercamer, D., Steurtewagen, B., Van den Poel, D., Vermeulen, F.: Predicting consumer load profiles using commercial and open data. IEEE Trans. Power Syst. **31**(5) (2015). https://doi.org/10.1109/tpwrs.2015.2493083
2. Li, R., Li, F., Smith, N.D.: Multi-Resolution load profile clustering for smart metering data. In: 2017 IEEE Power & Energy Society General Meeting, Chicago, IL, p. 1 (2016). https://doi.org/10.1109/pesgm.2017.8273828
3. Wang, Q., Yu, X., Chou, P., Savage, D., Zhang, X.: Power usage spike detection using smart meter data for load profiling. In: 2016 IEEE 25th International Symposium on Industrial Electronics (ISIE), Santa Clara, CA, pp. 732–737 (2016). https://doi.org/10.1109/isie.2016.7744980
4. Lu, Y., Zhang, T., Zeng, Z.: Adaptive weighted fuzzy clustering algorithm for load profiling of smart grid customers. In: 2016 IEEE/CIC International Conference on Communications in China (ICCC), Chengdu, pp. 1–6 (2016). https://doi.org/10.1109/iccchina.2016.7636874
5. Ghuman, M.K., Singh Mann, B.: Profiling customers based on their social risk perception: a cluster analysis approach. https://doi.org/10.1177/0972622518768679. 2018 Indian Institute of Management. SAGE Publications, Lucknow. http://sagepub.in/home.nav. http://journals.sagepub.com/home/met
6. Rai, R., Chettri, P., Chettri, L.: E-commerce and its software (2018). https://doi.org/10.4018/978-1-5225-3646-8.ch007
7. Sriram, A., Rao, A.P.: Security Issues in E-commerce (2018). https://doi.org/10.4018/978-1-5225-3646-8.ch008
8. Soto-Acosta, P., Popa, S., Palacios-Marqués, D.: E-business, organizational innovation and firm performance in manufacturing SMEs: an empirical study in Spain. Technol. Econ. Dev. Econ. **22**(6), 885–904 (2015). https://doi.org/10.3846/20294913.2015.1074126

9. Chuanga, S., Lin, H.: Performance implications of information-value offering in e-service systems: examining the resource-based perspective and innovation strategy. J. Strat. Inf. Syst. **26**(1), 22–38 (2017). https://doi.org/10.1016/j.jsis.2016.09.001

10. Sun, T., Watanabe, W.C.: The study of critical success factors of cross-border e-commerce freight forwarder from China to Thailand. In: 2017 IEEE International Conference on Industrial Engineering and Engineering Management (IEEM), Singapore, pp. 1848–1852. https://doi.org/10.1109/ieem.2017.8290211

11. Adebanjo, D.: Classifying and selecting e-CRM applications: an analysis-based proposal. Manag. Decis. **41**(6), 570–577 (2015). https://doi.org/10.1108/00251740310491517

12. Kourtesopoulou, A., Kehagias, J., Papaioannou, A.: Evaluation of E-service quality in the hotel sector: a systematic literature review. In: Springer Proceedings in Business and Economics (2018). https://doi.org/10.1007/978-3-319-67603-6_13

13. Fanga, J., Lia, J., Prybutok, V.R.: Posting-related attributes driving differential engagement behaviors in online travel communities. Telemat. Inform. **35**(5), 1263–1276 (2018). https://doi.org/10.1016/j.tele.2018.02.008

14. Rouyendegh, B.D., Topuz, K., Oztekin, A.D.: An AHP-IFT integrated model for performance evaluation of E-commerce web sites. Inf. Syst. Front. (2018). https://doi.org/10.1007/s10796-018-9825-z

15. Wu, C., Tsai, S.: Using DEMATEL-Based ANP model to measure the successful factors of E-commerce. J. Glob. Inf. Manag. **26**(1), 120–135 (2018). https://doi.org/10.4018/JGIM.2018010107

16. Qin, J., Liu, X., Pedrycz, W.: A multiple attribute interval type-2 fuzzy group decision making and its application to supplier selection with extended LINMAP method. Soft. Comput. **21**, 3207 (2017). https://doi.org/10.1007/s00500-015-2004-y

17. Zaim, H., Ramdani, M., Haddi, A.: Splitting method for decision tree based on similarity with mixed fuzzy categorical and numeric attributes. In: Tabii, Y., Lazaar, M., Al Achhab, M., Enneya, N. (eds.) Big Data, Cloud and Applications, BDCA 2018. Communications in Computer and Information Science, vol. 872. Springer, Cham (2018)

18. Zaim, H., Ramdani, M., Haddi, A.: Multi-criteria analysis approach based on consumer satisfaction to rank B2C E-commerce websites. In: 11th International Conference on Intelligent Systems: Theories and Applications, SITA 2016 (2016). https://doi.org/10.1109/sita.2016.7772260

19. Zaim, H., Ramdani, M., Haddi, A.: Fuzzy-based mining framework of browsing behavior to enhance E-commerce website performance: case study from Kelkoo.com. In: 12th International Conference on Intelligent Systems: Theories and Applications, SITA 2018 (2018, in press)

Search Convenience and Access Convenience: The Difference Between Website Shopping and Mobile Shopping

Ibrahim Almarashdeh[1](✉), Kamal Eldin Eldaw[1], Mutasem AlSmadi[1],
Usama Badawi[1], Firas Haddad[1], Osama Ahmed Abdelkader[1],
Ghaith Jaradat[2], Ayman Alkhaldi[3], and Yousef Qawqzeh[4]

[1] College of Applied Studies and Community Service,
Imam Abdurrahman Bin Faisal University, Al-Dammam, Saudi Arabia
`ibramars@gmail.com, iaalmarashdeh@iau.edu.sa`
[2] Department of Computer Science, Faculty of Computer Science and
Information Technology, Jerash University, Jerash 26150-311, Jordan
[3] Management Information Systems Department,
Hail University, Hail, Saudi Arabia
[4] College of Science – Zulfi, Majmaah University,
Al Zulfi, Kingdom of Saudi Arabia

Abstract. Now a day we use online platforms for everything in our daily life. Using shopping online via mobile application and shopping via online website become a typical procedure in our daily life. People like to do shopping online, but we still did not know the different between shopping via mobile apps and website adoption in term of access convenience and search convenience. This study aimed to investigate the different between shopping via mobile apps and website adoption in term of access convenience and search convenience. This study used a sample of 143 participant to measure the adoption of shopping online via both website and mobile shopping. The collected data analyzed using SPSS. The results show that mobile apps is more adopted in term of accessibility but website shopping is more adopted in term searchability. This study did not find significant difference it term of total adoption of both website and mobile shopping. The future researchers could focus in measuring the difference in term of usability and security which we assume could bring a valid result.

Keywords: Access convenience · Search convenience ·
Behavioural intention · Online shopping

1 Introduction

The rise of mobile apps and smartphones has changed the way we live, purchase, communicate, and the way we do our business. The number of internet users has been increased up to 1,052% from the year 2000 to 2018 (IWS 2018). Consumers increasingly use various Internet-enabled devices for online shopping (Mosteller et al. 2014). Online shopping enables consumers to purchase products and services at any point of time and wherever they are located. Online shopping allows consumers to save

© Springer Nature Switzerland AG 2020
A. M. Madureira et al. (Eds.): SoCPaR 2018, AISC 942, pp. 33–42, 2020.
https://doi.org/10.1007/978-3-030-17065-3_4

money, effort, and time when purchasing products (Al-Debei et al. 2015). The emergence of e-commerce has led to the establishment of a number of online purchasing portals both as e-commerce as well as m-commerce ventures. Consumers have tuned to sites such as flipkart.com, amazon.com, ebay.com, jabong.com, myntra.com and many more for their discounts and shopping convenience (Dey et al. 2015). However, it is important to identify the reasons that customers choose to visit an online store (Pappas et al. 2017).

Previous research suggested that it is important for managers and future researchers to investigates outcomes of using an app, emphasizing the relevance of the identified dimensions (Schmitz et al. 2016). To succeed in the rapidly growing and highly competitive e-commerce environment, it is important to understand the continued usage behaviour of online shopping customers as they relate to enhancing customer conversion and retention (Lin et al. 2018). Although, there is generally a lack of research on the implementation of smartphone apps in the service delivery process (Schmitz et al. 2016).

Continued usage and adoption of online shopping might be measured by how convenience is costumer with online shopping. Even though, convenience has received relatively little attention in marketing literature, and efforts to develop a valid and comprehensive measure of it have been limited (Seiders et al. 2007). As discussed above, the internet users increased rabidly, and the online shopping. On the overhand, we can see that the online market now is huge, and people start ordering everything online including daily grocery needs. Since the usage is increased and the marketing and online business became more popular, we still need an answer for a question of what costumer like to adopt more: shopping from online website or shopping from mobile apps? From here, this study aims to understand the costumer's usage behaviour of website and mobile apps shopping by measuring the different between website shopping and mobile shopping in term of customer service convenience. This study aims to make a comparison in term of user adoption of website and mobile shopping.

2 Related Work and Research Model

Based on the literature review, there is a need to investigate the changes in consumers' attitudes toward using online shopping over time (Al-Debei et al. 2015). There are many researchers interested in understanding the effects of convenience on consumer behaviour, and recent empirical studies indicate that convenience influences critical marketing consequences, including customer evaluation and purchase behaviour (Seiders et al. 2007). Consistent representation and measurement is especially germane in service contexts, where convenience is difficult to standardize and deliver.

The Technology acceptance model (TAM) has been used commonly for several years to measure the adoption of new technology (Almarashdeh and Alsmadi 2016; Almarashdeh and Alsmadi 2017; Almarashdeh et al. 2010; Almarashdeh et al. 2011; Davis et al. 1989; Shareef et al. 2016; Saxena 2017). The TAM aims to predict the user acceptance based on tow factors; perceived ease of use and perceived usefulness (Davis et al. 1989; Almarashdeh 2016). Furthermore, Unified Theory of Acceptance and Use of Technology (UTAUT) aimed to measure the user acceptance based on the user expectation, social influence and facilitating condition (Almarashdeh 2016; Venkatesh and Zhang 2010).

This study adopted two major dimensions from Jiang et al. (2013). Access convenience and search convenience are the first factors that the previous research measured to predict the adoption of online shopping. This dimension has turned out to be the foremost driver of overall online shopping convenience. Online consumers have the advantage of shopping at any time and are able to make multiple economies of time. They can also purchase products from such locations as home and office, rather than at physical stores. These two types of flexibility – time and place – in turn provide psychological benefits by avoiding crowds, reducing waiting time, and expending less effort in traveling to physical stores. Consumers enjoy the benefits of accessibility to products, brands, and stores that are not available in the location where they reside or work. Accessing product over the internet associated potential issues categorised to Availability of products and brands, time flexibility, space flexibility, accessibility of web sites and energy used (Jiang et al. 2013).

Theoretically, the search convenience is measuring how online customers can research products and compare costs without physically visiting multiple locations to find their desired products. According to Jiang et al. (2013), consumers regard search inconvenience as a major obstacle to efficient and convenient online shopping. All the product search over the internet associated potential issues categorised to product classification, download speed, search function, and web site design (Jiang et al. 2013).

Empirical studies indicate that convenience influences a variety of consequences, including consumers' behavioural intentions (Jiang et al. 2013; Seiders et al. 2007; Szymanski and Hise 2000). In term of mobile shopping, Jiang et al. (2013), identifies five key convenience dimensions of online shopping including transaction, access, evaluation, search, and possession/post-purchase convenience, as convenience has been one of the principal motivations underlying customer inclinations to adopt online shopping. According to Seiders et al. (2007), the service convenience is related positively to behavioural intentions to use the services (Seiders et al. 2007). Based on the literature review there are limited studies offers an in-depth, systematic studies related into dimensions of online shopping convenience and their specific components of each dimension (Jiang et al. 2013; Colwell et al. 2008; Bednarz and Ponder 2010; Palacios 2016; Nuryakin and Farida 2016).

In e-commerce, service convenience dimension has turned out to be the foremost driver of overall online shopping convenience. Online consumers have the advantage of shopping at any time and are able to make multiple economies of time. They can also purchase products from such locations as home and office, rather than at physical stores. These two types of flexibility – time and place – in turn provide psychological benefits by avoiding crowds, reducing waiting time, and expending less effort in traveling to physical stores. Consumers enjoy the benefits of accessibility to products, brands, and stores that are not available in the location where they reside or work. Accessing product over the internet associated potential issues categorised to Availability of products and brands, time flexibility, space flexibility, accessibility of web sites and energy used (Jiang et al. 2013). Behavioural intention is the main predictor of the service adoption, it can predict the future behaviour of the users to reuse the service again in the future and the willingness to recommended to others (Almarashdeh and

Alsmadi 2017). This study adopted two major dimensions from Jiang et al. (2013). Access convenience and search convenience are the first factors that the previous research measured to predict the adoption of online shopping. Table 1 below conclude the suggested hypothesis to achieve the aim of this study.

Table 1. Suggested hypotheses

No	Hypothesis
H1	Access convenience has significant effect on behavioural intention to shop via mobile apps
H2	Search convenience has significant effect on behavioural intention to shop via mobile apps
H3	Access convenience has significant effect on behavioural intention to shop via online website
H4	Search convenience has significant effect on behavioural intention to shop via online website

3   Research Methodology

Data for this study were collected from 130 online shopping users in Saudi Arabia. The survey data was obtained online by using google forms. Scales from prior research were adjusted to the online shopping context. All items were measured on a five-point Likert scale ranging from 1 = strongly disagree to 5 = strongly agree. The collected date verified by two professors from the department of MIS (Management information system). The survey data test using SPSS and the initial reliability test for all items illustrate the collect data acceptable level of reliability .871 which is higher that the recommended level above .70 (Sekaran 2003). The Cronbach's Alpha for the measuring the access convenience for shopping via website (ACW) is .763, search convenience for shopping via website (SCW) .711, behavioural intention to use shopping via website (BIW) .770. he Cronbach's Alpha for the measuring the access convenience for shopping via mobile apps (ACM) is .809, search convenience for shopping via mobile apps (SCM) .713, behavioural intention to use shopping via mobile apps (BIM) .871. these results illustrated that all items used in the study have stable consistency.

3.1   Sample Characteristic

From the total 143 participants, the highest participation of the study comes from the age 35–39 years old which represent 38.5% of the sample size. The smallest participation comes from the age group above 50 years old which represent 3.5% from the total sample size. This sample represent a 53.1 male and 46.9 females. The income of this sample is medium from 1000$ to 3000$ per month. Sample of the study illustrated in Table 2.

Table 2. Sample characteristics

		N	%
Age	18–24	8	5.6
	25–34	41	28.7
	35–39	55	38.5
	40–49	34	23.8
	Above 50	5	3.5
Gender	Male	76	53.1
	Female	67	46.9
Income	Less than 1000$	38	26.6
	1000$–3000$	47	32.9
	More 3000$–less 5000$	37	25.9
	More than 5000$	21	14.7
Total		143	100

3.2 Descriptive Statistics

Table 3 below illustrated the mean and standard deviation of the items used in this study to measure the adoption of shopping via website and mobile shopping.

Table 3. Exploring the items used in this study

Constructs	Code	Mean	SD	Constructs	Code	Mean	SD
Access convenience via shopping website	ACW	3.82	.892	Access convenience via mobile apps	ACM	4.11	.769
When I use online shopping via website, I could shop anytime I wanted	ACW1	3.74	.998	When I use online shopping via mobile apps, I could shop anytime I wanted	ACM1	4.01	.884
The web site is always accessible	ACW2	3.87	.788	The mobile apps is always accessible	ACM2	4.19	.927
When I use online shopping via website, I could order products wherever I am	ACW3	3.85	.841	When I use online shopping via mobile apps, I could order products wherever I am	ACM3	4.14	.885

(*continued*)

Table 3. (*continued*)

Constructs	Code	Mean	SD	Constructs	Code	Mean	SD
Search convenience via shopping website	SCW	3.84	.658	Search convenience via mobile apps	SCM	4.17	.752
The web site is user-friendly for making purchases	SCW1	3.73	.858	The mobile shopping apps is user-friendly for making purchases	SCM1	4.20	.939
The web site is easy to understand and navigate	SCW2	3.74	.967	The mobile shopping apps is easy to understand and navigate	SCM2	4.29	.924
The web site is very attractive	SCW3	3.99	.746	The mobile shopping apps is very attractive	SCM3	4.17	.661
When I use online shopping via website, I am able to find desired products quickly	SCW4	4.03	.921	When I use online shopping via mobile apps, I am able to find desired products quickly	SCM4	4.36	.835
When I use online shopping via website, the product classification is intuitive and easy to follow	SCW5	3.77	.979	When I use online shopping via mobile apps, the product classification is intuitive and easy to follow	SCM5	4.09	.941
When I use online shopping via website, I am able to find the same product using a variety of online search options	SCW6		.922	When I use online shopping via mobile apps, I am able to find the same product using a variety of online search options	SCM6	4.13	.777

(*continued*)

Table 3. (*continued*)

Constructs	Code	Mean	SD	Constructs	Code	Mean	SD
Behavioural intentions to shop using website	BIW	4.17	.763	Behavioural intentions to shop using mobile apps	BIM	4.06	.959
I will continue to shop online using website shopping	BIW1	4.07	.819	I will continue to shop online using mobile shopping	BIM1	4.14	.898
I encourage others to shop online using website	BIW2	4.25	.953	I encourage others to shop online using mobile apps	BIM2	4.03	.797
I will use shopping website to do shopping more often in the future	BIW3	4.20	.988	I will use mobile shopping more often in the future	BIM3	4.02	.858

From the Table 3 above, we illustrate that the mean for access convenience for shopping via website is 3.82 while the access convenience for shopping via mobile application is 4.20. Furthermore, the mean value for search convenience via website is 3.84 and for mobile application shopping is 4.11. In the end we can illustrate the intention for shopping via website, based on the mean value of 4.17 is higher than costumer's intention to shop using mobile application with mean value of 4.06.

4   Discussion

4.1   Results of Regression Test

The regression test is used in this study to measure the effect of independents variables on the dependent variables. Table 4 illustrate that all research hypotheses were accepted. As shown in the Table 4 below, the effect access convenience on behavioural intention to use mobile apps is 3.59 which is higher than the access convenience for shopping via online website (t = 2.73). In the other hand, the effect of search convenience via website shopping (t = 4.81) is higher than search convenience via mobile apps (t = 2.10).

Table 4. Regression test for hypotheses

No	Hypothesis	t	Sig.	Indicator
Ha	Access convenience has significant effect on behavioural intention to shop via mobile apps	3.59	.000	Accepted
H2	Search convenience has significant effect on behavioural intention to shop via mobile apps	2.10	.037	Accepted
H3	Access convenience has significant effect on behavioural intention to shop via online website	2.73	.007	Accepted
H4	Search convenience has significant effect on behavioural intention to shop via online website	4.81	.000	Accepted

4.2   Results of Paired Samples Test

The Paired Samples Test is used in this study to measure the different between both dependent variables behavioural intention to use online shopping via website (BIW) and behavioural intention to use online shopping via mobile apps (BIM).

Table 5. One-sample test

	Paired differences					t	df	Sig. (2-tailed)
	Mean	Std. dev	Std. error mean	95% confidence interval of the difference				
				Lower	Upper			
BIW - BIM	.1072	.97592	.08161	−.05410	.26856	1.314	142	.191

Table 5 above conclude the Paired Samples Test which was run to determine whether there are a significant different in mean between both BIW and BIM. The mean for BIM is 4.06 and the mean for BIW is 4.17. The Confidence Interval (CI) shows the lowers value of −.0541 and the upper CI value is .268. The average difference between the BIW and BIM is .107. The paired T test statistic (t) value is 1.314 which statistically not significant with p value of .191 which less than 0.05.

In summary, this study aims to investigate the different between shopping online via mobile apps and website shopping. The result of this study conclude that costumers find that using mobile apps for online shopping is more acceptable (t value 3.59) than shopping using online website (t value 2.73). But in the other hand, costumers find that the search convenience (t value 4.18) via shopping website is more acceptable than searching via mobile application (t value 2.10). Furthermore, there is no significant different between shopping via mobile apps and website shopping. The implication of this study helps the managers to increase accessibility of shopping website as costumers to make it as efficient as shopping via mobile application. Meanwhile, the mangers should consider the search convenience of mobile application still not sufficient as shopping via online website. This study contributes the field of marketing and application developers to enhance the searchability and accessibility of the e-commerce platforms. The future researchers could investigate other factors that might influence the adoption of shopping online or can make a different between shopping via mobile applications or via shopping website. Factors such as security and usability could bring valid results.

5 Conclusion

The purpose of this paper is to investigate the different between shopping via mobile application and website shopping in term of adoption. This study adopted two main factors namely access convenience and search convenience as a predictors of online shopping acceptance. The data collected from 143 participants and analysed using SPSS. The results of data analysis illustrated that the mobile application acceptance is more effect by the access convenience than search convenience, while using website to shop is more effected by search convenience than access convenience. This study did not find significant difference between the adoption of shopping via website or mobile application. This study limited to two factors only, future researchers could investigate the different between both mobile apps and website shopping in term of security and usability.

References

Al-Debei, M.M., Akroush, M.N., Ashouri, M.I.: Consumer attitudes towards online shopping: the effects of trust, perceived benefits, and perceived web quality. Internet Res. **25**(5), 707–733 (2015)

Almarashdeh, I.: Sharing instructors experience of learning management system: a technology perspective of user satisfaction in distance learning course. Comput. Hum. Behav. **63**, 249–255 (2016). https://doi.org/10.1016/j.chb.2016.05.013

Almarashdeh, I., Alsmadi, M.: Investigating the acceptance of technology in distance learning program. In: International Conference on Information Science and Communications Technologies (ICISCT), Tashkent Uzbekistan, pp. 1–5. IEEE (2016)

Almarashdeh, I., Alsmadi, M.K.: How to make them use it? Citizens acceptance of M-government. Appl. Comput. Inform. **13**(2), 194–199 (2017). https://doi.org/10.1016/j.aci.2017.04.001

Almarashdeh, I., Sahari, N., Zin, N.M., Alsmadi, M.: Instructors acceptance of distance learning management system. In: International Symposium on Information Technology 2010 (ITSim 2010), Kuala Lumpur, vol. 1, pp. 1–6. IEEE (2010)

Almarashdeh, I.A., Sahari, N., Zin, N.A.M., Alsmadi, M.: Acceptance of learning management system: a comparison between distance learners and instructors. Adv. Inf. Sci. Serv. Sci. **3**(5), 1–9 (2011)

Bednarz, M., Ponder, N.: Perceptions of retail convenience for in-store and online shoppers. Mark. Manag. J. **20**(1), 49–65 (2010)

Colwell, S.R., Aung, M., Kanetkar, V., Holden, A.L.: Toward a measure of service convenience: multiple-item scale development and empirical test. J. Serv. Mark. **22**(2), 160–169 (2008)

Davis, F., Bagozzi, R., Warshaw, P.: User acceptance of computer technology: a comparison of two theoretical models. Manag. Sci. **35**(8), 982–1003 (1989)

Dey, R., D'souza, R., D'souza, J.: Relationship between Brick and Mortar stores and m-commerce facilitates customers towards shopping online. Educ. Quest **6**(3), 189 (2015)

IWS: Internet world stats: usage and population statistics (2018). Retrieved from Internet World Stats: http://www.internetworldstats.com/stats.htm

Jiang, L., Yang, Z., Jun, M.: Measuring consumer perceptions of online shopping convenience. J. Serv. Manag. **24**(2), 191–214 (2013)

Lin, C., Wei, Y.-H., Lekhawipat, W.: Time effect of disconfirmation on online shopping. Behav. Inf. Technol. **37**(1), 87–101 (2018)

Mosteller, J., Donthu, N., Eroglu, S.: The fluent online shopping experience. J. Bus. Res. **67**(11), 2486–2493 (2014)

Nuryakin, N., Farida, N.: Effects of convenience online shopping and satisfaction on repeat-purchase intention among students of higher institutions in Indonesia. J. Internet Bank. Commer. **21**(2) (2016)

Palacios, S.: Examining the Effects of Online Shopping Convenience, Perceived Value, and Trust on Customer Loyalty. New Mexico State University, Las Cruces (2016)

Pappas, I.O., Kourouthanassis, P.E., Giannakos, M.N., Lekakos, G.: The interplay of online shopping motivations and experiential factors on personalized e-commerce: a complexity theory approach. Telematics Inform. **34**(5), 730–742 (2017)

Saxena, S.: Enhancing ICT infrastructure in public services: factors influencing mobile government (m-government) adoption in India. The Bottom Line (just-accepted), 00-00 (2017)

Schmitz, C., Bartsch, S., Meyer, A.: Mobile app usage and its implications for service management-empirical findings from german public transport. Procedia-Soc. Behav. Sci. **224**, 230–237 (2016)

Seiders, K., Voss, G.B., Godfrey, A.L., Grewal, D.: SERVCON: development and validation of a multidimensional service convenience scale. J. Acad. Mark. Sci. **35**(1), 144–156 (2007)

Sekaran, U.: Research Methods for Business: A Skill-Building Approach, 4th edn. Wiley, New York (2003)

Shareef, M.A., Dwivedi, Y.K., Laumer, S., Archer, N.: Citizens' adoption behavior of mobile government (mGov): a cross-cultural study. Inf. Syst. Manag. **33**(3), 268–283 (2016). https://doi.org/10.1080/10580530.2016.1188573

Szymanski, D.M., Hise, R.T.: E-satisfaction: an initial examination. J. Retail. **76**(3), 309–322 (2000)

Venkatesh, V., Zhang, X.: Unified theory of acceptance and use of technology: U.S. vs. China. J. Glob. Inf. Technol. Manag. **13**(1), 5–27 (2010)

Automatic Classification and Segmentation of Low-Grade Gliomas in Magnetic Resonance Imaging

Marta Barbosa[1], Pedro Moreira[1], Rogério Ribeiro[1], and Luis Coelho[1,2(✉)]

[1] Polytechnic Institute of Porto, ISEP-IPP, 4200-072 Porto, Portugal
{1160374,1160357,1160358,lfc}@isep.ipp.pt
[2] CIETI - Center for Innovation in Engineering and Industrial Technology,
4200-072 Porto, Portugal

Abstract. In this article a new methodology is proposed to tackle the problem of automatic segmentation of low-grade gliomas. The possibility of knowing the limits of this type of tumor is crucial for effectively characterizing the neoplasm, enabling, in certain cases, to obtain useful information about how to treat the patient in a more effective way. Using a database of magnetic resonance images, containing several occurrences of this type of tumors, and through a carefully designed image processing pipeline, the purpose of this work is to accurately locate, isolate and thus facilitate the classification of the pathology. The proposed methodology, described in detail, was able to achieve an accuracy of 87.5% for a binary classification task. The quality of the identified regions had an accuracy of 81.6%. These are promising results that may point the effectiveness of the approach. The low contrast of the images, as a result of the acquisition process, and the detection of very small tumors are still challenges that bring motivation to further pursue additional results.

Keywords: Low-grade glioma · Image segmentation · MRI

1 Introduction

1.1 Framework

Cancer is the major cause of death in economically developed countries and the second major in developing countries [1]. The tumors of the central nervous system are defined by their origin cell and by their histopathological characteristics, which predict their behavior. Gliomas are neuroepithelial tumors originating from glial cells supporting the central nervous system. They are the most common type of primary brain tumor, comprising about 50% of malignant brain tumors in adults. They are responsible for 189.000 new cases and 142.000 deaths annually (1.7% of new cancer cases and 2.1% of cancer deaths), presenting themselves as one of the deadliest cancers [1, 2]. Since 2016 World Health Organization (WHO) divides the glioblastomas into isocitrate dehydrogenase (IDH)-wildtype (about 90% of cases), which normally corresponds to primary treatment and predominates in patients over 55 years of age, and in

© Springer Nature Switzerland AG 2020
A. M. Madureira et al. (Eds.): SoCPaR 2018, AISC 942, pp. 43–50, 2020.
https://doi.org/10.1007/978-3-030-17065-3_5

IDH-Mutant (about 10% of cases), which corresponds to the secondary glioblastoma that arises in younger patients. They may also be classified in "not otherwise specified" (NOS) glioblastoma for those tumors for which the IDH assessment cannot be carried out [2].

The classification of tumors can encompass a wide range of dimensions: (a) the pathogenic dimension, that provides a dynamic classification, based on both the phenotype and genotype; (b) the prognosis dimension, which groups tumors that share similar prognostic markers; and (c) the patient dimension perspective, that guides therapies (conventional or directed) to biological and genetically similar entities [2]. The WHO classification system is based on histopathological characteristics, such as atypia cytological, anaplasia, mitotic activity, microvascular proliferation or necrosis, which categorizes gliomas from Grade 1 (lowest degree) to Grade 4 (highest degree) [3]. Low-grade gliomas, particularly, are a diverse group of primary brain tumors that usually arise in healthy, younger patients. Treatment options are individualized, based on the location of the tumor, histology, molecular profile and patient characteristics and include observation, surgery, radiation, chemotherapy or a combined approach [3].

Screening tests separate people who are apparently healthy but have a disease or a risk factor for a disease, from those that do not have disease. An ideal screening test should take only a few minutes to perform, and requires minimal patient preparation, causing minimal disruption, not relying on special scheduling, and is low cost and easily applicable. The results of a screening test should be valid, reliable, and reproducible. In summary, the validity of an instrument for screening or diagnosis is measured by its ability to do what it proposes, that is, to adequately categorize individuals with symptoms of the disease as a positive test and those subjects without symptoms of the disease as negative test. The relationships between disease and non-disease and positive test, or negative test are expressed through sensitivity, specificity and predictive values of the tests. A good screening test should be highly sensitive not to lose cases of disease present in the population tested, as well as high specificity in order to reduce the number of people with false positive results who need further investigation. Thus, one of the major challenges in the treatment of neuroepithelial images is the significant intra-observer variability, as well as an interinstitutional heterogeneity of the parameters, depending on the magnetic field and the types of sequences obtained [4, 5]. The introduction of automatic objective evaluation tools, that could consistently provide fast and unequivocal results, could be a major help on improving the response of the screening processes.

1.2   State of the Art

Currently, the standard procedure for definitive diagnosis of a glioma is the histological examination of tissue. Computed tomography (TC) and nuclear magnetic resonance (NMR) are the supplementary procedures used to define the stage of the disease [6]. In 2012, over 66.000 primary tumors of the central nervous system were diagnosed in the USA, of which about 30% were gliomas. In adults (20–34 years), the gliomas represent 32% of all central nervous system primary tumors, 17% of which are astrocytomas and 28% of these are glioblastomas. Incidence rates of this type of tumors are more than two times higher in Caucasian than in African descent; and are 40% more common in men than in women. The reason for this racial and sexual discrepancy is uncertain. Several environmental risk factors have been examined to highlight a link between environmental exposures and an increased risk of brain tumor formation. The only factor that has been correlated with an increased risk of secondary tumors is exposure to radiation doses. Other environmental exhibitions have been investigated, with no convincing evidence to support their role in the formation of brain tumors [7].

Low-grade Gliomas are heterogeneous tumors that present poorly defined margins and generally do not present a contrast enhancement after gadolinium injection. When the highlight occurs, which happens in between 10 to 50% of the cases, depending on the series, is moderate and only in a small portion of the tumor [6]. Being a relatively common oncological pathology, there is a lot of effort put in developing techniques and software capable of sensing and targeting gliomas from medical images, which allows to analyze the progression and extension of the glioma, providing valuable information about the best approach to treatment. The first step is always to get cranial images of the patient. Low-Grade Gliomas are notoriously complex to be viewed in images obtained by magnetic resonance (MR), due to its late anaplastic transformation. With the purpose to obtain imagens with sufficient quality, it's common to associate the MR with spectroscopy and with dynamic perfusion [6].

After obtaining the required medical images, it is possible to use several software programs that allow to identify and segment the oncological structure. One of the most prominent applications for this type of study is the GLioma Image SegmenTation and Registration (GLISTR) [8]. This software package uses a semi-automated approach to label and map both healthy brain tissue and the tumor also allowing to estimate a growth model for each tumor. Additionally it can estimate 10 distinct tissue types: cerebrospinal fluid, gray matter, white matte, vessel, edema, necrosis, enhancing tumor, non-enhancing tumor, cerebellum, and background [9].

GLISTRBoost, a newer software based on GLISTR, is a hybrid generative-discriminative method developed for brain glioma segmentation and atlas registration in multimodal MRI, while incorporating tumor growth modelling via reaction-diffusion-advection. Requires images of prior patients with this pathology at different stages of development in memory to appropriately work and estimate a correct diagnosis. This software, in its first step makes a generative approximation of a segmentation-recognition set based on an expectation-maximization framework, allowing to obtain a growth model of the glioma, and allows to segment, label and map medical images (in a similar way as GLISTR); after this step, GLISTRBoost makes a descriptive classification of the pathology, estimating the phase of the pathology, based on the images in memory; finally, the software refines and finalizes the tumor segmentation using a Bayesian strategy. Compared to GLISTR, GLISTRboost can shape and label tissues more effectively and consequently obtain better diagnosis for each case. GLISTRboost, due to its complexity, is not yet incorporated in brain-CaPTk (later explained) but it is available for public use through the inline Image Processing Portal of Centre for Biomedical Image Computing and Analytics (CBICA), which provides access to performance computing resources [9–11]. Despite these possibilities, GLISTRBoost requires that all input MRI volumes (i.e. T1, T1-Gad, T2, T2-FLAIR) must be skull-stripped, co-registered and in the LPS coordinate system, for the method to produce meaningful results.

The segmentation of brain structures based on the Geodesic Distance Transform (GDT), a patient- based method fully incorporated in brain-CaPTk, is highly adaptable to several imaging modalities. Its "simplistic" nature supports its potential applicability in ample routine clinical settings for delineation of anatomical regions [9].

Brain-CaPTk is a dynamically growing software platform and can consider any anatomical site and image type, it currently focuses on analysing multimodal magnetic resonance imaging (MRI). Furthermore, Brain-CaPTk also supports visualization of diffusion images derivative measurements such as the tensor's apparent diffusion coefficient, axial diffusivity, radial diffusivity and fraction anisotropy, as well as parametric maps extracted of relative cerebral blood volume, peak height and percentage signal recovery [9]. This tool provides two segmentation methods, namely GLISTRboost and GDT, but both require initialization. This is an important limitation because it implies a semi-automated process that represent a greater time to obtain results and prevents the possibility of batch processing multiple images.

1.3 Motivation and Objectives

Currently, the diagnosis of these tumors is based on imaging studies and, for a more accurate diagnosis, tumor tissue is required for a histological analysis. This tissue is obtained through the biopsy or resection which poses a risk when the gliomas are in difficult-to-reach areas. Apart from the difficulty in tissue collection, these zones may present resistance to radio and chemotherapy and it may not be possible to perform a removal surgery. For these reasons, it is extremely important to find new strategies that allow for an early diagnosis, infallible treatment and more effective monitoring. The motivation behind this project is the necessity to develop an efficient image analysis tool, that is able to receive the necessary cranial images, and thus provide immediate information about the position and extent of the glioma thus allowing a reduction of diagnose time, which consequently leads to a greater likelihood of being able to make an appropriate treatment for each situation. In this paper we propose a new methodology that can efficiently provide solutions for the described problem and conditions.

2 Low-Grade Glioma Classification and Segmentation

2.1 Materials and Tools

The main resource that was used to support the technical developments was the TCIA database of carcinogenic images [12] composed by 17360 MR images from 159 patients. All the images are good resolution and include annotations, in particular, segmentation information. Depending on the position of the subject during the image acquisition process, the images can show several degree of rotation in relation to an imaginary vertical axis.

The main image processing tasks were implemented using the Matlab software and the respective image processing toolbox. Some experiments, for evaluating intermediate processes and results, were also conducted using Python and the open source OpenCV image processing library.

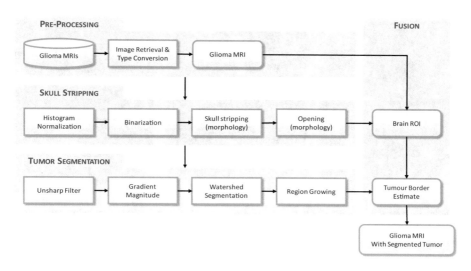

Fig. 1. Segmentation pipeline for the proposed method.

2.2 Methodology

The pipeline that was followed for the implementation of the proposed methodology is depicted in Fig. 1, where we can observe the organization in four stages and the related main steps: (A) Pre-processing: The original images were in DICOM format and required some operations to make them more compatible with the tools that were used. After retrieving the image from the database, we have started by maximizing contrast and performing a conversion to double format for improved precision and simplified processing. The resulting image exhibits an enhanced color distribution and the location of the glioma, when present, can be clearly identified, which can be a valuable information to make critical assessments to the results during the development of the algorithm; (B) Skull-Removal: In this stage we want to eliminate the skull walls section that surrounds the brain. We have started by normalizing the image histogram and then performing a binarization with a 1/3 threshold (on a 0 to 1 range). The bone sections are almost completely removed since their colors exist in a distinct range of the brain. A clear border operation is then performed followed by an image opening. After this we have delimited the region of interest (ROI) containing the area of the brain that will be useful for the following stage; (C) Tumor segmentation: Considering the estimated ROI in the original image we have started by applying an unsharp mask (Fig. 2a) to boost high frequency components, which allowed to emphasize the tumor borders. Then we have calculated the gradient magnitude (Fig. 2b) along the image to prepare the watershed segmentation (Fig. 2c). With this operation we were able to identify and enhance all the zones with the adequate brightness for our purpose, which in the case of our type of images was clearly limited. We have created a list of all the independent zones and have calculated their respective areas. After identifying the highest value, we determine the main properties of the region: center, orientation, highest vertical range and highest vertical range. With this information we apply a specific region growing

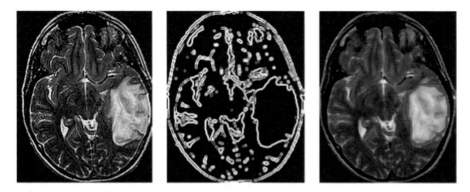

Fig. 2. Intermediate results of the segmentation process: (a) Result after application of unsharp filter (left); (b) Gradient magnitude from original image (middle); (c) Glioma border after watershed over original image (right).

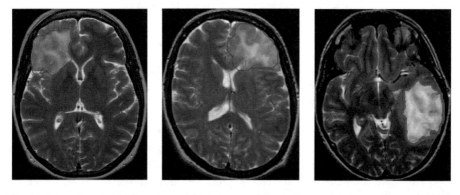

Fig. 3. Examples of results, at the end of the processing pipeline: (a) glioma with fairly regular borders and contrast (left); (b) glioma with not so good contrast (middle); (c) glioma with good contrast and irregular border, based on the Fig. 2(c) (right).

algorithm considering a maximum color variation equal to half of the maximum magnitude determined during gradient calculation. (D) Information fusion and presentation: Finally, using the original image, the brain ROI and the glioma segmentation information we create color masks and adjust the alpha channel in order to combine all information in a final image for presentation.

3 Evaluation and Results

Figure 3 shows examples of results of the application of the proposed methodology. The performance evaluation was made using sets of 32 random images, from different patients, and always considering similar brain sections. The first test that was performed consisted on evaluating if the algorithm was able to correctly detect the

Table 1. Confusion matrix for the detection of gliomas on a given MRI, considering the results of 10 runs of the proposed methodology, for a 95% confidence interval. The top labels are the true conditions while the left labels represent the output of the proposed methodology.

	True	False	Total
Positive	(TP) 50.0%	(FP) 0.0%	50.0%
Negative	(FN) 12.5%	(TN) 37.5%	50.0%
Total	62.5%	37.5%	(ALL) 100%

presence of a glioma on a given image. For this binary classification task, for a 95% confidence interval, we have obtained the results presented in Table 1. Considering widely used metrics for this task, we have obtained: accuracy = (TP + TN)/(ALL) = 87.5%, as an overall discrimination ability considering all the error possibilities; Sensitivity = TP/(TP + FN) = 80%, understood to be the likelihood of a test being positive in the presence of the disease; Specificity = TN/(TN + FP) = 100%, representing the ability to correctly reject healthy patients without a condition.

In a second evaluation phase, it was intended to evaluate the quality of the boundaries identified by the algorithm. By comparing, pixel by pixel on superimposed images, the annotations performed by a human expert with the results provided by the proposed methodology, we have obtained an accuracy value of 81.6%.

4 Conclusions

Low-grade gliomas are a diverse group of primary brain tumors. The prognostic for a successful treatment of the disease requires and accurate identification of the tumor boundaries in order to correctly evaluate its dimension and track its evolution during therapy. A clear definition of the affected tissues is also important when preparing a surgery. In this work we have proposed a new methodology for classification and segmentation of these type of tumors on magnetic resonance images. We have started by presenting the most important tools that are currently used for these purposes and have shown that they can be improved. The proposed methodology was then described and evaluated using widely accepted metrics. The success rate for the classification task was 87.5%, the sensitivity rate was 80% and the specificity rate was 100%. The quality of the identified regions had an accuracy of 81.6%. A good screening test should have high sensitivity to not lose cases of disease present in the tested population, as well as high specificity to reduce the number of people with false-positive results who need further investigation. In this way, we can conclude that the obtained results can support the effectiveness of the proposed method, despite some open challenges. The obtained rates can be improved if the contrast of the original images is enhanced during the acquisition process. It is also envisioned to introduce advances that can bring a higher robustness in the presence of very small tumors.

References

1. Gaspar, B.M.: Biomarcadores em gliomas: conhecimento atual e perspetivas futuras (2016). http://hdl.handle.net/10316/46899
2. Louis, D.N., Perry, A., Reifenberger, G., von Deimling, A., Figarella-Branger, D., Cavenee, W.K., Ohgaki, H., Wiestler, O.D., Kleihues, P., Ellison, D.W.: The 2016 World Health Organization classification of tumors of the central nervous system: a summary. Acta Neuropathol. (Berl.) **131**, 803–820 (2016)
3. Forst, D.A., Nahed, B.V., Loeffler, J.S., Batchelor, T.T.: Low-grade gliomas. Oncologist. **19**, 403–413 (2014)
4. de Goulart, B.N.G., Chiari, B.M.: Testes de rastreamento x testes de diagnóstico: atualidades no contexto da atuação fonoaudiológica. Pró-Fono Rev. Atualização Científica **19**, 223–232 (2007)
5. Bø, H.K., Solheim, O., Jakola, A.S., Kvistad, K.-A., Reinertsen, I., Berntsen, E.M.: Intra-rater variability in low-grade glioma segmentation. J. Neurooncol. **131**, 393–402 (2017)
6. Guillevin, R., Herpe, G., Verdier, M., Guillevin, C.: Low-grade gliomas: the challenges of imaging. Diagn. Interv. Imaging **95**, 957–963 (2014)
7. Ostrom, Q.T., Gittleman, H., Xu, J., Kromer, C., Wolinsky, Y., Kruchko, C., Barnholtz-Sloan, J.S.: CBTRUS statistical report: primary brain and other central nervous system tumors diagnosed in the United States in 2009-2013. Neuro-Oncology **18**, v1–v75 (2016)
8. Gooya, A., Pohl, K.M., Bilello, M., Cirillo, L., Biros, G., Melhem, E.R., Davatzikos, C.: GLISTR: glioma image segmentation and registration. IEEE Trans. Med. Imaging **31**, 1941–1954 (2012)
9. Rathore, S., Bakas, S., Pati, S., Akbari, H., Kalarot, R., Sridharan, P., Rozycki, M., Bergman, M., Tunc, B., Verma, R., Bilello, M., Davatzikos, C.: Brain cancer imaging phenomics toolkit (brain-CaPTk): an interactive platform for quantitative analysis of glioblastoma. In: Brainlesion: Glioma, Multiple Sclerosis, Stroke and Traumatic Brain Injuries, Third International Workshop, BrainLes 2017, Held in Conjunction MICCAI 2017, Quebec City, QC, Canada, 14 September 2017, Revised Selected Papers, vol. 10670, pp. 133–145 (2018)
10. Bakas, S., Zeng, K., Sotiras, A., Rathore, S., Akbari, H., Gaonkar, B., Rozycki, M., Pati, S., Davatzikos, C.: GLISTRboost: combining multimodal MRI segmentation, registration, and biophysical tumor growth modeling with gradient boosting machines for glioma segmentation. In: Brainlesion: Glioma, Multiple Sclerosis, Stroke and Traumatic Brain Injuries, First International Workshop, Brainles 2015, Held in Conjunction with MICCAI 2015, Munich, Germany, 5 October 2015, Revised Selected Papers, vol. 9556, pp. 144–155 (2016)
11. Bakas, S., Akbari, H., Sotiras, A., Bilello, M., Rozycki, M., Kirby, J.S., Freymann, J.B., Farahani, K., Davatzikos, C.: Advancing the cancer genome atlas glioma MRI collections with expert segmentation labels and radiomic features. Sci. Data **4**, 170117 (2017)
12. Akkus, Z., Ali, I., Sedlář, J., Agrawal, J.P., Parney, I.F., Giannini, C., Erickson, B.J.: Predicting deletion of chromosomal arms 1p/19q in low-grade gliomas from MR images using machine intelligence. J. Digit. Imaging **30**, 469–476 (2017)

Enhancing Ensemble Prediction Accuracy of Breast Cancer Survivability and Diabetes Diagnostic Using Optimized EKF-RBFN Trained Prototypes

Vincent Adegoke[(✉)], Daqing Chen, Ebad Banissi, and Safia Barsikzai

Computer Science and Informatics, School of Engineering,
London South Bank University, London, UK
{adegokev, chenq, banisse, barikas}@lsbu.ac.uk

Abstract. We are in a machine learning age where several predictive applications that are life dependent are made by machines and robotic devices that relies on ensemble decision making algorithms. These have attracted many researchers and led to the development of an algorithm that is based on the integration of EKF, RBF networks and AdaBoost as an ensemble model to improve prediction accuracy. Firstly, EKF is used to optimize the slow training speed and improve the efficiency of the RBF network training parameters. Secondly, AdaBoost is applied to generate and combine RBFN-EKF weak predictors to form a strong predictor. Breast cancer survivability and diabetes diagnostic datasets used were obtained from the UCI repository. Results are presented on the proposed model as applied to Breast cancer survivability and Diabetes diagnostic predictive problems. The model outputs an accuracy of 96% when EKF-RBFN is applied as a base classifier compare to 94% when Decision Stump is applied and AdaBoost as an ensemble technique in both examples. The output accuracy of ensemble AdaBoostM1-Random Forest and standalone Random Forest models is 97% in both cases. The study has gone some way towards enhancing our knowledge and improving the prediction accuracy through the amalgamation of EKF, RBFN and AdaBoost algorithms as an ensemble model.

Keywords: AdaBoost · Breast cancer · Diabetes diagnosis · EKF · Ensemble · RBFN · Optimization · RMSE

1 Introduction

Ensemble models play crucial roles in many applications and related devices that are automated with the use of decision control mechanisms. Many ensemble algorithms are essentially iterative, and their results are inconsistent and not as accurate as it should be. Therefore, the need to develop enhanced predictive ensemble models are very important to the acceptability of such devices in the health and other industrial sectors that relies on them. However, the way algorithms are designed and trained plays a major role in machine learning prediction's accuracy and reliability. In the past three decades the need for this have attracted many researchers that led to the development of

© Springer Nature Switzerland AG 2020
A. M. Madureira et al. (Eds.): SoCPaR 2018, AISC 942, pp. 51–65, 2020.
https://doi.org/10.1007/978-3-030-17065-3_6

wide range of approaches and variants of ensemble algorithms. The selection and the diversity of the selected hypotheses also plays important role in prediction accuracy and reliability of ensemble models. However, the potential improvement in ensemble prediction through merging of the existing predictive models to improve classification accuracy has not been fully studied. In general, ensemble method combines several hypotheses (weak learners) to produce a strong classifier instead of the traditional standalone algorithms that are based on a single classifier.

One of the main objectives of ensemble machine learning algorithms as addressed in this research is to build and combine multiple weak learners on the same task to stabilize the prediction accuracy and achieve a better generalization result. For example, the accuracy prediction of breast cancer survival and diabetes diagnosis using data mining techniques based on historical records of patients can save lives by assisting doctors and policy makers in managerial decisions. It can also reduce the overhead cost of healthcare and other public service provisions. The proposed model is an extension of AdaBoost algorithm for forming committee of decision makers. The rationale behind this is that it takes the advantage of AdaBoost's high bias RBFN's (Radial Basis Function Network) noncomplex design. It also has good generalization, strong tolerance to input noise and EKF's (Extended Kalman Filter) quicker convergence during iterations in addressing complex estimation problems.

This paper therefore annexed this problem by proposing a concept that implements the process of integrating EKF, RBFN and AdaBoost algorithms as an ensemble model for binary classification tasks. A substantial additional output of this paper is the creation of working computerised EKF-RBFN-AdaBoost ensemble models. The models were evaluated and used as a computer assisted diagnosis device for early prediction of breast cancer and diabetic diagnostic diseases. The rest of the paper is arranged in the following format: In Sect. 2 we provided an overview background of the problem. In Sect. 3 we present an outline of algorithms that were integrated in the model proposed in this paper. Section 4 covers the results of our investigation and discussion of our findings. Finally, in Sect. 5 we present conclusion of the model we proposed in this study and further work to be carried out in the future.

2 Background and Problem Overview

Ensemble algorithms are essentially iterative, their results are inconsistent and not as accurate as it should be. For example, the application of algorithms in early prediction of breast cancer and diabetes which are two common diseases that affects a lot of peoples requires algorithms with high prediction accuracy and reliability.

Breast cancer is one of the most common causes of cancer related death among women in the world in the past years. In the USA alone in 2015 an estimated 231,840 new cases of invasive breast cancer were diagnosed among women and 60,290 additional cases of in-situ breast cancer (Society American Cancer 2015; Adegoke et al. 2017). Similarly, in the UK over 55,222 women were diagnosed with new cases of the disease in 2014 which amounted to 11, 433 deaths (UK 2018) and the ailment reached 25.2% of women worldwide (Kwon and Lee 2016). The disease is also a looming epidemic in the developing countries where advanced techniques for early detection and treatments are not readily available (Formenti et al. 2012; Adegoke et al. 2017).

Similarly, "Diabetes is a chronic progressive disease that is characterized by elevated levels of blood glucose. Diabetes of all types can lead to complications in many parts of the body and can increase the overall risk of dying prematurely" (WHO 2016). According to the British Heart Foundation "the increasing number of people suffering from the epidemic could trigger a 29% rise in the number of heart attacks and strokes linked to the condition by 2035" (BHF 2018; ITV 2018). Currently, about four million people in the UK have diabetes with condition accounting for 10% of all NHS spending (BBC 2018).

2.1 Related Work

Even though considerable research has been carried out in data mining using different ensemble techniques in predicting probable events based on historical datasets. One of the key challenges is the choice of the base classifier and appropriate loss function that goes with it. The goal of any ensemble algorithm is to minimize error rate to achieve accuracy and improve reliability. Despite the successful research efforts and application of ensemble methods (Adegoke et al. 2017), recent work shows that the problem with prediction accuracy, speedy and computational cost are still puzzling tasks. Therefore, the development of reliable ensemble models that can be applied for efficient medical diagnosis, incidents management and execution of automated technologies that are decision based and in some cases life dependent are highly essential. To address the issue of prediction accuracy/reliability and to extend the applications of ensemble algorithms, we propose a new model that bridges the potentials of RBFN, EKF and AdaBoost algorithms.

2.2 Breast Cancer Survivability Models

Medically, breast cancer can be detected early during screening examinations through mammography or after a woman notices an unusual lump (Society American Cancer 2015) in her breast. Owing to advancement in technology and availability of patient medical records, computer aided diagnosis cancer detection systems have been developed to detect and thus control the spread of the disease (Adegoke et al. 2017). However, such systems rely on pattern recognition algorithms that are used to process and analyse medical information of images obtained from mammograms for diagnostic and decision making (Weedon-Fekjær et al. 2014; Sapate and Talbar 2016).

Different algorithms have also been proposed to extract relevant patterns from patients' breast cancer datasets; for instance Yang et al. (2013) came up with a genetic algorithm that identify the relationship between genotypes that can lead to cancer cases using mathematical analysis. Similarly, McGinley *et al.* (2010) and Adegoke et al. (2017) applied Spiking Neural Networks algorithm as a novel tumour classification method in classifying cancer tumours as either benign or malignant. In another approach (Pak et al. 2015) proposed a breast cancer detection and classification in digital mammography based on Non-Subsampled Contourlet Transform (NSCT) and Super Resolution was proposed to improve the quality of digital mammography images. The authors then applied AdaBoost algorithm to determine the probability of a disease being a benign or malign cancer. Likewise, in breast mass cancer classification (Xie et al. 2015) the authors used computer-aided diagnosis (CAD) system for the processing and diagnosis of breast cancer. In their work, Adegoke *et al.* proposed standalone and ensemble predictive models using AdaBoost as a technique and several base/standalone classifiers. The authors found that the topology and complexity of the algorithms does not necessarily improve the prediction and performance accuracy of the models (Adegoke et al. 2017).

2.3 Diabetes Diagnostic Models

In their study Alghamdi *et al.* using SMOTE and ensemble techniques carried out experimental work applying a number of algorithms to establish and compare their performances in predicting diabetes based data obtained from patients' medical history (Alghamdi et al. 2017). The model comprises of ensemble-based predictive method that uses 13 out of the 62 available classified attributes. The selected attributes for the model depends on clinical importance, multiple linear regression (MLR) and the Information Gain (IG). The authors reported an accuracy of 89% for G1/G2 attributes and accuracy (AUC) of 0.922 for the ensemble method. Similarly, in (Zheng et al. 2017), the authors proposed a framework that identifies type 2 diabetes using patient's medical data. They utilized various classification models that extract features to predict identification of T2DM in datasets. According to the authors, the average results of the framework was 0.98 (UAC) compare with other algorithms at 0.71. To validate whether there is a connection between diabetes mellitus and glaucoma chronic diseases, in their work (Apreutesei et al. 2018) applied a simulation technique constructed using artificial neural networks on clinical observations datasets. According to the authors the model was able to predict an accuracy of 95%. In another study (Barakat et al. 2010) the authors proposed a multi-purpose model for the diagnosis and prediction of diabetes using support vector machines algorithm. The results of the model show a prediction accuracy of 94%, precision of 94%, and sensitivity of 93%.

3 RBFN, EKF and AdaBoost Algorithms

Review on ensemble modelling reveals that there are substantial gaps in the literature where EKF algorithm could be used to train RBFN to form committee of EKF-RBFN model using AdaBoost. The performance of radial basis function network is based on how the network is trained and the training parameters are obtained. Though, EKF have been used for modelling and calibration of dynamic systems such as model-based engine control architecture, ballistic and other space-based projects (Csank and Connolly 2016) because of its performance even when noises are present. Despite the reliability and advanced application of EKF and RBFN and the benefits they offered individually they have not been integrated together with AdaBoost as an ensemble model.

3.1 Radial Basis Function (RBF) Network

RBF network is a special type of MLP (Multi-Layer Perceptron) artificial neural network for non-linear modelling (Nabney 2002). The commonly used activation function for the network is radial basis function, however other functions such as Multiquadric or Thin-plate spline can similarly be applied. The output of RBF network is a linear combination of the radial basis functions of the inputs and neuron parameters that form part of the training process of the network. It has only one hidden layer and can be trained in many ways, unlike the MLP that are typically trained with backpropagation algorithms. The structure of a typical RBF network is shown in Fig. 1. The output of the network can be expressed as in Eq. 1

$$y(x) = \sum_{j=1}^{M} w_j \phi_j + w_k \qquad (1)$$

$$\phi_j = \phi\left(\left\|x - c_j\right\|\right) \qquad (2)$$

where, w_j is the weight of jth centre, ϕ_j are the basis functions and w_k are the bias weights and $\left\|x - C_j\right\|$ as expressed in Eq. 2 is the Gaussian activation function.

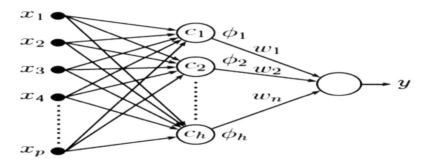

Fig. 1. The topology of a radial basis function network

3.2 Extended Kalman Filter

Theoretically, Kalman filter is a recurrence algorithm with a number of equation that can be used to estimate the state of a process that is based on series of measurements taken over time. The mean square error of the filter is minimised even when the measurements taken contains noises or missing data. The filter has been used in training neural network (Lima et al. 2017; Chernodub 2014). The derivation and application of EKF are widely available (Ribeiro 2004). It uses several measurements observed over time that contains noises and other inaccuracies. The filter consists of a number ensemble equation as illustrated in Fig. 2. The filter also produces estimates of unknown variables that is more precise than those based on a single measurement. It minimizes the estimated covariance error in a Gaussian environment. When the conditions of Gaussian are not met it has been found that Kalman Filter still outperforms other class of linear filters (Merwe et al. 2004). Extended Kalman Filter on the other hand is the nonlinear form of the Kalman filter that linearizes the estimate of the current mean and covariance. Due to EKF's simplicity and accuracy of predictions even in the presence of noise the algorithm has been widely accepted and used among scholars as a standard in the concept of GPS, state estimation and other related complex nonlinear problems (Moreno and Pigazo 2009).

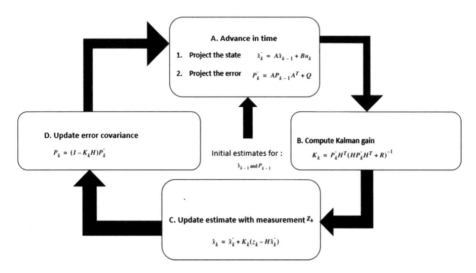

Fig. 2. Basic equations and process of Kalman Filter as a sequential ensemble method

3.3 AdaBoost as an Ensemble Technique

AdaBoost an ensemble technique forms a strong classifier by combining the outputs of the weak classifiers (Adegoke et al. 2017). It has many potential applications it has been successfully applied in many areas such as text classification, natural language processing, drug discovery and computational biology (Fan et al. 2015) vision and object recognition (Viola and Jones 2004; Lee et al. 2013), medical diagnosis (Abuhasel et al. 2015) and industrial chemical fault diagnosis (Karimi and Jazayeri-Rad 2014). The key objective of AdaBoost as a meta-classifier is to improve the accuracy of the base classifiers by constructing and combining multiple instances of a weak classifiers (Schapire and Freund 2014; Adegoke et al. 2017) then produces a strong classifier that performs better than arbitrary guessing. Each instance of the classifier is trained on the same training dataset with different weights assigned to each instance based on its classification accuracy. The final classifier $H(x)$ as illustrated in Fig. 3 is computed as a weighted majority of the weak hypothesis h_t by vote where each hypothesis is assigned a weight α_t can be expressed as illustrated in Eq. 3 below.

$$H(x) = sign\left(\sum_{t=1}^{T} \alpha_t h_t(x)\right) \qquad (3)$$

During the training procedure there is a difference between the predicted values $h_i(x)$, and the expected values, $y_i(x)$. This difference over the committee of classifiers can be stated mathematically as in Eq. 4:

$$E_{err} = \frac{1}{N} \sum_{i=1}^{N} [y_i(x) - h_i(x)]^2 \qquad (4)$$

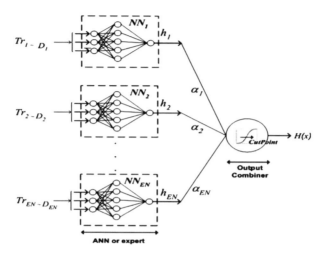

Fig. 3. An ensemble model showing committee of weak neural network predictors

3.4   Optimizing RBFN Training Parameters Using EKF

Kalman filter can be used to optimize weight matrix and error centre of RBFN as a least squares minimization problem (Simon 2002). Therefore, in this session emphasis is on how EKF algorithm can be applied to minimize errors when used in training RBFN to improve their performance following a similar approach (Adegoke 2018). Assuming a non-linear finite dimension discrete time system we can represent the state and measurements as:

$$\theta_{k+1} = f(\theta_k) + \omega_k \qquad (5)$$

$$y_k = h(\theta_k) + v_k \qquad (6)$$

where: the vector θ_k is the state of the system at time k, ω_k is the process noise, y_k is the observation vector, v_k is the observation noise, $f(\theta_k)$ and $h(\theta_k)$ are the non-linear vector functions of the state and process respectively (Adegoke 2018). If the dynamic models $f(\theta_k)$ and $h(\theta_k)$ in Eqs. 5 and 6 are assumed to be known. Then EKF can therefore be used as a standard technique of choice in approximating the maximum likelihood estimation of the state θ_k (Wan and Van Der Merwe 2000). The state and the output white noises ω_k and v_k have zero-correlation with covariance matrix Q_t and R_t respectively and can therefore be modelled as:

$$Q = E\left[\omega_k \omega_k^T\right] \tag{7}$$

$$R = E\left[v_k v_k^T\right] \tag{8}$$

$$MSE = E\left[e_k e_k^T\right] = P_k \tag{9}$$

where P_k is the error covariance matrix at time k.

The solution that Kalman filter provide is to find an estimate for $\widehat{\theta}_{n+1}$ from θ_{k+1} given y_j $(j = 0, \ldots, k)$. If the EKF model in Eqs. 5 and 6 are further assumed to be sufficiently smooth, then we can expand them and approximate them around the estimate θ_k using first-order Taylor expansion series such that:

$$f(\theta_k) = \left(\widehat{\theta}_k\right) + F_k * \left(\theta_k - \widehat{\theta}_k\right) + \text{ Higher orders} \tag{10}$$

$$f(\theta_k) = \left(\widehat{\theta}_k\right) + H_k^T * \left(\theta_k - \widehat{\theta}_k\right) + \text{ Higher orders} \tag{11}$$

where

$$F_k = \frac{\partial f(\theta)}{\partial(\theta)}\Big|_{\theta = \widehat{\theta}_k} \tag{12}$$

$$H_k^T = \frac{\partial h(\theta)}{\partial(\theta)}\Big|_{\theta = \widehat{\theta}_k} \tag{13}$$

Removing the higher order terms of the Taylor series and substitute Eqs. 10 and 11 and substitute them into Eqs. 5 and 6 respectively, then Eqs. 5 and 6 can be approximated as Eqs. 14 and 15 respectively

$$\theta_{k+1} = F_k \theta_k + \omega_k + \emptyset_k \tag{14}$$

$$y_k = H_k^T + v_k + \varphi_k \tag{15}$$

4   Experimental Setup and Discussion

In this section we briefly describe the integration of RBFN, EKF and AdaBoost algorithms that were applied to enhance the prediction accuracy of the ensemble models we proposed in our study.

4.1   Improving RBFN-EKF Model Predictions with AdaBoost

In our simulation we fit the weak classifier RBFN to a version of the dataset described in the previous section. We used EKF to train the RBFN at each iteration. The training process comprises of several training points (X_i, Y_i) where $X_i, \in X$ and $Y_i \in \{-1, +1\}$, on round t, where $t = 1, \ldots T$. Then we calculate the weighted misclassification rate of the learner and update the weighting measure used in the next *round t + 1*. During the training process AdaBoost called the base classifier *T* times, in this case 20 times. As AdaBoost trains RBF network at each round, RBFN layers are optimized using EKF to train and update the network training parameters, namely the: standard deviation (σ), mean (μ) and the weights (w).

The architectural flowchart of our model is as illustrated in Fig. 4 and the framework is as depict in Fig. 5. As shown in Fig. 5, it is possible to swap the dotted part (i.e. RBFN parameter optimization) of the framework with other optimization methods such as training the network with Decoupled Kalman filter or Particle Swarm Optimization (PSO).

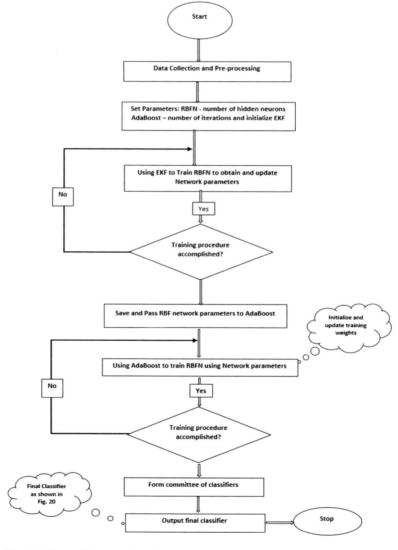

Fig. 4. The architectural flowchart of the proposed EKF-RBFN-AdaBoost model

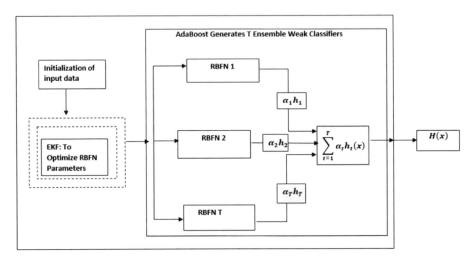

Fig. 5. The framework of the proposed ensemble model based on training RBFN with EKF showing the exchangeable node as illustrated in Fig. 4

4.2   Results and Analysis

Some of the results of applying the proposed model, *EKF-RBFN-AdaBoost* as described on Breast Cancer survivability and Diabetes Diagnostic datasets are presented in this section. To evaluate the prediction accuracy and performance of the proposed models, we compare their results with standalone and ensemble models on the same datasets. In doing this, the following evaluation measures were used: Overall Accuracy Error Rate, True Positive, False Positive and F-Measure; Sensitivity and Precision. Tables 1 and 2 depicts the performance of the proposed model on cancer survivability and diabetic diagnostic datasets when the network was trained with EKF algorithm, compare with ensemble and standalone models. As illustrated Fig. 7 the prediction accuracy of the model outperforms other ensemble models apart from AdaBoost-Random Forest. Figure 6 also shows the FPR, Precision and F-measure of the model are 0.03, 0.97 and 0.87 respectively. Figures 8 and 9 shows the performance of other standalone algorithms when tested on the same dataset. Figures 10 and 11 demonstrates the performance of the proposed model on diabetic dataset when the network EKF was used to train the network, compare with other ensemble models. It appears from Fig. 9 that prediction accuracy of the model was 0.76 and outperform other models apart from AdaBoost-Random Forest which was also 0.76. As Fig. 8 indicates, the TPR (True Positive Rate), FPR (False Positive Rate), Recall, Precision and F-measure of the model are 0.74, 0.34, 0.74, 0.74 and 0.74 respectively. Figures 12 and 13 illustrate the performance of other standalone algorithms when tested on the diabetes dataset.

Table 1. Prediction comparison of Wisconsin Cancer Survivability dataset

Algorithms/Measures	TPR	FPR	Recall/Sensitivity	Precision	F-Measure	Accuracy
Predictive Models of Ensemble Classifiers						
EKF-RBFN-AdaBoost	*0.93*	*0.03*	*0.80*	*0.97*	*0.87*	*0.96*
AdaBoostM1 with Decision stump	0.94	0.08	0.94	0.94	0.94	0.94
AdaBoostM1 with RBFN trained with K-Means	0.96	0.04	0.96	0.96	0.96	0.96
AdaBoostM1 with Random Forest	**0.97**	**0.04**	**0.97**	**0.97**	**0.97**	**0.97**
AdaBoostM1 with Support Vector Machine	0.97	0.04	0.96	0.96	0.96	0.96
Predictive Models of Standalone Classifiers						
Random Forest	**0.97**	**0.04**	**0.97**	**0.97**	**0.97**	**0.97**
Support Vector machine	0.97	0.03	0.97	0.97	0.97	0.96
K-NN	0.96	0.06	0.96	0.96	0.96	0.96
ANN	0.96	0.04	0.96	0.96	0.96	0.96
Naïve Bayes	0.96	0.03	0.96	0.97	0.96	0.96

Table 2. Prediction comparison on Diabetes Diagnostic dataset

Algorithms/Measures	TPR	FPR	Recall/Sensitivity	Precision	F-Measure	Accuracy
Predictive Models of Ensemble Classifiers						
EKF-RBFN-AdaBoost	*0.74*	*0.34*	*0.74*	*0.74*	*0.74*	*0.76*
AdaBoostM1 with Decision stump	0.74	0.35	0.74	0.74	0.74	0.74
AdaBoostM1 with RBFN trained with K-Means	0.74	0.34	0.74	0.74	0.74	0.74
AdaBoostM1 with Random Forest	**0.76**	**0.32**	**0.76**	**0.76**	**0.76**	**0.76**
AdaBoostM1 with Support Vector Machine	0.65	0.65	0.65	0.42	0.51	0.65
Predictive Models of Standalone Classifiers						
Random Forest	**0.76**	**0.31**	**0.76**	**0.75**	**0.76**	**0.76**
Support Vector machine	0.65	0.65	0.65	0.42	0.79	0.65
K-NN	0.65	0.65	0.65	0.42	0.51	0.65
ANN	0.75	0.31	0.75	0.75	0.75	0.75
Naïve Bayes	0.76	0.31	0.76	0.76	0.76	0.76

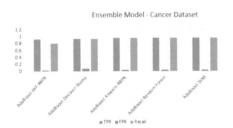

Fig. 6. TPR, FPR and Recall

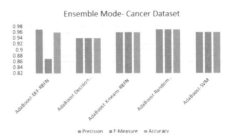

Fig. 7. Precision, F-Measure and Accuracy

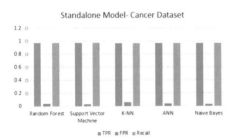

Fig. 8. TPR, FPR and Recall

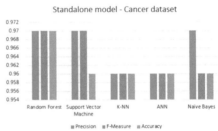

Fig. 9. Precision, F-Measure and Accuracy

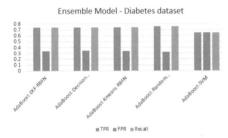

Fig. 10. TPR, FPR and Recall

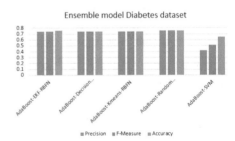

Fig. 11. Precision, F-Measure and Accuracy

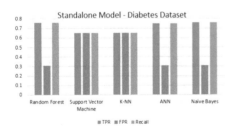

Fig. 12. TPR, FPR and Recall

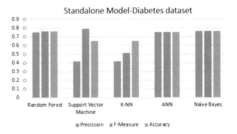

Fig. 13. Precision, F-Measure and Accuracy

5   Conclusion and Further Work

This paper demonstrates how to train RBF networks with EKF algorithm to optimize the RBFN training parameters and improve ensemble prediction accuracy. We used AdaBoost to generate committee of weak learners of EKF-RBFN that are combined to form the final prediction. A performance comparison was carried out using Breast Cancer Survivability and Diabetes Diagnostic datasets that were obtained from the UCI repository. The result shows a good prediction outcome and fast convergence rate compared with other standard standalone models and similar ensemble RBFN models trained with K-means algorithm or Support Vector Machine. The proposed model improves the generalization and prediction accuracy of ensemble models. It also minimizes overfitting problems and improve rate of convergence during training compare to other models used during our study. In the near future, further research will be focused on the application of the proposed model on complex/imbalance datasets, effect of diversity and algorithmic settings on prediction accuracy.

References

Abuhasel, K., Iliyasu, A., Fatichah, C.: A combined AdaBoost and NEWFM technique for medical data classification. Inf. Sci. Appl. **339**, 801–809 (2015)

Adegoke, V.F.: Research report. London South Bank University, Computer Science informatics, School of Engineering, London, UK (2018)

Adegoke, V.F., Chen, D., Barikzai, S., Banissi, E.: Predictive ensemble modelling: experimental comparison of boosting implementation methods. In: European Modelling Symposium (EMS). IEEE, Manchester (2017). https://doi.org/10.1109/EMS.2017.13

Adegoke, V., Chen, D., Banissi, E.: Prediction of breast cancer survivability using ensemble algorithms. In: International Conference on Smart Systems and Technologies (SST). IEEE, Osijek (2017)

Alghamdi, M., Al-Mallah, M., Keteyian, S., Brawner, C., Ehrman, J., Sakr, S.: Predicting diabetes mellitus using SMOTE and ensemble machine learning approach: the Henry Ford ExercIse Testing (FIT) project. PLoS One **12**(7) (2017). https://doi.org/10.1371/journal.pone. 0179805

Anil, K., Duin, R., Mao, J.: Statistical pattern recognition: a review. IEEE Trans. Pattern Anal. Mach. Intell. **22**(1), 4–37 (2000). https://doi.org/10.1109/34.824819

Apreutesei, N.A., Tircoveanu, F., Cantemir, A., Bogdanici, C., Lisa, C., Curteanu, S., Chiseliţă, D.: Predictions of ocular changes caused by diabetes in glaucoma patients. Comput. Methods Programs Biomed. **154**, 183–190 (2018)

Barakat, N., Bradley, A.P., Barakat, M.N.: Intelligible support vector machines for diagnosis of diabetes mellitus. Trans. Inf. Technol. Biomed. **14**(4), 1114–1120 (2010)

BBC UK: Labour's Tom Watson 'reversed' type-2 diabetes through diet and exercise. British Broadcasting Corporation, London (2018). http://www.bbc.co.uk/news/uk-politics-45495384. Accessed 12 Sept 2018

BHF: CVD Statistics – BHF UK Factsheet. BHF (British Heart Foundation) (2018). https://www. bhf.org.uk/-/media/files/research/heart-statistics/bhf-cvd-statistics—uk-factsheet.pdf. Accessed 23 Aug 2018

Chernodub, A.: Training neural networks for classification using the extended Kalman filter: a comparative study. Opt. Mem. Neural Netw. **23**(2), 96–103 (2014)

Csank, J.T., Connolly, J.W.: Model-based engine control architecture with an extended Kalman filter. The American Institute of Aeronautics and Astronautics, San Diego, California (2016). NASA STI Program. https://ntrs.nasa.gov/archive/nasa/casi.ntrs.nasa.gov/20160002248.pdf. Accessed 06 Feb 2018

Fan, M., Zheng, B., Li, L.: A novel Multi-Agent Ada-Boost algorithm for predicting protein structural class with the information of protein secondary structure. J. Bioinform. Comput. Biol. **13**(5) (2015). https://doi.org/10.1142/S0219720015500225

Formenti, S., Arslan, A., Love, S.: Global breast cancer: the lessons to bring home. Int. J. Breast Cancer (2012)

Freund, Y., Schapire, R.: A decision-theoretic generalization of on-line learning and application to boosting. J. Comput. Syst. Sci. **55**, 119–139 (1997). http://citeseerx.ist.psu.edu/viewdoc/summary?doi=10.1.1.32.8918

Haykin, S.: Adaptive Filter Theory, 3rd edn. Prentice-Hall, Upper Saddle River (1996)

ITV: Surge in heart attacks and strokes predicted as diabetes epidemic takes its toll. ITV Report, London (2018). https://www.itv.com/news/2018-08-23/surge-in-heart-attacks-and-strokes-predicted-as-diabetes-epidemic-takes-its-toll/. Accessed 06 Aug 2018

Karimi, P., Jazayeri-Rad, H.: Comparing the fault diagnosis performances of single neural networks and two ensemble neural networks based on the boosting methods. J. Autom. Control. **2**(1), 21–32 (2014)

Kwon, S., Lee, S.: Recent advances in microwave imaging for breast cancer detection. Int. J. Biomed. Imaging (2016). https://doi.org/10.1155/2016/5054912

Lee, Y., Han, D., Ko, H.: Reinforced AdaBoost learning for object detection with local pattern representations. Sci. World J. **2013**, 14 (2013). https://doi.org/10.1155/2013/153465

Lima, D., Sanches, R., Pedrino, E.: Neural network training using unscented and extended Kalman filter. Robot. Autom. Eng. J. **1**(4) (2017)

McGinley, B., O'Halloran, M., Conceicao, R., Morgan, F., Glavin, M., Jones, E.: Spiking neural networks for breast cancer classification in a dielectrically heterogeneous breast. Prog. Electromagn. Res. C **17**, 74–94 (2010). https://doi.org/10.2528/PIERC10100202

Merwe, R., Nelson, A., Wan, E.: An introduction to Kalman filtering. OGI School of Science & Engineering Lecture (2004)

Moreno, V.M., Pigazo, A.: Kalman Filter: Recent Advances and Applications. I-Tech Education and Publishing KG, Vienna (2009)

Nabney, I.: NETLAB Algorithms for Pattern Recognition (Ed. by M. Singh). Springer, London (2002)

Pak, F., Kanan, H., Alikhassi, A.: Breast cancer detection and classification in digital mammography based on Non-Subsampled Contourlet Transform (NSCT) and super resolution. Comput. Methods Programs Biomed. **122**, 89–107 (2015)

Ribeiro, M.: Kalman and extended Kalman filters: concept, derivation and properties. CiteSeer (2004)

Sapate, S., Talbar, S.: An overview of pectoral muscle extraction algorithms applied to digital mammograms. In: Studies in Computational Intelligence (2016). https://doi.org/10.1007/978-3-319-33793-7_2

Schapire, R., Freund, Y.: Boosting: Foundations and Algorithms, 2nd edn. MIT Press, Cambridge (2014)

Simon, D.: Training radial basis neural networks with the extended Kalman filter. Neurocomputing **48**, 455–457 (2002)

American Cancer Society: Breast Cancer Facts & Figures 2015-2016 (Ed. by DeSantis, R. Siegel, A. Jemal) (2015). American Cancer Society https://www.cancer.org/content/dam/cancer-org/research/cancer-facts-and-statistics/breast-cancer-facts-and-figures/breast-cancer-facts-and-figures-2015-2016.pdf. Accessed 05 May 2017

Cancer Research UK: Cancer Research UK (2018). http://www.cancerresearchuk.org/health-professional/cancer-statistics/statistics-by-cancer-type/breast-cancer. Accessed 20 Aug 2018

Viola, P., Jones, M.J.: Robust real-time face detection. Int. J. Comput. Vis. **57**(2), 137–154 (2004)

Wan, E., Van Der Merwe, R.: The unscented Kalman filter for nonlinear estimation. In: Symposium on Adaptive Systems for Signal Processing Communications and Control, pp. 153–158 (2000). https://doi.org/10.1109/ASSPCC.2000.882463

Weedon-Fekjær, H., Romundstad, P., Vatten, L.: Modern mammography screening and breast cancer mortality: population study. BMJ **348**, g3701 (2014). https://doi.org/10.1136/bmj.g3701

WHO: Global report on diabetes - World Health Organization. WHO Library Cataloguing-in-Publication Data, Geneva (2016). http://apps.who.int/iris/bitstream/handle/10665/204871/9789241565257_eng.pdf;jsessionid=DFE5616C3480A8F293D9970CC0FA4EF1?sequence=1

Xie, W., Li, Y., Ma, Y.: Breast mass classification in digital mammography based on extreme learning machine. Neurocomputing **73**(3), 930–941 (2015). https://doi.org/10.1016/j.neucom.2015.08.048

Yang, C.-H., Lin, Y.-U., Chuang, L.-Y., Chang, H.-W.: Evaluation of breast cancer susceptibility using improved genetic algorithms to generate genotype SNP barcodes. IEEE/ACM Trans. Comput. Biol. Bioinf. **10**(2), 361–371 (2013). https://doi.org/10.1109/TCBB.2013.27

Zheng, T., Xie, W., Xu, L., He, X., Zhang, Y., You, M., Chen, Y.: A machine learning-based framework to identify type 2 diabetes through electronic health records. Int. J. Med. Inform. **97**, 120–127 (2017)

Improving Audiovisual Content Annotation Through a Semi-automated Process Based on Deep Learning

Luís Vilaça[1], Paula Viana[1,2(✉)], Pedro Carvalho[2], and Teresa Andrade[2,3]

[1] School of Engineering, Polytechnic of Porto, Porto, Portugal
1121405@isep.ipp.pt
[2] INESC TEC, Porto, Portugal
{paula.viana,pedro.carvalho,mandrade}@inesctec.pt
[3] Faculty of Engineering, University of Porto, Porto, Portugal

Abstract. Over the last years, Deep Learning has become one of the most popular research fields of Artificial Intelligence. Several approaches have been developed to address conventional challenges of AI. In computer vision, these methods provide the means to solve tasks like image classification, object identification and extraction of features.

In this paper, some approaches to face detection and recognition are presented and analyzed, in order to identify the one with the best performance. The main objective is to automate the annotation of a large dataset and to avoid the costy and time-consuming process of content annotation. The approach follows the concept of incremental learning and a R-CNN model was implemented. Tests were conducted with the objective of detecting and recognizing one personality within image and video content.

Results coming from this initial automatic process are then made available to an auxiliary tool that enables further validation of the annotations prior to uploading them to the archive.

Tests show that, even with a small size dataset, the results obtained are satisfactory.

Keywords: Content annotation · Computer Vision ·
Machine Learning · Deep Learning · Object detection ·
Facial detection · Facial recognition

1 Introduction

Machine Learning (ML) has been applied in problems that require a high degree of manual parametrization or the manipulation of large sets of data. When combined with Computer Vision (CV), where the goal is to extract useful information from digital images, ML has demonstrated good results when comparing with traditional algorithms. Object detection, facial detection and recognition, are examples of problems that have been addressed.

A. M. Madureira et al. (Eds.): SoCPaR 2018, AISC 942, pp. 66–75, 2020.
https://doi.org/10.1007/978-3-030-17065-3_7

The main objective of this paper is to semi-automate the process of image and video content annotation, by using ML algorithms for face detection and recognition. In this first version, a single person of interest was selected. One of the restrictions for the solution to be implemented is that the data set used for training should be small to avoid the need of manually annotation the content for creating the ground truth.

The proposed solution creates a pipeline of two applications: an application for automatically generating the pre-annotations after an initial short training phase and a validation application enabling cleaning any noisy annotations and creating valid information to be further used to improve the accuracy.

The remaining of this paper is structured as follows. A brief review of the related work is presented in Sect. 2. The experiments, evaluation parameters, results and decision taking is presented in Sect. 3. Section 4 presents the proposed solution, along with its functionalities. Finally, the final section concludes this document with some discussion and conclusions.

2   Related Work

Searching and browsing large collections of video assets depends greatly on the capacity of describing this content. Several approaches have been proposed in the literature to enhance the accuracy of search queries. Examples based on image, video, audio or text analysis as well as on semantic related approaches can be found applied to several areas of application [3,17]. Other methods have been exploiting user contributions and implementing methods that automatically filter noisy information [14,19,20,30].

Approaches based on image processing date from mid-1960s and have been evolving progressively. In 2001, Viola and Jones developed a facial detection algorithm, later called "Viola-Jones Algorithm" [31], that is usually regarded as an important milestone. It was the first time a facial detection software showed acceptable results for real-time analysis. Some facial features, such as the eyes, mouth, nose, and the relationships between them, were manually coded and then, the information was submitted in a binary classifier, with the purpose of classifying the face as true or false. This algorithm was distinguished by its fast detection capability when compared to SoA solutions. Its main limitation was the decrease of accuracy with the non-frontality of the faces.

In 2005 Dadal and Triggs developed a more efficient technique the "Histograms of Oriented Gradients" [5]. This algorithm, still used nowadays, was applied to face detection as well as to other problems such as pedestrians detection. Pixels in the image are compared to its neighbors, enabling the detection of maximum gradients. Classification was achieved by comparison with pre-coded facial features and the decision was based on a confidence threshold.

Deep Learning (DL) matures in 2010 assuming the neural networks, a computational method inspired by the brains biological networks, as its core. In the same year, at ImageNet, Krizhevsky et al. used a Convolutional Neural Network (CNN) [12] that outperformed all other algorithms. This was possible due to the amount of audiovisual information made available for training and to the ever-increasing graphic processing power.

Although these results are promising, they still have some drawback that limit their application to real use cases as images used for training were pre-processed and only contained the target objects. Moreover, this mechanism is quite computationally demanding and trying to overcome this problem, in 2015, the concept of a Regional Convolutional Neural Networks (R-CNN) [7] was proposed. This approach uses a selective search process to create windows of different sizes within the original image and each one groups adjacent pixels with similar characteristics ("Region Proposals"). After this initial pre-processing phase, each window is inserted into a trained CNN for classification. Several adaptation approaches have been developed to improve the efficiency of this method: Fast R-CNN [6], Faster R-CNN [23] and Mask R-CNN [8].

Although this learning approach provides good results, it relies heavy on the training sets. Alternative techniques optimized for facial detection and recognition, that might not rely on learning processes, have also been exploited.

Four main categories can be identified for facial detection: (1) Knowledge-based models - use hand coded features, defined as rules from the human knowledge of what best characterize a face; (2) Feature-based approaches - Aims to find structural features that exist even with the variation of the external conditions; (3) Template matching - Uses standard patterns of the face and detects them through correlation evaluation; (4) Appearance-based models: Through a set of examples, the representative variability of the facial appearance is captured.

Within the Knowledge-based model approach, [32] uses a multiresolution hierarchical method to detect faces that, although did not achieve high detection rates, was used in later works due to its simplicity [11]. In contrast to these methods, the feature-based approaches aim to find invariant features of faces for detection. Sirohey [24] proposed a solution to segment a face from a cluttered background, achieving 80% accuracy on 48 images. It uses an edge map [4] and heuristics for grouping and selecting edges, to find the contour of the face. Recently, methods that combine several facial features, like the one used in [33], which is based on structure, color and geometry, have been proposed.

An example of Template matching was described by Tsukamoto et al. in [27,28]. It uses a qualitative model for face pattern (QMF), dividing the sample image into blocks, estimating the qualitative features for each one and defining a template. In the Appearance-based model approach, the template is learned from the training examples using one of the traditional ML approaches.

Facial recognition approaches can be split into two main classes: (1) Template-based - compute the correlation between the template and the face submitted; (2) Geometry feature-based - analyze the local features and their geometric relationships.

For the Template models for facial recognition, statistical tools are used to compute an adequate set of templates: Support Vector Machines [18,29], Linear Discriminant Analysis [2], Principal Component Analysis [25] and Neural Networks [10,15] have been used for this purpose. For Geometry-based solutions, approaches like the one presented by Cootes et al. [13], that uses a built flexible model for face recognition, can be identified.

3   ML Frameworks and Services for Facial Detection and Recognition

The scientific community as well as the industry have been making available ML frameworks that can be used and adapted for several purposes. This section describes the most relevant cloud-based services and software libraries available to implement facial detection and recognition algorithms. The first set of solutions enable detecting faces and returning their location within the image, usually as a surrounding box (BBOX). The second set of algorithms adds the capability of identifying a person from a set of possible options.

Tests were conducted using the following cloud-based services: Microsoft Azure, Google Vision API, Clarifai and Amazon Rekognition. For the architecture based on local software libraries Tensorflow and YOLO [21,22] (Darknet) were tested. For YOLO, Darkflow [1], the translated version of Darknet for Tensorflow, was used. Two different Tensorflow models were tested - SSD Mobilenet [9,16] and Faster RCNN Inception v2 [23,26], while for the YOLO the Tiny model version was the one used. This was due to limitations on the available hardware used for the training phase.

The goal of this testbed is to identify the most efficient facial recognition and detection services available, taking into account the small size of the training dataset.

3.1   Experiment Setup

Given that the application is expected to be used in several scenarios that include the capture of the content in non-controlled environments that comprise indoor and outdoor situations, different illumination and individual as well a group pictures, one of the aspects under analysis is the robustness of the approach with the non-frontality of the faces. Failure of the detection (false negatives), wrong detections (false positives) and the interval between the minimum positive values and the maximum negative values, so that a threshold can be identified, will be considered. In the cloud-based APIs, due to limitations imposed by the frameworks, available pre-trained networks were used. The only exception was in using the Clarifai web-service for which training with our training dataset was done. For the others, the experiment used the functionality of providing a reference image.

For testing purpose, all the services will be analyzed using the same dataset that includes seven images of Prof. Marcelo Rebelo de Sousa and a video in which he also appears. Several other people are also present in the content and the target appears in different locations and positions. Given the limitations imposed by some of the frameworks, the video was fragmented into frames, taken every second, each being analyzed individually. This resulted in 52 additional images in the test dataset. The training dataset consisted of 393 images of Prof. Marcelo Rebelo de Sousa. Each one was annotated manually to generate the ground truth required for the training process. A Supervised Learning method using

Transfer Learning, where a pre-trained model is reused to detect new classes, was implemented. Figure 1 depicts the improvement in respect to the total loss metric returned by the frameworks and that include several parameters that reflect the error of the model (an example is the difference between the ground truth and the returned BBox location). These results were used to define the number of iterations to be used during the training process in order to guarantee an optimal generalization. Training and testing were performed in the same system with the following characteristics: Intel Core i7 6700K, 4.3 GHz; 16 Gb of Ram, at 2400 MHz; Gigabyte GeForce GTX 1060 6 Gb, 1700 MHz.

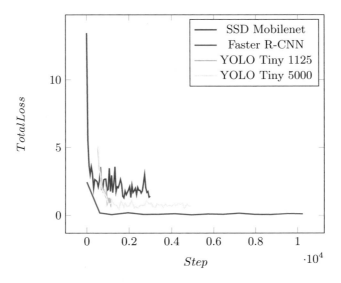

Fig. 1. DL models training evolution

3.2   Results

Figure 2 presents the results obtained by the Cloud-based APIs. Both Clarifai and Google Vision obtained 100% success, locating the target in all the test images. However, it is worth mentioning that among those, Clarifai was the only one that was trained with our training dataset. Additionally, some of these frameworks return extra information providing context information of the image that include image properties, text, emotions and approximate age in facial analysis. Clarifai, Google Vision and Amazon Rekognition provide also semantic aware information returning tags that identify a given scenario. Examples of tags include Sport, People, Parking Lot, Car, Intersection, Urban, Nature, Sunrise and Sunset, Urban Area, Street, Road, City, Skyline, Daytime, Dusk, Evening and Night. Figures 3 and 4 show the results obtained for each of the models of the local frame-works. Testing was performed using the described test dataset.

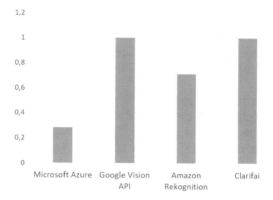

Fig. 2. Detection results with the cloud-based APIs

The results in Fig. 3 present the number of true and false positives for the total number of Bbox generated (one of the test images may contain several Bbox) while in Fig. 4 we can see the true and false positives when having as purpose being able to state if the target person is on an image.

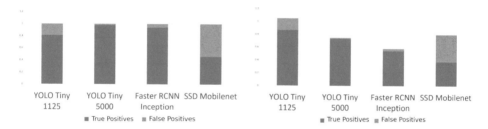

Fig. 3. Percentage of true and false positives per total of BBOXes generated

Fig. 4. Percentage of true and false positives per number of images analyzed

3.3   Discussion and Conclusions

The results presented in the previous figures show that the "Faster RCNN Inception" model is the one resulting in more correct identifications. Results have however been achieved at the expense of additional computational costs for the inference process (sec/image): (1) Yolo-Tiny: 0.095; (2) Yolo-Full: 0.287; (3) SSD-Mobilenet: 3.046; (4) Faster-RCNN: 4.366. These results are a direct consequence of the architecture used for each approach and of the pre-processing phase methodology. While in F-RCNN two neural networks are used – one for generating the *region proposals* and the other for classifying those regions – on SSD a neural network is eliminated as pre-processing relies on a random process to generate the initial BBoxes. For the Yolo model, a simpler network enables reducing the processing time and this is further exploited in the Tiny version by reducing the number of layers.

Tests also show that Faster-RCNN returns a higher confidence value than the others, although it still fails to detect 45% of the target person appearing in the images.

By analyzing the results obtained in Fig. 2 and taking into account that Google Vision API does not allow to train a network with a local dataset, the best choice of implementation is with the Clarifai Web Service. This API will be implemented with the general-purpose network to generate tags about the context of the image.

4   Proposed Solution

The main goal, as stated before, is to speed up the annotation process of an audiovisual archive. The proposed solution includes a two-stage process that was implemented as two independent software applications developed in Python: the Annotation Generator and the Annotation Validator.

Annotation Generator: This application uses the model selected in the previous chapter to perform the detection of Prof. Marcelo Rebelo de Sousa. A confidence value higher than 60% was defined as a requirement. The same image is submitted to the "General" model of the Clarifai service, to obtain additional tags relative to the image's semantics. The output of this process is a XML file having the structure presented in Listing 1.1.

```
<annotation>
<folder>images</folder>
<filename>image1.jpg</filename>
<size>
    <width>1619</width>
    <height>1080</height>
</size>
<tags>administration, outfit, politician</tags>
<object>
    <name>Person</name>
    <bndbox>
        <xmin>794</xmin>
        <ymin>41</ymin>
        <xmax>1618</xmax>
        <ymax>1075</ymax>
    </bndbox>
</object>
</annotation>
```

Listing 1.1. XML File Structure

Annotations resulting from this process are them made available for customization and validation through the Annotation Validator: this application reads the information generated in the previously described step and provides a GUI with editing functionalities. Figure 5 shows the main screen where the tags and the Bbox generated are presented and made available either for validation or for editing. Any action will be reflected on the XML file, updating the descriptions. The full process is depicted in Fig. 6.

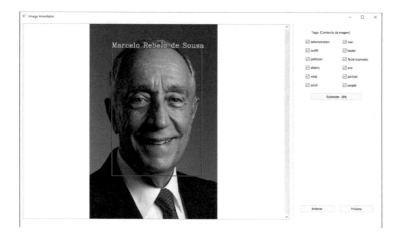

Fig. 5. Overview of the UI in the annotation validation application

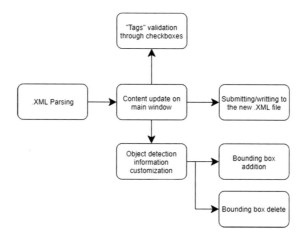

Fig. 6. Functionalities of the annotation validation application

5   Conclusions

This paper analyses the performance of several ML approaches for detecting and recognizing a target person considering the small size of the training dataset. A prototype that enables preliminary automatic annotation of the multimedia content followed by a simple and intuitive application that enables removing noisy information or adding additional data is also presented. Initial tests show that the performance of the system enables accelerating the annotation process. The solution that best fits the requirements is the YOLO Tiny and the Faster R-RCNN Inception, this last one outperforming all for the collected dataset. Test showed also that the user validation application could indeed hasten the annotation process while comparing to the fully manual introduction of the same data. Future work includes testing the solution with the YOLO full model and to repeat the tests with training datasets of different sizes.

Acknowledgments. The work presented was partially supported by the following projects: FourEyes, a Research Line within project "TEC4Growth: Pervasive Intelligence, Enhancers and Proofs of Concept with Industrial Impact/NORTE-01- 0145-FEDER-000020" financed by the North Portugal Regional Operational Programme (NORTE 2020), under the PORTUGAL 2020 Partnership Agreement, and through the European Regional Development Fund (ERDF); FotoInMotion funded by H2020 Framework Programme of the European Commission.

References

1. Darkflow repository. https://github.com/thtrieu/darkflow. Accessed 09 July 2018
2. Belhumeur, P.N., Hespanha, J.P., Kriegman, D.J.: Eigenfaces vs. fisherfaces: recognition using class specific linear projection. Technical report, Yale University, New Haven, United States (1997)
3. Bertini, M., Del Bimbo, A., Torniai, C.: Automatic video annotation using ontologies extended with visual information. In: Proceedings of the 13th Annual ACM International Conference on Multimedia, MULTIMEDIA 2005, pp. 395–398. ACM, New York (2005)
4. Canny, J.: A computational approach to edge detection. IEEE Trans. Pattern Anal. Mach. Intell. **PAMI-8**(6), 679–698 (1986)
5. Dalal, N., Triggs, B.: Histograms of oriented gradients for human detection, vol. 1, pp. 886–893, June 2005
6. Girshick, R.: Fast R-CNN. In: Proceedings of the IEEE International Conference on Computer Vision, pp. 1440–1448 (2015)
7. Girshick, R.B., Donahue, J., Darrell, T., Malik, J.: Rich feature hierarchies for accurate object detection and semantic segmentation. CoRR abs/1311.2524 (2013)
8. He, K., Gkioxari, G., Dollár, P., Girshick, R.: Mask R-CNN. In: 2017 IEEE International Conference on Computer Vision (ICCV), pp. 2980–2988. IEEE (2017)
9. Howard, A.G., Zhu, M., Chen, B., Kalenichenko, D., Wang, W., Weyand, T., Andreetto, M., Adam, H.: Mobilenets: efficient convolutional neural networks for mobile vision applications. CoRR abs/1704.04861 (2017)
10. Howell, A.J., Buxton, H.: Invariance in radial basis function neural networks in human face classification. Neural Process. Lett. **2**(3), 26–30 (1995)
11. Kotropoulos, C., Pitas, I.: Rule-based face detection in frontal views. In: Proceedings International Conference on Acoustics, Speech and Signal Processing, vol. 4, pp. 2537–2540 (1997)
12. Krizhevsky, A., Sutskever, I., Hinton, G.E.: ImageNet classification with deep convolutional neural networks, pp. 1097–1105 (2012)
13. Lanitis, A., Taylor, C.J., Cootes, T.F.: Automatic interpretation and coding of face images using flexible models. IEEE Trans. Pattern Anal. Mach. Intell. **19**(7), 743–756 (1997)
14. Larson, M., Soleymani, M., Serdyukov, P., Rudinac, S., Wartena, C., Murdock, V., Friedland, G., Ordelman, R., Jones, G.J.F.: Automatic tagging and geotagging in video collections and communities. In: Proceedings 1st ACM International Conference on Multimedia Retrieval, ICMR 2011, pp. 51:1–51:8. ACM, New York (2011)

15. Lawrence, S., Giles, C.L., Tsoi, A.C., Back, A.D.: Face recognition: a convolutional neural-network approach. IEEE Trans. Neural Netw. **8**(1), 98–113 (1997)
16. Liu, W., Anguelov, D., Erhan, D., Szegedy, C., Reed, S., Fu, C.Y., Berg, A.C.: SSD: single shot multibox detector. In: European Conference on Computer Vision, pp. 21–37. Springer (2016)
17. Moxley, E., Mei, T., Hua, X., Ma, W., Manjunath, B.S.: Automatic video annotation through search and mining. In: 2008 IEEE International Conference on Multimedia and Expo, pp. 685–688, June 2008
18. Osuna, E., Freund, R., Girosit, F.: Training support vector machines: an application to face detection. In: Proceedings of IEEE Computer Society Conference on Computer Vision and Pattern Recognition, pp. 130–136, June 1997
19. Pinto, J.P., Viana, P.: TAG4VD: a game for collaborative video annotation. In: Proceedings of the 2013 ACM International Workshop on Immersive Media Experiences, ImmersiveMe 2013, pp. 25–28. ACM, New York (2013)
20. Pinto, J.P., Viana, P.: Using the crowd to boost video annotation processes: a game based approach. In: Proceedings of the 12th European Conference on Visual Media Production, CVMP 2015, pp. 22:1–22:1. ACM, New York (2015)
21. Redmon, J., Divvala, S., Girshick, R., Farhadi, A.: You only look once: unified, real-time object detection. In: Proceedings of the IEEE Conference on Computer Vision and Pattern Recognition, pp. 779–788 (2016)
22. Redmon, J., Farhadi, A.: YOLO9000: better, faster, stronger. CoRR abs/1612.08242 (2016)
23. Ren, S., He, K., Girshick, R., Sun, J.: Faster R-CNN: towards real-time object detection with region proposal networks (2015)
24. Sirohey, S.A.: Human face segmentation and identification. Technical report (1993)
25. Sirovich, L., Kirby, M.: Low-dimensional procedure for the characterization of human faces. J. Opt. Soc. Am. A **4**(3), 519–524 (1987)
26. Szegedy, C., Vanhoucke, V., Ioffe, S., Shlens, J., Wojna, Z.: Rethinking the inception architecture for computer vision. CoRR abs/1512.00567 (2015)
27. Tsukamoto, A., Lee, C.W., Tsuji, S.: Detection and pose estimation of human face with synthesized image models. In: Proceedings of 12th International Conference on Pattern Recognition, vol. 1, pp. 754–757, October 1994
28. Tukamoto, A.: Detection and tracking of human face with synthesized templates. In: Proceedings of the ACCV 1993, pp. 183–186 (1993)
29. Vapnik, V.: The Nature of Statistical Learning Theory. Springer, New York (2013)
30. Viana, P., Pinto, J.P.: A collaborative approach for semantic time-based video annotation using gamification. Hum.-Centric Comput. Inf. Sci. **7**(1), 13 (2017)
31. Viola, P., Jones, M.: Rapid object detection using a boosted cascade of simple features, vol. 1, pp. I-511–I-518 (2001)
32. Yang, G., Huang, T.S.: Human face detection in a complex background. Pattern Recognit. **27**(1), 53–63 (1994)
33. Yang, M.H., Ahuja, N.: Detecting human faces in color images. In: Proceedings of the International Conference on Image Processing, ICIP 1998, vol. 1, pp. 127–130, October 1998

Subject Identification Based on Gait Using a RGB-D Camera

Ana Patrícia Rocha[1(✉)], José Maria Fernandes[1],
Hugo Miguel Pereira Choupina[2], Maria do Carmo Vilas-Boas[2],
and João Paulo Silva Cunha[2]

[1] Institute of Electronics and Informatics Engineering of Aveiro (IEETA),
and Department of Electronics, Telecommunications and Informatics,
University of Aveiro, Aveiro, Portugal
{aprocha,jfernan}@ua.pt
[2] Institute for Systems Engineering and Computers – Technology and Science
(INESC TEC), and Faculty of Engineering (FEUP),
University of Porto, Porto, Portugal
hugo.m.choupina@inesctec.pt,
mcarmo.vilasboas@gmail.com, jcunha@ieee.org

Abstract. Biometric authentication (i.e., verification of a given subject's identity using biological characteristics) relying on gait characteristics obtained in a non-intrusive way can be very useful in the area of security, for smart surveillance and access control. In this contribution, we investigated the possibility of carrying out subject identification based on a predictive model built using machine learning techniques, and features extracted from 3-D body joint data provided by a single low-cost RGB-D camera (Microsoft Kinect v2). We obtained a dataset including 400 gait cycles from 20 healthy subjects, and 25 anthropometric measures and gait parameters per gait cycle. Different machine learning algorithms were explored: k-nearest neighbors, decision tree, random forest, support vector machines, multilayer perceptron, and multilayer perceptron ensemble. The algorithm that led to the model with best trade-off between the considered evaluation metrics was the random forest: overall accuracy of 99%, class accuracy of $100 \pm 0\%$, and F_1 score of $99 \pm 2\%$. These results show the potential of using a RGB-D camera for subject identification based on quantitative gait analysis.

Keywords: Subject identification · Gait · RGB-D camera · Kinect v2 · Biometrics · Machine learning

1 Introduction

Biometric authentication, i.e., the verification of a subject's identity relying on biological characteristics, is being increasingly used in the area of security for access control or smart surveillance [1]. Biometric authentication methods involve the analysis of the face, fingerprint, iris, vital signs, voice, or gait [1, 2]. Human identification based on gait using markerless vision-based sensors has the advantage of not being as

© Springer Nature Switzerland AG 2020
A. M. Madureira et al. (Eds.): SoCPaR 2018, AISC 942, pp. 76–85, 2020.
https://doi.org/10.1007/978-3-030-17065-3_8

intrusive as the other methods, since it does not require the attention or cooperation of the subject being identified. Furthermore, gait is difficult to hide, steal or fake.

In the past years, several studies on subject identification based on gait recognition have been performed using vision-based sensors [3–15]. The first studies mainly used the model-free approach relying on the subject's binary silhouette, which was obtained from color or infrared image sequences provided by RGB or infrared cameras [4–7]. More recently, many contributions have relied on the model-based approach based on body segment and/or joint information provided by RGB-D cameras, such as the Kinect camera [8–15]. The latter approach typically has the advantage of being viewpoint and scale invariant [2]. Most of the studies relying on 3-D body joint data considered both anthropometric measures (e.g., length of body segments, and subject's height) and gait parameters [8, 11–14].

In this contribution, we present a study on subject identification using machine learning techniques, and features extracted from gait data provided by a RGB-D camera (Kinect v2). We obtained a dataset including 400 instances (gait cycles), and 25 features (3 anthropometric measures, and 22 spatiotemporal and kinematic gait parameters). The features were computed over 3-D body joint data acquired from 20 healthy subjects while they walked towards the camera.

To obtain a predictive model for subject identification, we explored the six different machine learning algorithms. For each algorithm, we validated the model by using the 10-fold cross validation approach, and obtaining several performance evaluation metrics (overall accuracy, mean and standard deviation for class accuracy and class F_1 score, as well as for training and prediction time). For the model with the best trade-off between the considered metrics, we also compared the three types of features (anthropometric measures, spatiotemporal gait parameters, and kinematic gait parameters) for subject identification.

2 Related Work

Several studies on subject identification were carried out using features extracted from gait data provided by the Kinect v1 [11–15]. Preis et al. obtained a success rate of 91% when using the Naïve Bayes algorithm (side-view, 9 subjects, 4 anthropometric measures) [14]. Ball et al. relied on the k-means algorithm for unsupervised clustering, achieving a mean accuracy of 44% (arbitrary walking paths, 4 subjects, 18 kinematic gait parameters) [15].

Sinha et al. reported a mean F_1 score of 62% (side-view, 10 subjects) when using the adaptive neural network algorithm [13]. The used features were selected from the ones proposed by Preis et al. [14] and Ball et al. [15], and 14 new features (related with body area, and distance between each joint and upper body centroid). In the study performed by Andersson et al., the best result was achieved when using k-NN: overall accuracy of 88% (140 subjects, walking in a semi-circle, 20 anthropometric measures and 60 spatiotemporal and kinematic gait parameters) [11].

Jiang et al. proposed the use of the nearest neighbor approach, together with 5 anthropometric features and 4 angle measures [12]. For each anthropometric feature, the Euclidean distance was used as a distance measure between two examples. For each angle measure, the distance between two time series was computed using dynamic time warping (DTW). When using both types of features, a single score of similarity between two feature vectors corresponding to gait cycles was obtained by computing the mean of the normalized distance score for each feature. For data acquired from 10 subjects (side-view), an overall accuracy of 82% was reported.

A similar approach to the one proposed by Jiang et al. was used by Nambiar et al. in a study relying on the Kinect v2 [8], with the difference that the single score of similarity is the sum of the normalized Euclidean distance for each feature. A re-identification rate of 92% (20 subjects, 7 anthropometric measures and 35 gait parameters, best match) was achieved for frontal direction (63 to 72% for other directions).

Also relying on the Kinect v2, Ahmed et al. proposed the use of DTW as the similarity measure between two examples of a measure, and rank-level fusion based on majority voting to identify a subject [9]. The considered measures computed for each gait cycle included the distance and angle for all joint pairs (subset selected using a genetic algorithm). They achieved a recognition rate of 92% (20 subjects, best match).

Rahman et al. used the mean and variance of the distance for each joint pair (760 features), computed for each gait cycle, for subject identification [10]. An overall accuracy of 93% (k-NN) was achieved when using the Kinect v1 and the k-NN algorithm (30 subjects), while a slightly higher accuracy of 95% (k-NN and SVM) was reported when using the Kinect v2 (20 subjects).

3   Materials and Methods

3.1   Subjects

An experiment was carried out at Porto Biomechanics Laboratory (LABIOMEP) with the participation of twenty healthy subjects (ten male, age: 31 ± 8 years, height: 1.71 ± 0.11 m, body mass: 67.9 ± 15.3 kg). The subjects were recruited from the university community, and the only exclusion criterion was the existence of a disease or injury affecting the gait. The experiment was approved by the Ethics Committee of Santo António Hospital (Porto, Portugal), and all subjects signed an informed consent form.

3.2   Experimental Setup and Protocol

The experimental setup included a single RGB-D camera (second version of the Kinect camera – Kinect v2), connected to a laptop. The camera was mounted on a tripod, at a height of 1 m, with a tilt angle of −10 degrees.

The experimental protocol included the following gait task: walking towards and away from the Kinect for a total of 14 m, at a self-selected pace. The turn was performed at around 1.2 m from the camera (outside its practical depth range). Each subject carried out the defined task ten times (200 gait trials in total).

3.3   Data Acquisition and Pre-processing

Data provided by the Kinect were acquired at approximately 30 Hz, using our *KiT* software application [16]. The data included infrared, depth and 3-D body joint data. Each frame of the latter includes the 3-D position of 25 joints.

Using the automatic gait analysis solution developed by us [17], the data corresponding to walking towards the camera were automatically selected, and the different gait cycles were identified. Additionally, we detected the toe off instants by finding the maximum of the absolute difference between the left and right shank angles between consecutive heel strikes.

For each gait cycle, 3 anthropometric measures and 22 spatiotemporal and kinematic gait parameters were computed. The anthropometric measures included the trunk, arm and leg length. The trunk/arm/leg length corresponds to the mean of the sum of the distances between adjacent trunk/arm/leg joints for all gait cycle frames. The arm/leg length is the mean between left and right arms/legs.

The spatiotemporal parameters included: stride, step, stance, swing, single support and double support duration; stride and step length; step width; gait speed, and gait speed variability; foot and arm swing velocity. The kinematic parameters included: neck, spine shoulder, and spine middle angle; elbow and knee angle minimum and maximum; hip and ankle angle range. These parameters were computed based on the detected gait events and the 3-D joint data provided by the Kinect v2.

3.4   Subject Identification

The possibility of identifying subjects based on their gait was studied by investigating the use of machine learning techniques to build a predictive model for subject identification using the dataset described above. Since the original dataset was imbalanced with regards to the number of instances (or gait cycles) per subject, we firstly obtained a balanced dataset by randomly selecting without replacement the same number of gait cycles for each subject (20 gait cycles, which was the minimum number of gait cycles available per subject, when considering all subjects). Therefore, the balanced dataset included 400 gait cycles in total.

We explored different machine learning algorithms, including the k-nearest neighbors (k-NN), decision tree, random forest, support vector machines (SVM), multilayer perceptron (MLP), and multilayer perceptron ensemble (MLPE). For k-NN, we used a weighted version [18, 19], considering 7 nearest neighbors, the Euclidean distance, and the "optimal" kernel function [20] to obtain the weight of each nearest neighbor.

For the decision tree, we used an implementation [21, 22] of the recursive partitioning method for building classification and regression trees (CART) [23]. We considered the Gini index for computing the impurity of a node, the class frequency in the training set as the class prior probability, and a minimum of 20 instances or an improvement by a factor of 0.01 (complexity parameter) for attempting a split [21, 22].

For the random forest, we used an implementation [24] of Breiman's random forest [25], considering 500 grown trees, and a size of $\lfloor \sqrt{n} \rfloor$ for the feature subset selection for each node, where n is the number of features in the training set.

For SVM, we used an implementation [26, 27] of the C-SVM formulation, with a cost parameter value of 1, and the Gaussian radial basis function as the kernel function. For multi-class problems, the used implementation relies on the one-against-one approach.

For MLP, we used the implementation described in [28], where the activation function of the hidden neurons is the logistic function. For multi-class problems, the output layer has a linear neuron per class. We considered 10 hidden neurons, and a maximum of 100 iterations for finding the best set of weights.

Since the training of a MLP model is not optimal, to avoid the dependence of the final solution on the choice of starting weights (chosen at random) [29], three different MLP models were built. Then, the one with the lowest value of the fitting criterion was selected. In the case of MLPE, all built MLPs were used, where the output is the mean between the individual predictions.

For each algorithm, the performance of the corresponding predictive model was evaluated by relying on a stratified 10-fold cross validation approach [30], and the following metrics: overall accuracy (1), class accuracy (2), and class F_1 score (5).

$$\text{Overall accuracy} (\%) = \frac{\text{no. of correctly classified instances}}{\text{no. of total instances}} \times 100 \tag{1}$$

$$\text{Class accuracy} (\%) = \frac{TP + TN}{TP + TN + FP + FN} \times 100 \tag{2}$$

$$\text{Class sensitivity} (\%) = \frac{TP}{TP + FN} \times 100 \tag{3}$$

$$\text{Class precision} (\%) = \frac{TP}{TP + FP} \times 100 \tag{4}$$

$$\text{Class } F_1 \text{ score} = 2 \times \frac{\text{class precision} \times \text{class sensitivity}}{\text{class precision} + \text{class sensitivity}} \tag{5}$$

In these equations, TP, TN, FP and FN correspond to (each subject is a class):

- True positives (TP): the number of instances correctly classified as belonging to the considered class;
- True negatives (TN): the number of instances correctly classified as belonging to a class other than the one considered;
- False positives (FP): the number of instances incorrectly classified as belonging to the considered class;
- False negatives (FN): the number of instances incorrectly classified as belonging to a class other than the one considered.

For each model, we also obtained the training time, i.e., the time required for training the model, as well as the prediction time, i.e., the time it takes to identify the subject for each instance (gait cycle). These results were obtained on a computer with an i7-4600U CPU (dual-core, 2.1 GHz), and 8 GB RAM. Training time should be as low as possible, since it will be necessary to re-train the model every time there is data from a new subject. A low prediction time is also desirable for online subject identification.

The dataset balancing, as well as model training and testing, were carried out in the R environment [31]. The "rminer" package [29] was used to train the models and obtain the models' predictions for each gait cycle, and the "caret" package [32] was used to compute the performance evaluation metrics.

4 Results

The results obtained for the model evaluation metrics are presented in Table 1, for each machine learning algorithm. For overall accuracy, the results are presented as the mean value for the 10 iterations of cross validation. For class accuracy and F_1 score, the results are presented as the mean and standard deviation values for all classes (or subjects) computed over the mean value per class for the 10 iterations. The training and prediction time are presented as the mean and standard deviation values for 10 runs. The prediction time for each run is the mean prediction time for a gait cycle.

Table 1. Subject identification evaluation results for different machine learning algorithms. The best result for each metric is indicated in bold.

Algorithm	k-NN	Decision tree	Random forest	SVM	MLP	MLPE
Overall accuracy (%)	94.8	91.5	**98.8**	96.5	84.8	92.5
Accuracy (%)	99.5 ± 0.4	99.2 ± 0.6	**99.9 ± 0.2**	99.7 ± 0.4	98.5 ± 0.6	99.2 ± 0.5
F1 score (%)	94.8 ± 4.4	91.3 ± 6.7	**98.7 ± 1.7**	96.5 ± 3.6	84.7 ± 6.3	92.5 ± 4.8
Training time (s)	**0.0 ± 0.0**	0.1 ± 0.0	1.0 ± 0.2	0.4 ± 0.1	1.3 ± 0.4	1.1 ± 0.1
Prediction time per gait cycle (ms)	12.0 ± 0.8	**4.7 ± 1.8**	11.4 ± 2.5	54.2 ± 10.5	6.3 ± 1.7	11.2 ± 1.9

From Table 1, we can see that the random forest model achieved the best results for the overall accuracy, as well as the class accuracy and F_1 score. The k-NN had the lowest training time (0 s). However, training time was still relatively low for the remaining algorithms (mean of 0.1 to 1.3 s). Regarding the prediction time, the decision tree model was the fastest for performing subject identification (mean of 5 ms). However, the mean prediction time was relatively low for all other models (mean between 6 and 12 ms) except for the SVM model (54 ms). Therefore, we considered that the random forest model has the best trade-off between all considered evaluation metrics.

Table 2 presents the same results as Table 1, but for the random forest algorithm only, when considering the feature subsets corresponding to the three types of features (anthropometric measures, spatiotemporal gait parameters, and kinematic parameters). The results for all gait parameters and all features are also included in the table.

Table 2. Subject identification evaluation results for different feature subsets, when using the random forest algorithm. The best result for each metric is indicated in bold.

Features	3 anthropometric measures	13 spatiotemporal gait parameters	9 kinematic parameters	22 gait parameters	All 25 features
Overall accuracy (%)	88.2	63.2	76.5	88.5	**98.8**
Accuracy (%)	98.8 ± 1.1	96.3 ± 1.8	97.7 ± 0.8	98.8 ± 0.6	**99.9 ± 0.2**
F1 score (%)	88.2 ± 12.0	63.2 ± 20.9	76.5 ± 11.9	88.5 ± 7.6	**98.7 ± 1.7**
Training time (s)	**0.2 ± 0.1**	0.5 ± 0.1	0.4 ± 0.1	0.8 ± 0.0	1.0 ± 0.2
Prediction time per gait cycle (ms)	**5.9 ± 1.3**	9.3 ± 1.9	8.0 ± 1.8	9.9 ± 2.1	11.4 ± 2.5

As can be seen from Table 2, the model's performance was better when using all three types of features. When comparing each different feature type, the anthropometric measures had the best performance, followed by the kinematic gait parameters, and finally the spatiotemporal gait parameters. Model training and prediction time was the lowest for the smallest feature set, as expected. However, the differences between the different feature sets are not considerable.

5 Discussion

From the different explored machine learning algorithms, the random forest led to the model with highest overall accuracy (99%), mean class accuracy (100%), and mean class F_1 score (99%). The overall accuracy for the remaining models varied between 85% (MLP) and 97% (SVM). As expected, the random forest achieved a better performance than the decision tree, since it relies on a large number of decision trees instead of a single tree. When comparing MLPE with MLP, the performance was better for MLPE, since it relies on three rather than a single MLP model.

Although the SVM also had a relatively good performance, it had the highest prediction time (54 ms). The decision tree model had the lowest prediction time and one of the lowest training times, but it presented one of the worst performances. Regarding interpretability, k-NN and decision tree models are easier to understand. When taking into account all these aspects, we considered the random forest algorithm as the most appropriate for subject identification.

The results for the models built with each one of the three types of features, and the random forest algorithm, showed that the overall accuracy is relatively low for the spatiotemporal or kinematic gait parameters (63% and 77%, respectively). When considering all 22 gait parameters, the performance was much better (89%). A similar performance was achieved when using only the 3 anthropometric measures (88%). Nevertheless, the best performance was achieved when using all features (overall accuracy of 99%). Therefore, all three feature types seem to be important for subject identification. Both training and prediction time were higher when a greater number of features was used, as expected, but the difference between using 3 or 25 features is not substantial (0.2 s or 1.0 s for training, and 6 or 11 ms for predicting).

When comparing our study with other similar studies, our results (overall accuracy of 99%, mean accuracy of 100%, and mean F_1 score of 99%) are better than or similar to those reported in other contributions. Ball et al. achieved a mean accuracy of 44% (frontal view), when using 18 kinematic parameters [15] (we achieved 77% with 9 kinematic parameters). Sinha et al. reported a mean F_1 score of 62% [13], and Jiang et al. obtained an overall accuracy of 82% [12]. Preis et al. achieved an overall accuracy of 91% with 4 anthropometric measures [14], while we obtained a slightly lower result (88%) with 3 anthropometric measures. However, the dataset of Preis et al. included only 9 subjects, while ours corresponded to 20 subjects.

Andersson et al. reported an overall accuracy of ≈96% for 20 subjects [11], which is similar to our result. However, we used a much lower number of features (25 instead of 80 features). The results obtained by Rahman et al. are also similar to ours (overall accuracy between 93% and 95% [10]), but they used a much larger number of features (760 features). Ahmed et al. [9] and Nambiar et al. [8] both reported an overall accuracy of 92%. We obtained a better result (99%) with a lower number of features (25 features, versus 42 or 40 features/measures, respectively).

6   Conclusion and Future Work

In this contribution, we presented a study on subject identification using features extracted from gait data provided by a single low-cost RGB-D camera (Kinect v2). From data acquired from twenty healthy subjects while they walked towards the camera several times, we computed 3 anthropometric measures and 25 spatiotemporal and kinematic parameters for each detected gait cycle (400 gait cycles in total). From the explored machine learning algorithms, the random forest led to the predictive model with the best performance regarding subject identification based on a single gait cycle: overall accuracy of 99%, class accuracy of $100 \pm 0\%$, and F_1 score of $99 \pm 2\%$. When compared with other similar contributions (Kinect v2, frontal view, 20 subjects) [8–10], we achieved a better performance using a lower number of features. In the future, we will further investigate if our results can be improved by carrying out automatic feature selection to choose the most relevant features.

The results of our study show the potential of using a RGB-D camera to perform biometric authentication based on the subject's gait, which can be useful for smart surveillance and access control. However, the gait cycles used for model validation belonged to the same data acquisition as the ones used for model training. Given that a subject's gait can change overtime, in the future we will perform different acquisitions with each subject (a few months apart, for example) and carry out a new study to verify if the solution proposed in this contribution can be used to identify a given subject at different points in time.

Acknowledgments. This work was supported by EU funds through Programa Operacional Factores de Competitividade (COMPETE), and by national funds through Fundação para a Ciência e a Tecnologia (FCT), in the context of scholarship SFRH/DB/110438/2015, and projects UID/CEC/00127/2013, Incentivo/EEI/UI0127/2014, FCOMP-01-0124-FEDER-028943, and FCOMP-01-0124-FEDER-029673. It was also partially funded by national funds through North Portugal Regional Operational Programme (NORTE 2020) in the context of projects NORTE-01-0145-FEDER-000016 and NORTE-01-0145-FEDER-000020. The authors wish to thank the subjects that participated in the study, and the LABIOMEP staff who assisted in the data acquisitions.

References

1. Boulgouris, N.V., Hatzinakos, D., Plataniotis, K.N.: Gait recognition: a challenging signal processing technology for biometric identification. IEEE Sig. Process Mag. **22**(6), 78–90 (2005). https://doi.org/10.1109/MSP.2005.1550191
2. Wang, J., She, M., Nahavandi, S., Kouzani, A.: A review of vision-based gait recognition methods for human identification. In: International Conference on Digital Image Computing: Techniques and Applications (DICTA), Sydney, NSW, Australia, 1–3 December 2010, pp. 320–327. IEEE (2010)
3. Nixon, M.S., Tan, T., Chellappa, R.: Human Identification Based on Gait, vol. 4. Springer, Heidelberg (2010)
4. Collins, R.T., Gross, R., Jianbo, S.: Silhouette-based human identification from body shape and gait. In: IEEE International Conference on Automatic Face and Gesture Recognition, 21 May 2002, pp. 366–371. IEEE (2002)
5. Cheng, M.-H., Ho, M.-F., Huang, C.-L.: Gait analysis for human identification through manifold learning and HMM. Pattern Recogn. **41**(8), 2541–2553 (2008). https://doi.org/10.1016/j.patcog.2007.11.021
6. Lam, T.H.W., Cheung, K.H., Liu, J.N.K.: Gait flow image: a silhouette-based gait representation for human identification. Pattern Recogn. **44**(4), 973–987 (2011). https://doi.org/10.1016/j.patcog.2010.10.011
7. Wang, L., Tan, T., Ning, H., Hu, W.: Silhouette analysis-based gait recognition for human identification. IEEE Trans. Pattern Anal. Mach. Intell. **25**(12), 1505–1518 (2003). https://doi.org/10.1109/TPAMI.2003.1251144
8. Nambiar, A., Bernardino, A., Nascimento, J.C., Fred, A.: Towards view-point invariant person re-identification via fusion of anthropometric and gait features from Kinect measurements. In: International Joint Conference on Computer Vision, Imaging and Computer Graphics Theory and Applications (VISIGRAPP), Porto, Portugal, 27 February–1 March 2017, pp. 108–119. SciTePress (2017)
9. Ahmed, F., Paul, P.P., Gavrilova, M.L.: DTW-based kernel and rank-level fusion for 3D gait recognition using Kinect. Vis. Comput. **31**(6), 915–924 (2015). https://doi.org/10.1007/s00371-015-1092-0
10. Rahman, M.W., Gavrilova, M.L.: Kinect gait skeletal joint feature-based person identification. In: IEEE International Conference on Cognitive Informatics and Cognitive Computing (ICCI*CC), 26–28 July 2017, pp. 423–430 (2017)
11. Andersson, V.O., Araujo, R.M.: Full body person identification using the Kinect sensor. In: IEEE International Conference on Tools with Artificial Intelligence, 10–12 November 2014, pp. 627–633 (2014)
12. Jiang, S., Wang, Y., Zhang, Y., Sun, J.: Real time gait recognition system based on Kinect skeleton feature. In: Asian Computer Vision - ACCV 2014 Workshops Conference on Computer Vision, Singapore, 1–5 November 2014, pp. 46–57. Springer (2014)

13. Sinha, A., Chakravarty, K., Bhowmick, B.: Person identification using skeleton information from Kinect. In: International Conference on Advances in Computer-Human Interactions, Nice, France, 24 February–1 March 2013, pp. 101–108 (2013)
14. Preis, J., Kessel, M., Werner, M., Linnhoff-Popien, C.: Gait recognition with Kinect. In: International Workshop on Kinect in Pervasive Computing, New Castle, UK, 18–22 June 2012, pp. P1–P4 (2012)
15. Ball, A., Rye, D., Ramos, F., Velonaki, M.: Unsupervised clustering of people from 'skeleton' data. In: Annual ACM/IEEE International Conference on Human-Robot Interaction, Boston, Massachusetts, USA, 5–8 March 2012, pp. 225–226. ACM (2012)
16. Cunha, J.P.S., Rocha, A.P., Choupina, H.M.P., Fernandes, J.M., Rosas, M.J., Vaz, R., Achilles, F., Loesch, A.M., Vollmar, C., Hartl, E., Noachtar, S.: A novel portable, low-cost Kinect-based system for motion analysis in neurological diseases. In: Annual International Conference of the IEEE Engineering in Medicine and Biology Society (EMBC), Orlando, Florida, USA, 16–20 August 2016, pp. 2339–2342. IEEE (2016)
17. Rocha, A.P., Choupina, H.M.P., Vilas-Boas, M.d.C, Fernandes, J.M., Cunha, J.P.S.: System for automatic gait analysis based on a single RGB-D camera. PLoS ONE **13**(8), e0201728 (2018). https://doi.org/10.1371/journal.pone.0201728
18. Hechenbichler, K., Schliep, K.: Weighted k-nearest-neighbor techniques and ordinal classification (2004)
19. Hechenbichler, K., Schliep, K.: kknn: Weighted k-nearest neighbors (R package version 1.3.1) (2016)
20. Samworth, R.J.: Optimal weighted nearest neighbour classifiers. Ann. Stat. **40**(5), 2733–2763 (2012). https://doi.org/10.1214/12-aos1049
21. Therneau, T.M., Atkinson, E.J.: An introduction to recursive partitioning using the RPART routines, Mayo Foundation (2015)
22. Therneau, T., Atkinson, B.: rpart: Recursive partitioning and regression trees (R package version 4.1-13) (2018)
23. Breiman, L., Friedman, J., Olshen, R.A., Stone, C.J.: Classification and Regression Trees. Chapman & Hall, (1984)
24. Liaw, A., Wiener, M.: Classification and regression by randomforest. R News **2**(3), 18–22 (2002)
25. Breiman, L.: Random forests. Mach. Learn. **45**(1), 5–32 (2001). https://doi.org/10.1023/A:1010933404324
26. Karatzoglou, A., Meyer, D., Hornik, K.: Support vector machines in R. J. Stat. Softw. **15**(9), 1–28 (2006). https://doi.org/10.18637/jss.v015.i09
27. Karatzoglou, A., Smola, A., Hornik, K., Zeileis, A.: kernlab - an S4 package for kernel methods in R. J. Stat. Softw. **11**(9), 1–20 (2004). https://doi.org/10.18637/jss.v011.i09
28. Venables, W.N., Ripley, B.D.: Modern applied statistics with S. Statistics and Computing, 4th edn. Springer, New York (2002)
29. Cortez, P.: Data mining with neural networks and support vector machines using the R/rminer tool. In: Perner, P. (ed.) Advances in Data Mining: Applications and Theoretical Aspects, pp. 572–583. Springer, Heidelberg (2010)
30. Han, J., Kamber, M., Pei, J.: Data Mining: Concepts and Techniques, 3rd edn. Morgan Kaufmann, Burlington (2012)
31. R Core Team R: A language and environment for statistical computing (2015)
32. Kuhn, M.: caret: Classification and regression training (R package version 6.0-80) (2018)

Leakage Detection of a Boiler Tube Using a Genetic Algorithm-like Method and Support Vector Machines

Young-Hun Kim[1], Jaeyoung Kim[1], and Jong-Myon Kim[2(✉)]

[1] Department of Electrical and Computer Engineering,
University of Ulsan, Ulsan, South Korea
dudgns0908l@gmail.com, kjy7097@gmail.com
[2] School of IT Convergence, University of Ulsan, Ulsan, South Korea
jmkim07@ulsan.ac.kr

Abstract. In this paper, we propose a method to detect boiler tube leakage using a genetic algorithm (GA)-like method and support vector machines (SVM). The GA-like method allows for selection of significant features, and the SVM detects a leak in boiler tubes using the selected features. Experimental results indicate that the proposed method outperforms a state-of-the-art principle component analysis (PCA) method in leakage detection.

Keywords: Boiler tube · Genetic algorithm · Support vector machine · Tube leakage detection

1 Introduction

The number of thermal power plants is increasing due to technological advances. The boiler, which is an important component of a thermal power plant, contains thousands of exchanger tubes to transfer the energy released from combustion to water, creating high temperature and pressure steam.

Boiler tubes may develop defects during the bending process. Leakage can be caused by overheating, foreign matter inflow, and reduction of thickness due to machining. Unexpected boiler tube leakage causes tremendous economy loss. This accidental boiler tube leakage occurs more than a dozen times a year in thermal power plants [1]. Therefore, detection of boiler tube leakage is essential. Tube leakage identified by the acoustic emission method is a dynamic inspection method that can measure the reaction of a structure due to growth of defects under structural stress [1].

To address the issue, principal component analysis (PCA) has been widely used, which is a statistical procedure that extracts principal components of a large data set while retaining most of the features. This is mainly used as a dimension reduction algorithm and can be applied to the fault diagnosis method [2–7] using an existing acoustic emission signal. However, since PCA considers all features, those that are not required may be included, reducing classification performance.

© Springer Nature Switzerland AG 2020
A. M. Madureira et al. (Eds.): SoCPaR 2018, AISC 942, pp. 86–93, 2020.
https://doi.org/10.1007/978-3-030-17065-3_9

In this paper, we propose a feature selection method using a genetic algorithm (GA)-like method that selects features useful for classification using data collected through an acoustic emission sensor. Selected data is then used to support vector machines (SVM) for detecting the leakage of a boiler tube.

This paper is organized as follows. Section 2 describes the related research, and Sect. 3 explains the proposed method and results. Finally, Sect. 4 concludes this paper.

2 Background

2.1 Principal Component Analysis (PCA)

The guiding principle of PCA is to reduce high dimensional data to low dimension. This approach tries to find a new orthogonal axis while preserving the variance of data and converts specimens of high dimensional space into low dimensional space without a linear relation. It removes unnecessary data among a large data set and transforms it into a combination of data that does not deviate greatly from the tendency of the present data [8].

2.2 Genetic Algorithm

The genetic algorithm (GA) method was developed by John Holland and is one of the techniques used to solve optimization problems. It is a representative technique of evolutionary computation that imitates biological evolution. In this algorithm, there are terms that imitate real evolutionary processes such as selection, crossover, and mutation. GA represents possible solutions to select informative features which are useful for detecting leaks of a boiler tube. Through gradual transformation, it creates more and better solutions. The data structure representing the solutions is a gene, and the process of creating a better solution through transformation represents evolution.

2.3 Support Vector Machine (SVM)

SVM is part of the field of machine learning and is a supervised learning model for pattern recognition and data analysis. It is a non-stochastic binary linear classification model that determines the category to which new data belong based on a given data set. In this classification model, the data is represented in space as a boundary, and the SVM algorithm finds the boundary with the largest width.

3 Proposed Method

In this paper, we propose a feature selection method based on a genetic algorithm and detect leakage method of boiler tubes using SVM. The following is the overall structure of the proposed method (Fig. 1).

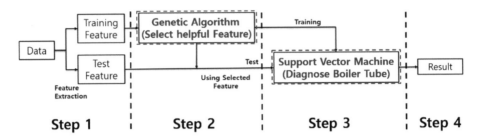

Fig. 1. Structure of the proposed method

In Step 1, we extract 22 features (see Table 1) for each channel using the data collected from four acoustic emission sensors. Since there are four channels, 88 features are ultimately extracted [2, 3] and are separated into learning features and test features for learning and testing.

Table 1. Feature types

No	Feature	No	Feature	No	Feature	No	Feature	No	Feature
1	Peak	6	Peak to peak	11	Shape factor	16	Normalize 6	21	Frequency spectrum energy
2	RMS	7	Kurtosis factor	12	Entropy	17	Shape factor 2	22	Hit
3	Kurtosis	8	Impulse factor	13	Energy	18	Frequency center		
4	Skewness	9	Margin factor	14	Clearance factor	19	RMS frequency		
5	SMR	10	Crest factor	15	Normalize 5	20	RV frequency		

In Step 2, since not all 88 extracted features contain useful information for boiler tube diagnostics, the GA-like method is performed to select useful features for boiler tube diagnostics. The use of the GA-like method reduces the number of features from 88 to three. The objective function of the GA-like method is to select features with useful information for detecting boiler tube leakage, and it is an evaluation index of discrimination between classes and density within a class.

$$C = \frac{1}{N_c \times N_s^2} \sum_{i=1}^{N_c} \sum_{j=1}^{N_s} \sum_{k=1}^{N_s} |S_i(j) - S_j(k)| \tag{1}$$

$$C = \frac{1}{N_c \times (N_c - 1) \times N_s^2} \sum_{i=1}^{N_c} \sum_{\substack{j=1 \\ j \neq i}}^{N_c} \sum_{k=1}^{N_s} \sum_{l=1}^{N_s} |S_i(k) - S_j(l)| \tag{2}$$

$$F = \frac{C}{S} \tag{3}$$

Equation 1 represents the density in a class, and Eq. 2 represents discrimination between classes. Since the density in a class should be high and the distinction between classes should be low, the objective function is constructed as shown in Eq. 3.

In Step 3, a classification model is created by training the SVM using three features selected by the genetic algorithm.

In Step 4, test features are selected in the same way as in the training process. Test results are obtained using the SVM, and the performance of the algorithm can be derived.

4   Experimental Environment and Results

In this study, four artificial pinholes were created in a boiler tube to obtain the acoustic emission signal for normal operation and during leakage. Pinhole diameters were 0.6 mm, 1 mm, and 2 mm. Figure 2 shows a side view of the boiler tube used in the actual experiments.

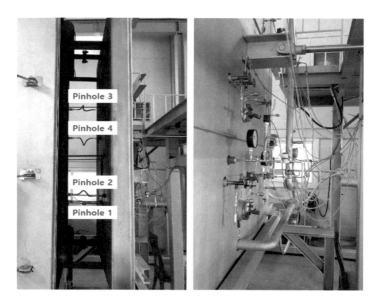

Fig. 2. Boiler tubes used in experiments

Figure 3 shows a floor plan of the boiler tube and the positions of the pinhole.

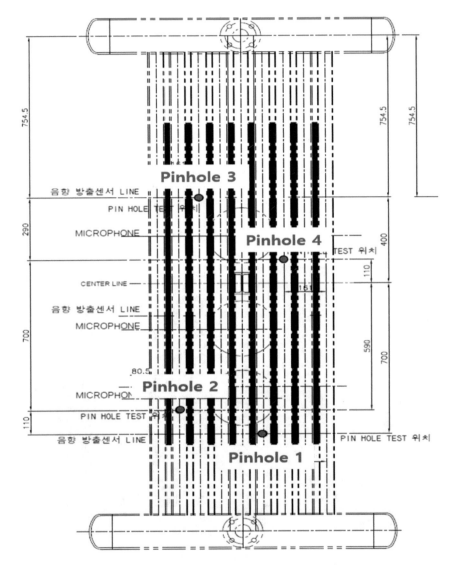

Fig. 3. Boiler tube drawings and pinhole locations

First, leakage was generated for each pinhole position and size, and an acoustic emission signal for each corresponding normal operation and leakage was obtained. Data was also obtained when pinholes 1 and 3 were generated simultaneously. In addition, each data was collected for 5 min. Pinholes were blocked for about 3 min and then exposed for 3 min to obtain data representing pre-leakage and post-leakage cases. Figure 4 shows the signal for an artificial pinhole having a diameter of 2 mm at pinhole position 4.

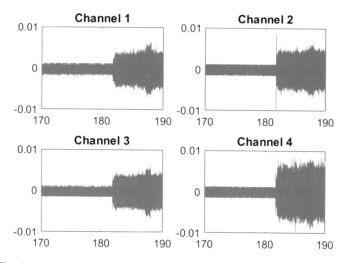

Fig. 4. Signal comparison before and after pinhole generation (pinhole 4)

In Fig. 4, the x-axis is seconds, and the y-axis is the signal. As shown in Fig. 4, the data before and after formation of the artificial pinhole show a marked difference upon initiation of the leak.

After extracting a total of 88 features from the acquired data, a GA-like method is applied to select features that contain information useful for detecting boiler tube leakage. Figure 5 shows the results of comparing GA and an existing PCA methods. As shown in Fig. 5, the GA-like method outperforms PCA in class separation. The selected features are then used as the input values of the SVM to detect leakage of the boiler tube.

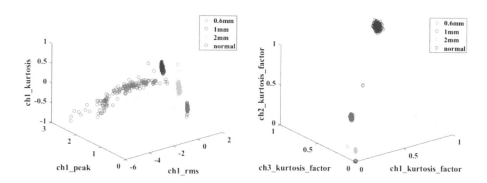

Fig. 5. Comparison of the PCA (left) and GA-like (right)

Table 2 shows the performance of the SVM for pinhole 4, where both normal and the three diameters are classified as 100% performance.

Table 2. Performance of SVM using a confusion matrix (pinhole 4)

	0.6 mm	1 mm	2 mm	Normal
0.6 mm	50	0	0	0
1 mm	0	50	0	0
2 mm	0	0	50	0
Normal	0	0	0	50

Table 3 shows the results of using the features selected by the proposed GA-like method and PCA as inputs to the SVM. As shown in Table 3, the results of selected features using GA show better classification performance than those of features selected using PCA.

Table 3. Performance comparison of the proposed GA-like method and PCA

Hole	Signal	
	Acoustic emission	
	Proposed	PCA
Pinhole 1	100%	99.5%
Pinhole 2	98.5%	97.5%
Pinhole 3	99.5%	99.5%
Pinhole 4	100%	99.5%
Pinhole 1, 3	100%	99%

5   Conclusions

In this paper, a GA-like method was used to select useful features for detection of boiler tube leaks. The selected features using the GA-like method was used as input of SVM to detect boiler tube leaks. Experimental results showed that the proposed GA-based method outperforms the PCA in classification accuracy.

Acknowledgement. This work was supported by the Korea Institute of Energy Technology Evaluation and Planning (KETEP) and the Ministry of Trade, Industry & Energy (MOTIE) of the Republic of Korea (No. 20161120100350, No. 20181510102160, No. 20162220100050).

References

1. Lee, S.B., Roh, S.M.: Developing an early leakage detection system for thermal power plant boiler tubes by using acoustic emission technology. J. Korean Soc. Nondestr. Test. **38**(2), 181–187 (2005)
2. Kim, J., Kim, J.: Methods and devices for diagnosing facility conditions, Patent Registration No. 10-1818394 (2018)

3. Kim, J., Kim, J.: Methods and devices for diagnosing machine faults, Patent Registration No. 10-1797402 (2018)
4. Kim, J., Kim, J.: Apparatus and method for machine fault diagnosis, Patent Registration No. 10-1745805 (2017)
5. Kim, J., Kim, J.: Machine fault diagnosis method, Patent Registration No. 10-1808390 (2017)
6. Kim, J., Kim, J.: Apparatus and method for monitoring machine condition, Patent Registration No. 10-1745805 (2017)
7. Kim, J., Kim, J.: Method and apparatus for predicting remaining life of a machine, Patent Registration No. 10-1808461 (2017)
8. Lee, K., Lee, B.W., Choi, D.-H., Kim, T.-O., Shin, D.: A study on fault detection monitoring and diagnosis system of CNG stations based on principal component analysis (PCA). J. Korean Inst. Gas **18**(3), 53–59 (2014)
9. Kang, M., Islam, M.R., Kim, J., Kim, J.-M., Pecht, M.: A hybrid feature selection scheme for reducing diagnostic performance deterioration caused by outliers in data-driven diagnostics. IEEE Trans. Ind. Electron. **63**(5), 3299–3310 (2016)
10. Kang, M., Kim, J., Wills, L.M., Kim, J.-M.: Time-varying and multiresolution envelope analysis and discriminative feature analysis for bearing fault diagnosis. IEEE Trans. Ind. Electron. **62**(12), 7749–7761 (2015)

Sentiment Analysis on Tweets for Trains Using Machine Learning

Sachin Kumar[(✉)] and Marina I. Nezhurina

College of IBS, NUST MISIS, Leninsky Prospect 4, 119049 Moscow, Russia
sachinagnihotri16@gmail.com

Abstract. Sentiment analysis is a popular theme in the natural language processing (NLP) domain. People at present share their stay experience in restaurants, shopping malls, hotels and their travel experience in taxis, buses, trains and airplanes. Online social media provide a platform for the people to share their experiences of stay and travel in the form of text, images and videos. Twitter is one of the popular and well known social media platforms across the world. In this study, we are using tweets data in respect to comfort services in Indian long route superfast trains. This tweet data is used to analyze the hidden sentiments using machine learning techniques such as support vector machines (SVM), Random forest (RF) and back propagation neural networks (BPNN). The results show that BPNN provides high accuracy with more training on the data. The results achieved from SVM and RF was also satisfactory but BPANN won the race with more training on the data.

Keywords: Classification · Support vector machines · Random forest · Twitter · Back propagation neural network

1 Introduction

Sentiment analysis is one of the trendy research domains in natural language processing (NLP). Sentiments or opinions can be defined as a mental situation or feeling of a person in certain circumstances and conditions [1]. These feelings may be a reflection of joy, sadness, discomfort or nervousness. Present world is full of technology and smart devices. Moreover, availability of fast internet and huge storage capability with social media platforms, people can share their feelings online. Twitter [2] is one of the biggest online social media platform followed by billions of people across the world. The personal account of individual on Twitter may be considered as a microblog [3] as it provides a limited text of 140 characters to share the views. There are a lot of studies on sentiment analysis which takes into consideration air travel, tourism, stay in hotels, restaurants, movies, politics etc. [4–8]. Trains are also one of the popular modes of transport which are time consuming (in comparison to air travel) but more convenient and comfortable for long routes.

India is a country with second largest population in the world after china [9]. In India, people mostly prefer trains to cover the long distances because of convenience of comfort, low fare etc. However, travel in trains may not be always comfortable due to various reasons i.e. food quality, cleanliness, number of people in the coach, quality of

A. M. Madureira et al. (Eds.): SoCPaR 2018, AISC 942, pp. 94–104, 2020.
https://doi.org/10.1007/978-3-030-17065-3_10

air conditioner or heating system and other reasons. These comfort and discomfort situations give rise to emotions or sentiments that illustrates the mental situation of a passenger during travel time. Present world is equipped with smart devices, high speed internet, Wi-Fi facilities and social media platforms. On social media platforms such as Facebook, Twitter or Instagram, people are sharing their opinions on different subjects every second. Facebook and Instagram are usually famous among common people around the world whereas Twitter is used by professionals, official government authorities, public sectors, multinational companies and celebrities etc. Therefore, people use Twitter as a platform to share their official and real opinion about anything as 51% of the contents are found generic opinions [10]. Figure 1 illustrates some tweets on Indian railways showing negative sentiments. It also shows a quick decision taken by respective authority to provide comfort.

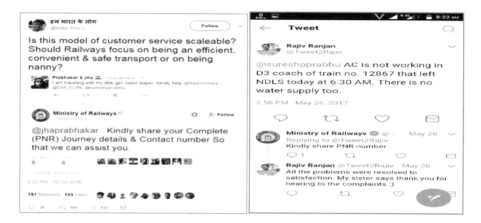

Fig. 1. A tweet criticizing the railway ministry

This study took twitter data in respect to travel in superfast trains, into consideration. The respective tweets from Twitter database have been extracted by using specific hash tags and further machine learning methods has been used to map these tweets into two categories of sentiments i.e. positive and negative. SVM, RF and deep neural network is deployed for this purpose.

The rest of the article is organized as follows: Sect. 2 provides a literature review on sentiment analysis in different domains. In Sect. 3, data set description and its preprocessing is discussed which is followed the methods used. Section 4 carries results and discussion. The study concludes in Sect. 5.

2   Related Work

Sentiment analysis or opinion mining has been a popular research domain for several years. In a study by Turney [11], movie reviews were classified as positive or negative using semantic orientation. They considered a review as positive if it has more than average positive nuance and negative for more than average negative nuance. Furthermore, it is also found that traditional topic based analysis of sentiments perform better than three standard machine learning algorithms namely Naïve Bayes (NB), maximum entropy classification, SVM [12]. They formed two class labels i.e. positive and negative for movie reviews data and performed binary supervised classification. They mentioned that performance of SVM was better than other two approaches but the difference was not big. However, another study [13] found machine learning techniques to be effective on sentiment analysis specifically for classification. They also mentioned that feature selection is powerful enough to strengthen the prediction accuracy of classification model. In a book chapter [14], Liu provides a comprehensive state of art literature regarding problems of sentiment analysis, different methods to investigate the hidden sentiments etc. Singh et al. [15] also took into consideration the movie reviews and performed an analysis on document level classification of sentiments using SentiWordNet approach with slight variations.

Besides movie reviews, sentiment analysis were performed for hotel industries as well. Shi and Li [16] used SVM to perform sentiment analysis on hotel reviews. They made use of unigram feature along with TF-IDF (term frequency-inverted document frequency). They mentioned that use of TF-IDF is more effective than just frequency of words. Furthermore Adebornaa and Siau [5] performed a sentiment analysis on airline reviews. They proposed a sentiment topic recognition model to investigate and detect the polarity and the sentiments from the given text. They used the airline reviews for AirTrans, Frontier and SkyWest airlines. Lacic et al. [17] investigated the traveler satisfaction level based on sentiment analysis of airline reviews. They extracted the reviews from Skytrax air travel review portal that provides four categories to review about airports, airlines, seats and lounges.In a different study by Zou et al. [18] suggested that it's a difficult task to analyze and process the microblog tweets due to massive amount and noisy information. They combined social and textual context to strengthen the sentiment analysis process.

In order to perform sentiment analysis, they proposed a method that makes use of social context in the sentences or tweets. They introduced new similarity index to define the structured similarity. They used Laplacian matrix to develop the model for sentiment polarity classification. Customer satisfaction has always been a primary objective for the hotels, lounges, airlines, trains, buses etc. In present high technology world, everyone is aware about that customer are habitual to write reviews after their visit or travel. Therefore, everyone is trying to provide facilities and maximum comfort to the best of their level to please the customer with a motive to receive positive reviews. These positive reviews are one of the important ways of online and genuine publicity which can result in a rise in business. The reviews are not only positive or negative but it contains a lot of hidden information inside. Several studies have attempted to investigate such hidden information for airplanes, hotels, online booking company, politics etc. There are no studies available that considered the sentiments of travelers in Indian trains. Therefore, this study presents a machine learning based approach to perform sentiment analysis on the reviews based on Indian long route trains.

3   Materials and Methods

3.1   Data Set Description

The respective tweets have been downloaded from Twitter database using tweepy package. The hash tags have been carefully chosen to extract tweets. The script to download tweets was written in Python 3.6 version. The tweets were downloaded from the duration between 02.08.2018 to 12.08.2018. The hash tags used for downloading tweets were #railminIndia, #indianrailways, #northernrailways, #railwayminister. Initially 55,499 tweets were extracted from the server. Further, a total of 15777 unique tweets were obtained after removing duplicate retweets.

3.2   Data Preprocessing

The tweets were extracted in JSON (JavaScript Object Notation) format. Therefore, once the tweets were extracted, they have been converted into CSV (Comma Separate Values) format. The dataset is prepared by providing a class label for each of the tweets. These labels were selected to be assigned as positive, negative or neutral. Furthermore the major reasons for the tweets were investigated based on the frequency of important words. These categories are illustrated in Fig. 2.

The percentage distribution of all the tweets for various categories in three sentiment classes is illustrated in Fig. 2 and all those categories are illustrated in Fig. 3. Bag of Words (BoW) is one of the known models in sentiment analysis to create feature vectors. The problem with BoW is that it does not consider the ordering of word in a given sentence. Therefore Doc2Vec approach is used that takes into consideration the word ordering and maps each sentence in a vector space.

3.3   Machine Learning Techniques

The level of sentiment analysis [22] has already achieved a significant rise using machine learning techniques. In this study, three machine learning techniques have been used to build a sentiment classification model for tweets extracted. Support Vector Machine, Random Forest and BPANN.

SVM is a popular classification method that develops a classification model by finding the margin threshold to distinguish between different classes (Fig. 4). SVM was initialized designed for binary classification problems where the target class values have two labels only. For multiclass classification problems, there versions of SVM were introduced [19–21]. These versions are one against all (OAA), one against one (OAO) and direct acyclic graph (DAG) SVM.

OAA-SVM was the earliest model used to solve multi-class classification problem. OAA-SVM creates n SVM models, where n is the number of target class. All these n models must be trained on the dataset with the respective class value as positive and all remaining classes as negative. Therefore, given a data set with k instances: $(x_1, y_1), (x_2, y_2), \ldots, (x_k, y_k), where x_i \in R^z, i = 1, 2, \ldots, k$ and $y_i \in \{1, 2, \ldots, n\}$ is the class value of x_i.

The zth SVM model solves the problem as follows:

$$\min_{w^z, b^z, \xi^z} \frac{1}{2}(w^z)^T w^z + C \sum_{i=1}^{k} \xi_i^z$$

$$(w^z)^T \phi(x_i) + b^z \geq 1 - \xi_i^z, \; if \; y_i = z, \tag{1}$$

$$(w^z)^T \phi(x_i) + b^z \leq -1 + \xi_i^z, \; if \; y_i \neq z$$

$$\xi_i^z \geq 0, i = 1, 2, \ldots, k$$

In the above Eq. 1, the training instance x_i has been mapped to higher dimension by function ϕ and C is the penalty.

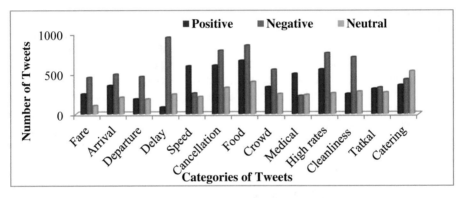

Fig. 2. Tweets categorization for different sentiment classes

Equation 1 tries to minimize $\frac{1}{2}(w^z)^T w^z$ in order to find the maximum margin between the positive and negative class present in the data. Solving Eq. 1 results in the n decision functions for all the class values as given in Eq. 2.

$$(w^i)^T \phi(x) + b^i, i = 1, 2, \ldots, n \tag{2}$$

The target class of a data sample x would be the given by the highest value of the decision function as follows:

$$Class_x \equiv \max_{z=1,\ldots n} ((w^z)^T \phi(x) + b^z) \tag{3}$$

OAO-SVM is another multiclass problem solving approach used in several studies. Unlike OAA-SVM, it starts constructing $n(n-1)/2$ classification models. Each of the classification models has been trained on all the data from two classes. The data from class c and d, the binary classification problem is solved as follows:

$$\min_{w^{cd},b^{cd},\xi^{cd}} \frac{1}{2}(w^{cd})^T w^{cd} + C \sum_{i=1}^{n} \xi_i^{cd}$$

$$(w^{cd})^T \phi(x_i) + b^{cd} \geq 1 - \xi_i^{cd}, \; if \; y_i = c \qquad (4)$$

$$(w^{cd})^T \phi(x_i) + b^{cd} \leq -1 + \xi_i^{cd}, \; if \; y_i \neq d,$$

$$\xi_i^{cd} \geq 0$$

Fig. 3. Various tweet categories

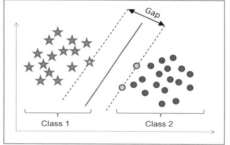

Fig. 4. Hyper plane separating two classes in SVM

After, the construction of all $n(n-1)/2$ classifiers, an instance x can be mapped to its corresponding class by following the voting strategy. Every data instance must be checked by all $n(n-1)/2$ classifiers to be a part of class c or d. Finally, a data instance would be assigned a class value with highest voting obtained by all the classification models.

DAG-SVM is the third version of SVM dealing with multi-class problems. DAG-SVM follows the same approach used by OAO-SVM by constructing $n(n-1)/2$ classifier for training the data. For testing purpose, it creates a rooted DAG with $n(n-1)/2$ internal nodes and n external nodes or leaves representing class values. Each internal node in DAG represents a binary SVM of class c and d. For every new sample in the test data, it solves a binary classification function to decide its further movement along the path. Based on the decision taken by the binary function, it proceeds along the respective branch and the process continues. Finally it ended with one leaf node which is the assigned class value for the data instance. Random forest is a classification technique based on the concept of ensemble learning. It starts by constructing several decision trees for the same data set. Furthermore, voting method is used to take into account for classification. The class voted by most of the decision trees models is to be assigned as the predicted class for particular data instance (Fig. 5). The main motive of RF is to enhance the prediction accuracy. Earlier decision tree itself has been a popular approach for classification. However, it was not guarantee to have an optimize tree with efficient prediction accuracy. Therefore, RF was proposed with a principle to encourage the diversity between trees. Different type of trees can be built for the same data set in a given model. Further, decision from every individual tree is to be combined to achieve the final classification. Finally, majority voting is used to map a given instance to a selected category or class values. RF has several advantages over decision trees i.e. it avoids over fitting by using ensemble learning approach and also improves stability and accuracy of the model.

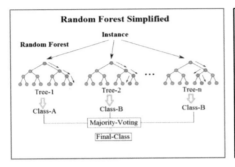 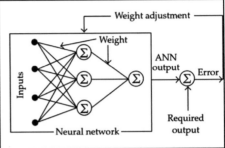

Fig. 5. Back-propagation ANN   **Fig. 6.** Classification in random forests

ANN is one of the popular prediction and classification technique for both numeric and categorical data. ANN usually has one input layer, one output layer and one or more hidden layers. Number of neurons in each layer depends on the type of data and number of features in the data set. There are respective weights on each layer that has to be multiplied with the input values and then pass on to the next layers. BPANN is another version of ANN which back propagates the prediction error (actual output – predicted output) to the previous layers (Fig. 6). This propagated error is used to improve the weights on each layer. The aim of this back propagation approach is to minimize the prediction error and improve the accuracy. The minimization of prediction error also depends on the number of times a network is trained upon particular dataset.

4   Results and Discussion

K-fold cross validation method with standard k value as 10 is used for training the classifiers. SVM and RF techniques were first used to train the models on the dataset. All 3 versions of SVM namely OAA, OAO and DAG were used for analysis. The effectiveness of the trained model is evaluated on the basis of precision, recall, F-score and area under curve (AUC). Table 1 provides the evaluation parameters for both the RF and three versions of SVM. It can be seen that both RF and SVM provides better prediction rate. RF provides 79.49% precision, 79.65% Recall and AUC value of 0.815 which is a good indicator. However, all three versions of SVM won the race with slightly better parameter values. SVM-DAG performed almost similar to SVM-OAO but still has the highest prediction rate as in Table 1. The highest AUC value is achieved by 0.845 by SVM DAG. On the other hand, Table 2 provides the performance of BPANN. As BPANN, tries to reduce the prediction error or improve the prediction accuracy by propagating the prediction error to the layers backwards. Further, weights on the hidden layers are modified accordingly and the process goes on until the number of iterations terminated. In this study, we used one hidden layer in our neural network to train the model. Initially, we started with 200 iterations for training the model on the data. Further, numbers of iteration have been increased from 200 to 1000. In very first experiment with just 200 iterations to train the model, the obtained value of AUC was 0.689, which is the worst performance among all the classifiers. Further, the number of iterations has been increased and with 800 iterations, BPANN model achieved AUC value 0.826, which is equal to the performance of SVM-OAA. Furthermore, on 1000 number of iterations we achieved the highest performance among all classifiers with AUC value 0.854. After, the model with 1000 number of training iterations, we did not achieved a better performance and model is considered to reach at saturation level.

Table 1. RF and SVM performance evaluation

Classifier	Precision	Recall	F-score	AUC
RF	79.49	79.65	79.5	0.815
SVM-OAA	81.54	82.73	82.13	0.824
SVM-OAO	82.78	82.89	82.83	0.834
SVM-DAG	82.81	82.94	82.87	0.845

Table 2. BPANN performance for different epoch

Number of epoch	Precision	Recall	F-score	AUC
200	65.79	65.82	65.80	0.689
500	74.45	75.46	74.95	0.745
800	81.63	81.95	81.79	0.826
1000	83.46	83.53	83.49	0.854

It can be seen in Table 2 that with increasing number of training iterations, the prediction rate improves. But after 1000 iterations, the results were found significantly similar (no further improvements). The error rate for BPANN is illustrated in Fig. 7. It is clear in Fig. 7 that error rate is reduced with an increasing number of training iterations up to a level of 1000. Further, we achieved only a straight line. It's clear that all classifiers namely RF and three versions of SVM and BPANN is able to build a good classification model for the selected tweets data. However, considering the experimental results BPANN with sufficient number of training iterations is suggested as the best classifier among all used in this study. The number of training iterations required for analysis may vary based on the size and nature of data. Therefore, it is suggested to train the model for a sufficient number of times until desired prediction accuracy is achieved. BPANN model is able to classify the tweets into respective categories i.e. positive, negative and neutral with good accuracy. Further, statistics as shown in Fig. 2 reveals that delay in train arrivals on stations was the top most reason for negative tweets. It can be seen as a genuine reason because very few positive and neutral tweets were in the category. Second cleanliness was another category with highest number of negative tweets and very few positive tweets. Next category for negative tweets was cancellation of trains, food quality, high-rates in the travel fare. There were some conflicts between positive and negative tweets under the cancellation of trains and food quality. Lot of tweets mentioned that cancellation of trains was not the government fault but it is due to bad weather. Besides this, speed of the train and medical facilities category has the comparatively higher positive tweets. Tatkal and Catering are the two categories that consist of similar number of positive and negative tweets. So it is hard to say if the services were fair or not.

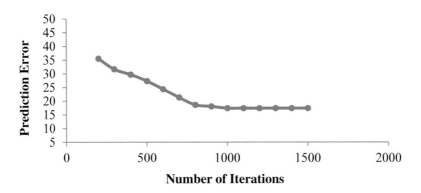

Fig. 7. Error rate vs number of training iterations for BPANN

5 Conclusion

In this study, we used machine learning techniques SVM, RF and BPANN to analyze the sentiments in the tweets in respect to Indian railways. The data were extracted from the Twitter database and then preprocessed to give it a suitable shape for analysis. Results shown that although RF and SVM provides good results but BPANN with 1000 training iterations provided better results. It is also identified that more training on data certainly increases the prediction accuracy but after a certain level there were no further improvement. Further, different categories of tweets with more positive or negative tweets have been discussed. The study can be extended with gathering more data and applying deep learning technique to seek if there are any further improvements or not.

Acknowledgement. The authors gratefully acknowledge the financial support of the Ministry of Education and Science of the Russian Federation in the framework of Increase Competitiveness Program of **NUST « MISiS »** (№ K4-2017-052).

References

1. Tiwari, P., Mishra, B.K., Kumar, S., Kumar, V.: Implementation of n-gram methodology for rotten tomatoes review dataset sentiment analysis. Int. J. Knowl. Disc. Bioinf. (IJKDB) **9**(1), 30–41 (2017)
2. Twitter: https://www.twitter.com. Accessed 05 July 2018
3. Microblogs: https://en.wikipedia.org/wiki/Microblogging. Accessed 20 July 2018
4. Elango, V., Narayanan, G.: Sentiment analysis for hotel reviews (2014). http://cs229.stanford.edu/projects2014.html
5. Adeborna, E., Siau, K.: An approach to sentiment analysis – the case of airline quality rating. In: PACIS 2014 Proceedings, Paper 363, Chengdu, 24–28 June (2014)
6. Pouransari, H., Ghili, S.: Deep learning for sentiment analysis of movie reviews (2015). https://cs224d.stanford.edu/reports/PouransariHadi.pdf
7. Doan, T., Kalita, J.: Sentiment analysis of restaurant reviews on yelp with incremental learning. In: 15th IEEE International Conference on Machine Learning and Applications (ICMLA), Anaheim, CA, pp. 697–700 (2016)
8. Ringsquandl, M., Petkovic, D.: Analyzing political sentiment on Twitter. In: Proceedings of AAAI Conference (2013)
9. Population in India: http://www.worldometers.info/world-population/population-by-country/. Accessed 23 July 2018
10. https://heidicohen.com/reliable-twitter-research/. Accessed 24 July 2018
11. Turney, P.: Thumbs up or thumbs down? Semantic orientation applied to unsupervised classification of reviews. In: Proceedings of the Association for Computational Linguistics, pp. 417–424 (2002)
12. Pang, B., Lee, L., Vaithyanathan, S.: Thumbs up? Sentiment classification using machine learning techniques. In: Proceedings of the Conference on Empirical Methods in Natural Language Processing (2002)
13. Michelle, A., Kondrak, G.: A comparison of sentiment analysis techniques: polarizing movie blogs. In: Advances in Artificial Intelligence, pp. 25–35. Springer, Heidelberg (2008)
14. Liu, B.: Sentiment analysis and subjectivity. In: Handbook of Natural Language Processing, pp. 627–666. CRC Press, Boca Raton (2010)
15. Singh, V.K., Piryani, R., Uddin, A., Waila, P.: Sentiment analysis of movie reviews: a new feature-based heuristic for aspect-level sentiment classification. In: 2013 International Mutli-Conference on Automation, Computing, Communication, Control and Compressed Sensing (iMac4 s), Kottayam, pp. 712–717 (2013)
16. Shi, H., Li, X.: A sentiment analysis model for hotel reviews based on supervised learning. In: 2011 International Conference on Machine Learning and Cybernetics, Guilin, pp. 950–954 (2011)
17. Lacic, E., Kowald, D., Lex, E.: High Enough?: Explaining and predicting traveler satisfaction using airline reviews. In: Proceedings of the 27th ACM Conference on Hypertext and Social Media, July 10–13, Halifax, Nova Scotia, Canada (2016)
18. Zou, X., Yang, J., Zhang, J.: Microblog sentiment analysis using social and topic context. PLoS ONE **13**(2), e0191163 (2018)
19. Bottou, L., Cortes, C., Denker, J.S., Drucker, H., Guyon, I., Jackel, L., LeCun, Y., Muller, U.A., Sackinger, E., Simard, P., Vapnik, V.: Comparison of classifier methods: a case study in handwriting digit recognition. In: International Conference on Pattern Recognition, pp. 77–87 (1994)

20. Knerr, S., Personnaz, L., Dreyfus, G.: Single-layer learning revisited: a stepwise procedure for building and training a neural network. In: Fogelman, J. (ed.) Neurocomputing: Algorithms, Architectures and Applications. Springer (1990)
21. Platt, J.C., Cristianini, N., Shawe-Taylor, J.: Large margin DAGs for multiclass classification. In: Advances in Neural Information Processing Systems, vol. 12, pp. 547–553. MIT Press (2000)
22. Mantyla, M.V., Graziotin, D., Kuutila, M.: The evolution of sentiment analysis-a review of research topics, venues, and top cited papers. Comput. Sci. Rev. **27**, 16–32 (2018)

A Genetic Algorithm for Superior Solution Set Search Problem

Ryu Fukushima[✉], Kenichi Tamura, Junichi Tsuchiya, and Keiichiro Yasuda

Department of Electrical and Electronic Engineering, Tokyo Metropolitan University,
1-1, Minamiosawa, Hachioji-shi, Tokyo 192-0397, Japan
fukushima-ryu@ed.tmu.ac.jp

Abstract. The superior solution set search problem contains parameters that provide constraints on evaluation value and distance. However, an optimization method explicitly incorporating these parameters has not yet been proposed.

There is a multi-objective optimization problem that is very similar to the superior solution set search problem. Studies on multi-objective optimization problems have been very active recently and solution applications to the superior solution set search problem are to be expected. Therefore, in this paper, we propose an evaluation indicator that is inspired by a method based on a dominance relationships in multi-objective optimization problems and includes the aforementioned parameters. We also propose a search method based on this indicator and perform numerical experiments on unique superior solution set search problems. The proposed method finds more superior solutions than the conventional single-objective optimization method, which confirms its usefulness.

1 Introduction

Typically, single-objective optimization aims at obtaining only a single optimal solution. In contrast, in most practical applications of optimization, the acquisition of multiple diverse solutions may be required [1–3]. In response to this shortcoming, authors have proposed the "superior solution set search problem" [2–4]. The goal of this problem is to obtain a set of superior multiple solutions. The superior solution set is comprised of a set of local optima with evaluation values above a defined level, where the distances between each local optimum are greater than a defined distance. Acquisition of this set enables a user to select a solution based on their own subjectivity.

There are multi-objective optimization problems and multi-modal optimization problems that are very similar to the superior solution set search problem. Research on multi-objective optimization problem and multi-modal optimization problems is very active and applications for developed methods are in high demand. In previous studies, solutions for the superior solution set search problem have been proposed, including novel methods that acquire a local optimal

© Springer Nature Switzerland AG 2020
A. M. Madureira et al. (Eds.): SoCPaR 2018, AISC 942, pp. 105–115, 2020.
https://doi.org/10.1007/978-3-030-17065-3_11

solution set containing a superior solution set, and the usefulness of these solutions has been evaluated through numerical experiments [2–5]. However, these methods do not directly focus on acquiring a superior solution set. Instead, they aim to acquire a set of local optimal solutions, which theoretically contain a superior solution set.

The remainder of this paper is organized as follows. First, the superior solution set search problem is analyzed in detail. Next, we propose a multi-objective-optimization-inspired evaluation indicator that provides fitness values for the superior solution set search problem. We then propose a novel search method based on this indicator is proposed. Finally, the usefulness of proposed method is evaluated through numerical experiments.

2   Superior Solution Set Search Problem

2.1   Definition of Superior Solution Set Search Problem

In this paper, we address a minimization problem using $X \subseteq R^n$ as the feasible region, $x \in X$ as the decision variable, and $f(x)$ as the evaluation function. In the following description, let $\delta \geq 0$ be the constraint on the evaluation values and $\varepsilon > 0$ be the constraint on the distance between solutions.

In order to meet the needs of an actual application, a superior solution set is defined as a set of local optimal solutions in which the differences from the evaluation value of the global optimal solution are less than δ and the distances between local optimum solutions are greater than ε. The superior solution set search problem is also defined in the reference [2] by authors. However, in this paper, in order to develop a novel search method for the superior solution set search problem, we define two novel relationships for the problem that are not mentioned in the referenced definition [2]. We then redefine the superior solution set search problem using these relationships. This definition differs from the referenced definition [2], but the resulting superior solution set is equivalent.

Definition of relationships using constraints on evaluation values: In order to consider constraints on the evaluation values, the following relationship is defined. When two solutions x_1 and $x_2 \in X$ satisfy the following condition, x_1 is superior to x_2 under the constraint on the evaluation value:

$$x_1 \prec_\delta x_2 \Leftrightarrow f(x_1) + \delta < f(x_2) \tag{1}$$

When $x \in X$ is not inferior to all feasible solutions to the above relationship (i.e., x is satisfactory $\forall y \in X, y \nprec_\delta x$), x is a solution in which the difference from the evaluation value of the global optimal solution is less than δ. In other words, x is a solution whose evaluation value differs from the global optimal by less than a defined value.

Definition of relationships using distance constraints: In order to consider the constraints on the distances between solutions, the following relationship is defined. Given a distance function d, the ε-neighborhood $\boldsymbol{B}(\boldsymbol{x}; \varepsilon)$ of $\boldsymbol{x} \in \boldsymbol{R}^n$ is defined by the following equation:

$$\boldsymbol{B}(\boldsymbol{x}; \varepsilon) = \{\boldsymbol{y} \in \boldsymbol{R}^n \mid d(\boldsymbol{x}, \boldsymbol{y}) < \varepsilon\} \tag{2}$$

When two solutions $\boldsymbol{x}_1, \boldsymbol{x}_2 \in \boldsymbol{X}$ satisfy the following condition, $\boldsymbol{x}_1$ is superior to $\boldsymbol{x}_2$ under the distance constraints:

$$\boldsymbol{x}_1 \prec_\varepsilon \boldsymbol{x}_2 \Leftrightarrow f(\boldsymbol{x}_1) < f(\boldsymbol{x}_2) \wedge \boldsymbol{x}_1 \in \boldsymbol{B}(\boldsymbol{x}_2; \varepsilon) \tag{3}$$

When $\boldsymbol{x} \in \boldsymbol{X}$ is not inferior to all feasible solutions based on the above relationship (i.e., $\boldsymbol{x}$ satisfies $\forall \boldsymbol{y} \in \boldsymbol{X}, \boldsymbol{y} \nprec_\varepsilon \boldsymbol{x}$), $\boldsymbol{x}$ is a local optimal solution, where no solution better than $\boldsymbol{x}$ in exists in a hypersphere with a radius ε centered on $\boldsymbol{x}$. In other words, $\boldsymbol{x}$ is a local optimal solution that is greater than a certain distance from any other local optimal solution. Hereafter, Euclidean distance is used as our distance metric.

Definition of superior solution set: Using the relationships defined above, we define a superior solution set $\boldsymbol{S}(\boldsymbol{X}; \delta, \varepsilon)$ as a set of solutions that are not inferior to all feasible solutions $\boldsymbol{x} \in \boldsymbol{X}$:

$$\boldsymbol{S}(\boldsymbol{X}; \delta, \varepsilon) = \{\boldsymbol{x}^\star \in \boldsymbol{X} \mid \forall \boldsymbol{x} \in \boldsymbol{X}, \boldsymbol{x} \nprec_\varepsilon \boldsymbol{x}^\star \wedge \boldsymbol{x} \nprec_\delta \boldsymbol{x}^\star\} \tag{4}$$

The optimal solution set $\boldsymbol{S}(\boldsymbol{X}; \delta, \varepsilon)$ is a set of local optimal solutions in which the differences from the global optimal solution are less than δ and the distances between solutions are greater than ε. We formulate the problem of finding a superior solution set $\boldsymbol{S}(\boldsymbol{X}; \delta, \varepsilon)$ as the superior solution set search problem.

2.2 Value of Superior Solution Set Search Problem

A superior solution set meets the requirements of the user and has diversity. In the paragraph, we consider the value of a superior solution set in relation to an automotive shape-design problem. We consider the value of a superior solution set in relation to an automotive shape-design problem. We consider a case in which aerodynamics is the evaluation function and the decision variables are width, height, and length. In typical single-objective optimization, the optimal solution is determined uniquely. Therefore, the solution is optimized only for aerodynamic properties. In contrast, in the superior solution set, finding a set of diverse solutions that all have excellent aerodynamic properties is the goal. Selecting a solution from the obtained solution set according to user preferences allows one to derive a final solution considering human subjectivity, which cannot be formulated. Therefore, the superior solution set problem has great value in the field of engineering because it can provide multiple solutions for complex problems.

2.3 Research Subjects in the Superior Solution Set Search Problem

The main research subjects in the superior solution set search problem are (1) utilization of the parameters δ and ε, and (2) utilization of knowledge in similar research fields.

We will first discuss subject (1). A superior solution set is a set of local solutions that satisfy user requirements. In contrast, because conventional single-objective optimization methods aim to find only a single global (quasi) optimal solution, it is difficult for such methods to derive multiple solutions that satisfy user requirements. The superior solution set search problem has unique parameters δ and ε. However, previous methods do not consider these parameters [2–5]. In other words, previous methods do not focus on finding a superior solution set. The main purpose of existing methods is to acquire multiple local solution sets that include a superior solution set. This is because previous studies focused only on basic examination of methods for solving the superior solution set search problem. Thus far, a method considering the parameters δ and ε has not been proposed. Therefore, development of a method considering these parameters is an important subject for the superior solution set search problem.

We will now discuss subject (2). There have been studies on niching methods (for multimodal function optimization) [6,7] and multi-objective optimization methods [8,9] for solving problems that are similar to the superior solution set search problem. Niching methods are optimization methods which aim at finding multiple global (or quasi) optimal solutions. Multi-objective optimization is an optimization that simultaneously considers multiple objectives. The study of these methods is very active and it is plausible that knowledge from these research fields can be utilized in research on the superior solution set search problem. The approaches in previous studies are most similar to niching methods because they search for multiple global optimal solutions. However, there are no previous studies that have utilized knowledge regarding multi-objective optimization for the superior solution set search problem. Therefore, this paper focuses on multi-objective optimization.

3 Search Method Based on Fitness of a Superior Solution Set

3.1 Dominance-Relationship-Based Evolutionary Multi-objective Optimization Method

In order to propose a search method based on fitness for the superior solution set search problem, we focus on multi-objective optimization. Multi-objective optimization methods are optimization methods that can consider multiple objectives simultaneously and are used to support user decision making. In this respect, the multi-objective optimization problem has a strong affinity with the superior solution set search problem.

Multi-objective optimization methods include methods for searching only one solution based on user preference information or metaheuristics searching for a set of diverse solutions. The evolutionary multi-objective optimization method falls into the latter category and is currently being studied as a powerful optimization method. This paper focuses on the evolutionary multi-objective optimization method from the viewpoint of the affinity between multi-objective optimization and the superior solution set search problem to find a set of diverse solutions. Evolutionary multi-objective optimization methods are roughly divided into three categories: dominance-relationship-based methods, division-based methods, and indicator-based methods. In this paper, we focus on dominance-relationship-based methods" Representative dominance-relationship-based methods include NSGA-II [8] and SPEA2 [9]. In these methods, an order relationships based on superiority and inferiority is assigned to the search points and survival selections are made based on these relationships to search for a non-inferior solution set.

With reference to this approach to multi-objective optimization, we define the ordering relationships for superiority and inferiority based on the definition of a superior solution set and propose a novel superior solution set search method. The proposed optimization method considers the parameters δ and ε to solve the superior solution set search problem.

3.2   Definition of Superior Solution Fitness

In order to propose a method that exploits knowledge regarding multi-objective optimization and problem-specific parameters (δ and ε), one must define superior solution fitness. The superior solution fitness value (fit), which is used to assign dominance relationships to solutions using problem-specific parameters (δ and ε) based on the definition of the superior solution set search problem, is defined below.

Superior Solution Fitness: $fit(x)$

1. Let $\boldsymbol{P}$ be the set of search points. Based on the definitions of the sets (definition of relationships using constraints on evaluation values and definition of relationships using distance constraints) for each search point $\boldsymbol{x} \in \boldsymbol{P}$, let fit be the number of search points $\boldsymbol{y} \in \boldsymbol{P}$ that are superior to $\boldsymbol{x}$.

$$fit(\boldsymbol{x}) = |\{\boldsymbol{y} \mid \boldsymbol{y} \prec_\delta \boldsymbol{x} \vee \boldsymbol{y} \prec_\varepsilon \boldsymbol{x}\}| \qquad (5)$$

However, if there are duplicate search points, duplicate search points excluding one search point are counted as $f(\boldsymbol{x}) = \infty$.

2. The smaller the value of fit, the better the solution. When the fit values of two solutions are equal, the solution with the smaller evaluation is chosen (in minimazation problem).

The $fit(\boldsymbol{x})$ value contains the superior solution set search problem-specific parameters (δ and ε) in its definition. Therefore, research knowledge regarding the utilization of parameters is incorporated into the proposed method. By using this $fit(\boldsymbol{x})$ value, dominance relationships can be defined between solutions.

3.3   The $fit(x)$ Based Optimization Method: GA-4S

We propose a novel superior solution set search method that uses $fit(\boldsymbol{x})$ based on a genetic algorithm (GA). The proposed method, which uses elite selection based on $fit(\boldsymbol{x})$ as the survival selection method for a GA, is called GA-4S. The main procedure for GA-4S is presented in Algorithm 1.

4   Specific Example of the Superior Solution Set Search Problem

To demonstrate the usefulness of the proposed method, we present a concrete example of the superior solution set search problem. The superior solution set

---

Algorithm 1. The Algorithmic Scheme for GA-4S

1: **procedure** GA-4S($m, p_c, p_m, \varepsilon, \delta, G_{max}$)
   Step 1: Initialization
2:     Give initial solutions $\mathcal{P}^0$ ($|\mathcal{P}^0| = m$), Set $G = 0$
   Step 2: Generation of New Solutions
3:     $\mathcal{Q}^G = \emptyset$
4:     **for** $i = 1 \ldots m$ **do**
5:         Choose randomly $\boldsymbol{x}_{r_1}, \boldsymbol{x}_{r_2} \in \mathcal{P}^G$ ($r_1 \neq r_2$)
6:         $\boldsymbol{q} =$ CROSSOVER-AND-MUTATION($\boldsymbol{x}_{r_1}, \boldsymbol{x}_{r_2}$)
7:         $\mathcal{Q}^G := \mathcal{Q}^G \cup \{\boldsymbol{q}\}$
8:     **end for**
   Step 3: Superior Solution Fitness Assignment
9:     $\mathcal{R}^G = \mathcal{P}^G \cup \mathcal{Q}^G$ ($|\mathcal{R}^G| = 2m$)
10:    **for** each $x \in \mathcal{R}^G$ **do**
11:        **for** each $y \in \mathcal{R}^G$ **do**
12:            $fit(\boldsymbol{x}) = |\{\boldsymbol{y} \mid \boldsymbol{y} \prec_\delta \boldsymbol{x} \vee \boldsymbol{y} \prec_\varepsilon \boldsymbol{x}\}|$
13:        **end for**
14:    **end for**
   Step 4: Superior Solution Fitness Based Selection
15:    Sort $\mathcal{R}^G$ in ascending order using $fit(x)$
16:    $\mathcal{P}^{G+1} := \mathcal{R}^G[1 : m]$
   Step 5: Termination
17:    **if** $G < G_{\max}$ **then**
18:        $G := G + 1$
19:        Go to **Step 2**
20:    **else**
21:        **return** Superior solutions in $\mathcal{P}^{G+1}$
22:    **end if**
23: **end procedure**

---

search problem is the problem of finding a set of local optimal solutions that have evaluation values within a defined value (δ) from the global optimal solution, where the distances between local optimum solutions is greater than a defined distance (ε). Therefore, the problem can be defined by determining the solution space, evaluation value constraint δ, and distance constraint ε.

In order to appropriately evaluate the performance of the superior solution set search method compared with the conventional single-objective optimization method, a solution space that can define various problems by constraints (δ and ε) is necessary. Therefore, it is desirable that the solution space of target problem whose solution set must be found containing local optimal solutions with various evaluation values and various distance in addition to a global optimum solution. Based on the above, we propose solution space and target problem.

5   Solution Space of Target Problem

The target function is defined as expression (6) in Table 1 and Fig. 1. This function has six local optimal solutions (◆) labelled A through F. The global optimal solution is B: $\boldsymbol{x} = [-3, -1.5]$. By specifying concrete δ and ε values, a superior solution set search problem can be defined.

$$f(\boldsymbol{x}) = -\sum_{i=1}^{6} \frac{c_i}{\exp(||\boldsymbol{x} - \boldsymbol{a}_i||)} \tag{6}$$

Table 1. Parameters of expression (6)

i	$\boldsymbol{a}_i$	c_i	Solution label	$f(\boldsymbol{x})$
1	$[-4, -1]$	90	A	-123
2	$[-3, -1.5]$	100	B	-130
3	$[-1, 4]$	40	C	-41.9
4	$[1, -4]$	40	D	-41.6
5	$[2, 1]$	60	E	-67.9
6	$[4, 2.5]$	80	F	-85.2

As a specific example, we define a superior solution set based on this function. The concrete values (δ and ε) and superior solutions defined by these values are presented in Fig. 2 and Table 2. The diamonds (◆) in Fig. 2 represent the superior solutions.

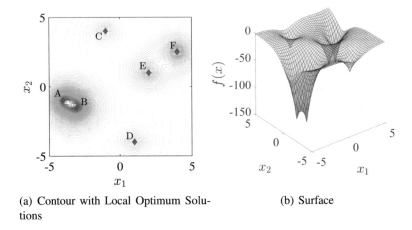

(a) Contour with Local Optimum Solutions

(b) Surface

Fig. 1. Solution space of target function

6 Numerical Experiments

We solved the aforementioned superior solution set search problem with a $fit(\boldsymbol{x})$-based elite selection GA called GA-4S. We evaluated the usefulness of GA-4S based on the number of superior solutions obtained using various settings and compared it to a basic GA.

6.1 Condition

In each experiment, we used Simulated Binary Crossover (SBX) [8,10] and Parameter-based Mutation (PBM) [8,10]. The parameters for these operations were crossover rate $p_c = 1$ and mutation rate $p_m = 0.5$. The value of the distribution adjustment variable was set to 25 different values of η_c and

Table 2. Example of superior solution set search problems

δ	ε	Superior solution set (◆ on Fig. 2)
30	1	A, B
	2	B
	3	B
70	1	A, B, E, F
	2	B, E, F
	3	B, F
100	1	A, B, C, D, E, F
	2	B, C, D, E, F
	3	B, C, D, F

$\eta_m = [2, 5, 10, 20, 50]$. The number of search points was 30 and the initial solution was randomized within a feasible region ($[-5, 5]^2$). The maximum number of generations is 100. We ran 50 trials with different initial solutions for each trials under the above conditions. Because it is difficult to acquire a strictly optimal solution, it was assumed that a superior solution is obtained when the Euclidean distance between each superior solution and the search point falls below 0.1.

6.2   Results and Consideration

The best experimental results are listed in Table 3 for all conditions. The experimental results reveal that GA-4S succeeds in acquiring more superior solutions than the basic GA. As an example demonstrating the usefulness of the proposed method, the search processes for each method for the problem defined by ($\delta = 100, \varepsilon = 3$) are shown in Figs. 3 and 4.

Table 3. Numerical experiment results: number of acquired superior solutions

		Function	GA				GA-4S			
δ	ε	Superior solution set	Mean	Best	Worst	S.D.	Mean	Best	Worst	S.D.
30	1	A, B	1.04	2	1	0.198	2	2	2	0
	2	B	0.98	1	0	0.141	1	1	1	0
	3	B	0.98	1	0	0.141	1	1	1	0
70	1	A, B, E, F	1.04	2	1	0.198	3.92	4	3	0.274
	2	B, E, F	0.98	1	0	0.141	2.88	3	2	0.328
	3	B, F	0.98	1	0	0.141	1.96	2	1	0.198
100	1	A, B, C, D, E, F	1.04	2	1	0.198	5.98	6	5	0.141
	2	B, C, D, E, F	0.98	1	0	0.141	4.96	5	4	0.198
	3	B, C, D, F	0.98	1	0	0.141	3.92	4	3	0.274

One can see from Fig. 3 that in the original GA, only the global optimal solution can be obtained. In contrast, the proposed method can properly obtain superior solutions defined by (δ, ε), not the only global optimal solution or local solutions.

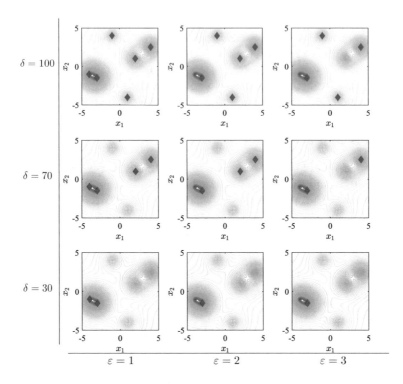

Fig. 2. Benchmark functions

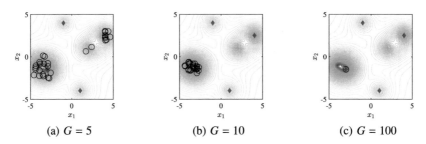

(a) $G = 5$ (b) $G = 10$ (c) $G = 100$

Fig. 3. Transition of search points in GA ($\delta = 100, \varepsilon = 3, \eta_c = 2, \eta_m = 10$)

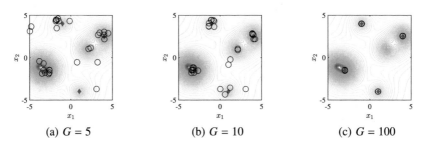

(a) $G = 5$ (b) $G = 10$ (c) $G = 100$

Fig. 4. Transition of search points in GA-4S ($\delta = 100, \varepsilon = 3, \eta_c = 2, \eta_m = 10$)

7   Conclusion

We proposed a superior solution set search method with evaluation an indicator $(fit(\boldsymbol{x}))$ that explicitly incorporates the parameters of the superior solution set search problem and demonstrated the usefulness of the proposed method.

This study applied the $fit(\boldsymbol{x})$ value to a GA. However, the application of $fit(\boldsymbol{x})$ is not limited to GAs. $fit(\boldsymbol{x})$ can also be applied to other evolutionary optimization methods. Additionally, it is also possible to develop optimization methods that perform functional specialization and control of diversification/intensification, as well as other operations, based on $fit(\boldsymbol{x})$.

References

1. Preuss, M., Burelli, P., Yannakakis, G.N.: Diversified virtual camera composition. In: Proceedings of European Conference on the Applications of Evolutionary Computation, EvoApplications 2012. Lecture Notes in Computer Science, vol. 7248, pp. 265–274 (2012)
2. Oosumi, R., Tamura, K., Yasuda, K.: Nobel Single-objective optimization problem and firefly algorithm-based optimization method. In: 2016 IEEE International Conference on Systems, Man, and Cybernetics, pp. 1011–1015 (2016)
3. Oosumi, R., Kumagai, W., Tamura, K., Yasuda, K.: A superior solution set search problem for single-objective optimization and a firefly algorithm. IEEJ Trans. Electron. Inf. Syst. **136**(10), 1947–1948 (2016). (in Japanese)
4. Oosumi, R., Kumagai, W., Tamura, K., Tsuchiya, J., Yasuda, K.: Proposal superior solution set search problem and firefly algorithm-based optimization method. In: Symposium on Evolutionary Computation 2016, vol. P1-03, pp. 12–20 (2016). (in Japanese)
5. Wang, H., Tamura, K., Tsuchiya, J., Yasuda, K.: Firefly algorithm using cluster information for superior solution set search. In: 2017 IEEE International Conference on Systems, Man, and Cybernetics, pp. 3695–3699 (2017)
6. Li., X., Engelbrecht, A., Epitropakis, M.G.: Benchmark functions for CEC'2013 special session and competition on niching methods for multimodal function optimization. Technical Report, RMIT University, Evolutionary Computation and Machine Learning Group, Australia (2013)
7. Singh, G., Deb, K.: Comparison of multi-modal optimization algorithms based on evolutionary algorithms. In: Proceedings of the 8th Annual Conference on Genetic and Evolutionary Computation (GECCO 2006), pp. 1305–1312 (2006)
8. Deb, K., Pratap, A., Agarwal, S., Meyarivan, T.: A fast and elitist multiobjective genetic algorithm: NSGA-II. IEEE Trans. Evol. Comput. **6**(2), 182–197 (2002)
9. Zitzler, E., Laumanns, M., Thiele, L.: SPEA2: improving the strength Pareto evolutionary algorithm. Technical report 103, Computer Engineering and Networks Laboratory (TIK), Department of Electrical Engineering, Swiss Federal Institute of Technology (ETH) Zurich (2001)
10. Deb, K., Agrawal, R.B.: Simulated binary crossover for continuous search space. Complex Syst. **9**(2), 115–148 (1995)

An Intelligent Tool for Detection of Phishing Messages

Marcos Pires and Petia Georgieva[✉]

Department of Electronics Telecommunications and Informatics,
University of Aveiro, Aveiro, Portugal
{marcosnetopires,petia}@ua.pt

Abstract. Phishing messages are a common attack on the web that results in the theft of user information. Finding a solution for this problem is a difficult task because phishers are very creative, and often it is hard even for a human to differentiate between legitimate and malign content. The goal of this project was to develop an intelligent tool for phishing detection that integrates only local information (the full content of the message) and does not rely on external (usually commercial) sources or black lists.

The major focus of this paper is the selection of appropriate features to discriminate between ordinary and phishing messages and the choice of an efficient classifier. The system can dynamically update the feature list and quickly adapt to new trends of phishing attacks. The proposed tool is suitable for implementation in email accounts or any other social network or communication channel. It is intended to reduce the workload on human experts that otherwise need to go through hundreds of messages everyday to verify their authenticity.

Keywords: Phishing messages · Text mining · Feature selection ·
Random Forest

1 Introduction

Phishing messages are a common attack on the web that results in the theft of confidential user information such as bank accounts, private data, personal logins or identity. These attacks have a heavy impact on the web and on the global economy, as billions of dollars are reported stolen every year [1].

Finding an automatic solution for this problem is a difficult task because phishers are very creative, and often it is hard even for a human to differentiate between legitimate and malign content. Most of the existing phishing detection systems are heavily based on URL analyses and the use of outside resources, such as blacklists or other APIs, [2]. Others use knowledge obtained from past attacks to create empirical rules that decide if a new message is phishing or not [3]. Over the last years machine learning (ML) algorithms gained also popularity as a mechanism to detect fraudulent messages [4,5].

© Springer Nature Switzerland AG 2020
A. M. Madureira et al. (Eds.): SoCPaR 2018, AISC 942, pp. 116–125, 2020.
https://doi.org/10.1007/978-3-030-17065-3_12

The objective of this project was to develop an intelligent tool for phishing detection, based on ML principles, that integrates only local information (the full content of the message) and does not rely on black lists or other external (expensive) sources.

The proposed solution has two stages. The most important stage is the extraction of reliable discriminative features. Two approaches were followed: (i) Binary features, often suggested in the literature; (ii) Mixed binary and real valued features.

The statistical analysis performed revealed that the most relevant features are related to the links and the correspondent URLs, the presence of HTML, scripts or forms in the message body, date and hour of sending, matching contents between the client info and the message subject. Additionally, a list of specific keywords from the content of the messages were extracted (bag of words) applying the TF-IDF text mining approach. The list is dynamic to quickly adapt to new trends of phishing attacks. The second stage is the choice of an efficient classification model. Among various classifiers (LR, SVM, K-NN, Tree-based models), Random Forest achieved the best performance.

The system is suitable for implementation in email accounts or any other social network or communication channel. It is intended to reduce the workload on human experts that go through hundreds of messages everyday to verify their authenticity, when only a small portion of them would actually need to be flagged as phishing.

2   Feature Extraction

A major issue for this project was the lack of labeled dataset with sufficiently high number of legitimate and fraudulent messages in order to build a trustful discrimination system. Based on two online available databases, we have created the testbed dataset. Around 7000 emails labeled as regular were randomly selected from over 600,000 emails available in the Enron Email Dataset [6], described in more details in [7]. Around 4000 phishing emails were obtained from the Fraudulent E-mail Corpus [8] uploaded on the website [9]. Data were separated in two sub-datasets, where 80% is used for training and validation, and 20% for testing.

2.1   TF-IDF Algorithm

Term Frequency-Inverse Document Frequency (TF-IDF) is a text mining method that searches for the most significant words in a collection of documents [10]. First the so called *stop words*, such as "the", "and", etc. are eliminated. Then the algorithm computes the *Term Frequency* (TF_{ij}) for every word i in document j. TF_{ij} is computed as the ratio between the occurrences of word i in document j (f_{ij}) and the number of occurrence of the word with the highest frequency in the same document. This gives a value between zero and one for every word in the same document.

$$TF_{ij} = \frac{f_{ij}}{MAX_k f_{kj}} \tag{1}$$

The *Inverse Document Frequency* is computed for every word i (IDF_i) (Eq. 2), where N is the total number of documents in the corpus and n_i is the number of documents where term i occurs

$$IDF_i = log_2(N/n_i) \tag{2}$$

The final score ($TF.IDF_{ij}$) for every word i in document j is given by

$$TF.IDF_i = TF_{ij} \times IDF_i \tag{3}$$

The higher this score, the most significant is this word in the document corpus.

Using the TF-IDF algorithm two distinct word lists were extracted. The words with the highest TF-IDF score, i.e. the most significant words for each class are stored. The dimensions of the extracted lists are hyper parameters that need to be carefully adjusted. For the phishing class the top 20 words were sufficient, whereas for the ordinary messages the top 100 words were extracted. All words present both in the phishing and the ordinary word lists were removed from the phishing word list. This way only the most characteristic words for phishing messages were kept. The sequence of steps of the TF-IDF based algorithm for keyword list extraction is schematically presented in Fig. 1.

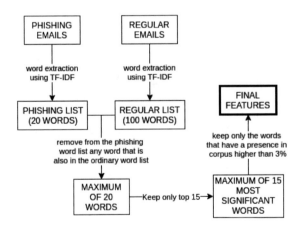

Fig. 1. TF-IDF based algorithm for keyword lists selection

2.2   Feature Structure 1 - Binary Features

Following the current trend in phishing detection systems we first extracted a list of binary features that consists of typical feature items suggested in the literature [11,12]. If the condition verifies, the feature value is set to 1, otherwise it is assigned to 0:

1. If the number of links present in the message are more than a *Hpar* (*Hpar* = 3)
2. If the URLs present in the message have different domains
3. If the average number of sub-domains of the URLs is greater than *Hpar*(*Hpar* = 1)
4. If any specific form exists in the message
5. If any specific script exists in the message
6. If the message is HTML formatted
7. Presence of IP in the URL
8. If the average number of dots in URLs exceeds *Hpar* (*Hpar* = 2)
9. If the average length of URLs is greater than *Hpar* (*Hpar* = 35)
10. If any "@" symbol is found in the URLs
11. If any "-" symbol is found in the URLs
12. If the client sent the message over the weekend
13. If the client sent the message outside working hours (between 8am and 6pm)
14. If the client created the account in less than a year
15. Presence of certain words in the message. This includes the words contained in the list obtained from the TF-IDF algorithm, where every word is considered as one feature.
16. Predefined sets of words/strings, where each set is one feature and if at least one of the words occurs in the message the feature is set to 1:
    (a) set1 - "Update" and "Confirm"
    (b) set2 - "User", "Customer" and "Client"
    (c) set3 - "Suspend", "Restrict" and "Hold"
    (d) set4 - "Verify", "Account"
    (e) set5 - "Login", "Username", "Password", "Click" and "Log"
    (f) set6 - "SSN", "Social Security", "Secur" and "Inconvinien".

2.3 Feature Structure 2 - Combination of Binary and Real Value Features

The binary features do not account for the frequency of occurrences of a condition. For example set5 gets value of 1 both if only one word or all words of the set occur in the email. To handle this problem we propose the following feature list that incorporates both binary and real valued features:

1. **n_link** (Real) - Sum of the links present in the message.
2. **n_domain** (Real) - Number of different domains. Every URL present has a domain, all existing domains in the message are compared. The total number of different domains is the value of this feature.
3. **n_subdomain** (Real) - Average number of sub-domains. For every URL present in the message the sub-domains are counted. The final value is the average of these values.
4. **form_script_html** (Real) - Sum of three elements: any existing forms; number of existing scripts; presence of HTML.

5. **ip_at_minus** (Real) - Sum of three elements relative to URLs: presence of IP; presence of "-" or "@" symbols.
6. **n_dots** (Real) - Average number of dots in the URLs present in the message.
7. **length** (Real) - Average length of the URLs present in the message.
8. **max_len_url** (Real) - Length of the longest URL.
9. **week_day** (Binary) - 1 if the message is sent over the weekend, otherwise is set to 0.
10. **working_hour** (Binary) - 1 if the message was sent outside working hours of the country of origin (from 8am to 6pm), otherwise is set to "0".
11. Predefined sets of words/strings, where each set is considered as one feature. The sum of occurrences of the words in the set is the value assigned to the feature:
    (a) **set1** (Real) - "Update" and "Confirm".
    (b) **set2** (Real) - "User", "Customer" and "Client".
    (c) **set3** (Real) - "Suspend", "Restrict" and "Hold".
    (d) **set4** (Real) - "Verify", "Account".
    (e) **set5** (Real) - "Login", "Username", "Password", "Click" and "Log".
    (f) **set6** (Real) - "SSN", "Secur" and "Inconvinien".
12. **Kwords_name_fuzzy** (Real) - Sum of occurrences of predefined keywords similar to the client's name. These words are a combination of previously seen attacks and brand names in phishing messages. Some examples are "bank", "card", "credit" and many brand names such as "visa", "yahoo", "wells". This keyword list is dynamic, i.e. words can be added or removed from the list (343 words were included in the studied feature structure). The phisher attackers often make small changes in one or few characters of the word so that it goes unnoticed to humans and exact matching between words. Therefore, the words do not need to have an exact matching with the client's name of the analyzed message, if the algorithm finds a similarity between the client's name and a keyword from the list it will count it as an occurrence.
13. **Kwords_subject** (Real) - This feature is similar to the Kwords_name_fuzzy, but now a similarity between the content of the message subject and the words from the predefined keywords list is searched.
14. **Kwords_name_exact** (Real) - similar to **Kwords_name_fuzzy**, but now exact matching between the client's name and the words from the predefined keywords list is searched. To improve the matching all letters are transformed into lower case. For example the word "Paypal" is first transformed into "paypal" and then is compared to the keyword "paypal". If all characters in the words match, the algorithm will count it as an occurrence.
15. **TF-IDF-based features** (Real) - maximum of 15 additional features are added corresponding to the sum of occurrences in the message of each of the words extracted by the TF-IDF algorithm.

2.4   Overall Message Processing

Each message was considered as a sample, and every sample went through an information extraction process regarding the fields of the message: the date, sender details, and body (the message itself). For messages with HTML content, the files were first parsed. All *href* tags were listed, to get the URLs in the message. The relevant URL information was extracted and combined: the number of links, the number of different domains, the average number of sub-domains, the existence of specific characters, average number of dots, average length and finally the length of the longest URL. Next, the existence of forms or scripts is checked. The features are finally normalized.

The email corpus preprocessing and feature extraction is schematically represented in Fig. 2. The implemented code is in Python 2.7 in order to be able to use the more comprehensive library of relevant functions compared to the new version of Python 3.6. The computer system used for this project, had Ubuntu 16.04 as an operating system, 12 gigabytes of RAM and a CPU with wight cores of model Intel®Core™i7-4720HQ and clock rate of 2.60 GHz.

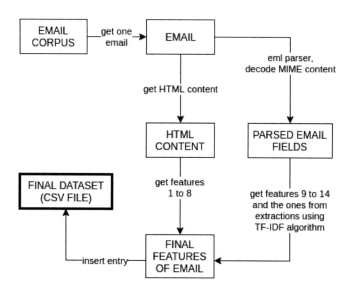

Fig. 2. Email corpus preprocessing and feature extraction process. Feature numbers correspond to Feature structure 2 (Sect. 2.3)

3   Classification Models

The most suitable classifier for this task was determined after evaluation of some of the most widely applied classifiers in diverse applications. The selected models were Logistic Regression, k-Nearest Neighbor (K-NN), Support Vector Machine (SVM with different kernels), Decision Tree (DT), Boosted Trees (BT), and Random Forest (RF). The structure of each classifier was optimized after tuning its most sensitive hyper parameter.

The performance of the K-NN classifier was assessed when the number of neighbors K varies in the range [1, 50]. The results (the error rate) are represented in Fig. 3, with $K = 6$ being determined as the optimal value. The DT model was evaluated varying the maximum depth of the tree in the range $[1, 50]$. The results depicted in Fig. 4 suggest 9 as the optimal value.

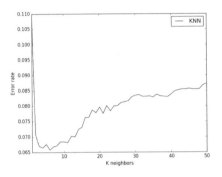 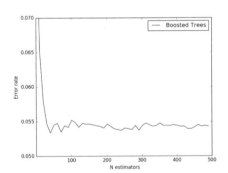

Fig. 3. K-NN model (variation of K)   **Fig. 4.** DT model (variation of max depth)

Random Forest and Boosted Trees are ensemble Tree-based classifiers where the number of estimators (trees) is varied in the range $[1, 500]$ as shown in Figs. 5 and 6. The optimal number of trees for RF was determined as 150 and for BT as 80, respectively.

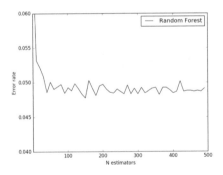

Fig. 5. RF model (variation of # estim.)   **Fig. 6.** BT model (variation of # estim.

SVM models with RBF (Radial Basis Function), linear, sigmoid and polynomial kernels were also studied. The comparative results of 5-folds cross-validation between all (nine) classifiers with binary features (Feature Structure 1) or combination of binary and real valued features (Feature Structure 2) are summarized in Tables 1 and 2. The testbed database built for this study is unbalanced to reflect a typical scenario of much less phishing messages than regular messages. Therefore besides the accuracy, the precision, recall and F1 score are also considered as performance metrics.

Table 1. Performance indicators of 5-fold cross-validation for Feature Structure 1

	Accuracy (%)	F1 Score (%)	Precision (%)	Recall (%)
Logistic Regression	93.05	90.21	92.17	88.33
SVM (Linear Kernel)	92.44	89.52	89.95	89.12
SVM (Polynomial Kernel)	90.42	85.60	93.99	78.60
SVM (RBF Kernel)	93.04	90.14	92.57	87.86
SVM (Sigmoid Kernel)	91.94	88.51	91.49	85.74
K-NN (K = 6)	92.54	89.23	93.60	85.27
Random Forest (150 Decision Trees)	**93.67**	**91.14**	**92.48**	**89.85**
Decision Tree (max depth = 9)	93.40	90.69	92.81	88.68
Boosted Trees (80 Decision Trees)	93.00	90.17	91.88	88.52

Table 2. Performance indicators of 5-fold cross-validation for Feature Structure 2

	Accuracy (%)	F1 Score (%)	Precision (%)	Recall (%)
Logistic Regression	91.80	88.27	92.15	84.71
SVM (Linear Kernel)	91.21	87.35	91.83	83.30
SVM (Polynomial Kernel)	66.53	17.18	87.35	9.53
SVM (RBF Kernel)	90.96	87.02	91.22	83.20
SVM (Sigmoid Kernel)	87.66	80.47	94.92	69.86
k-NN (K = 6)	93.10	90.24	93.09	87.57
Random Forest (150 Decision Trees)	**95.23**	**93.41**	**94.13**	**92.70**
Decision Tree (max depth = 9)	94.10	91.74	93.65	89.90
Boosted Trees (80 Decision Trees)	94.23	91.89	94.02	89.87

The results clearly suggest that the tree based models (RF, DT, BT) outperform the other classifiers. These results are in accordance to the reviewed literature [11]. Their performance is further improved when provided with Feature Structure 2 (mainly real valued features). Random Forest achieved state of the art performance of 95.23 % accuracy which is a very promising result.

4   Conclusions

The intelligent tool for detection of phishing messages proposed in this paper explores both the URLs information (if present) and the full content of the message (subject, date, sender details, and the body). It works only locally, and does not need to use on-line sources such as blacklists. The feature list can be dynamical updated if the message corpus is continuously supplied with new labeled data. Keywords can be deleted or new keywords added during the training process of the model. Better filtering of phishing messages is achieved with real-valued features where the frequency of occurrence of a certain condition is accounted.

The implementation of the proposed tool in commercial environments (digital marketing companies, web providers, social network services) would be favored if it is embedded into a platform for on-line message labeling. This platform allows authorized personnel of the company to label incoming messages and will feed continuously the system with new phishing examples. Periodic retraining with recently accumulated labeled data over a limited period of time would guarantee a smooth adaptation to new attack trends.

Acknowledgements. This Research work is funded by National Funds through the FCT - Foundation for Science and Technology, in the context of the project UID /CEC/00127/2013.

References

1. Jan, T.R.: Effectiveness and limitations of statistical spam filters. In: International Conference on New Trends in Statistics and Optimization (2009). https://pdfs. semanticscholar.org/85cc/8a68a7a822efcd24aa939170c03473c65846.pdf
2. Sharaff, A., Nagwani, N.K., Swami, K.: Impact of feature selection technique on email classification. Int. J. Knowl. Eng. **1**(1) (2015). http://www.ijke.org/vol1/10-E001.pdf
3. Zhang, J., Liu, Y.: Spam email detection: a comparative study. Tech. Data Min. J. (2013)
4. Awad, W.A., ELseuofi, S.M.: Machine learning methods for spam email classification. Int. J. Comput. Sci. Inf. Technol. **3**(1), 173–184 (2011)
5. Divya, S., Kumaresan, T.: Email spam classification using machine learning algorithm. Int. J. Innov. Res. Comput. Commun. Eng. **2**(1) (2014)
6. Enron corpus dataset. http://www2.aueb.gr/users/ion/data/enron-spam/
7. Klimt, B., Yang, Y.: The enron corpus: a new dataset for email classification research. Language Technologies Institute Carnegie Mellon University Pittsburgh, PA 15213-8213, USA. http://nyc.lti.cs.cmu.edu/yiming/Publications/ klimt-ecml04.pdf
8. Radev, D.: Clair collection of fraud emails (2008). http://aclweb.org/aclwiki
9. Fraudulent email corpus. https://www.kaggle.com/rtatman/fraudulent-email-corpus
10. Leskovec, J., Rajaraman, A., Ullman, J.D.: Mining of Massive Datasets (2011). on-line book

11. Akinyelu, A.A., Adewumi, A.O.: Classification of phishing email using random forest machine learning technique. J. Appl. Math. (2014). https://doi.org/10.1155/2014/425731

12. Basnet, R., Mukkamala, S., Sung, A.H.: Detection of phishing attacks: a machine learning approach. Soft Comput. Appl. Ind. (2008). https://doi.org/10.1007/978-3-540-77465-5_19

Discrete Wavelet Transform Application in Variable Displacement Pumps Condition Monitoring

Molham Chikhalsouk[✉], Balasubramanian Esakki, Khalid Zhouri, and Yassin Nmir

Abu Dhabi Men's College, Higher Colleges of Technology, Abu Dhabi, UAE
malsouk@hct.ac.ae

Abstract. Pumps performance and design can be detected through vibration signature comparison, which can detect malfunction and poor design aspects. In order to understand the vibration signature, there is a need to implement well-known techniques to obtain the required information. The most popular technique is Fourier Transform (FT). However, this technique and another frequency related technique focus on obtaining the frequency components and this is well accepted for stationery signatures. In condition monitoring process, it is very important to identify both the frequency and the time of occurrence, which is the nature of the transient signature. The most recent technique in identifying the useful information of transient signals is the Wavelet Analysis. In this study, the signals of healthy and defective control systems of the variable displacement pump are recorded and analyzed by using Wavelet Analysis. The study confirms the ability to apply the wavelet analysis in detecting the variable displacement pumps' defects.

Keywords: Vibration analysis · Fault diagnosis · Control unit ·
Variable displacement pumps · Wavelet analysis

1 Introduction

The recent trend of excellent performance, reliability, and safety of hydraulic systems has driven fault detection and condition monitoring to a higher level. With the recent breakthrough of sensors and transducers, the condition monitoring is becoming more popular and accurate especially in dynamic applications such as vibration displacement, dynamic force, noise levels, temperature, etc. Recently, vibration signature analysis is one of the most accurate techniques implemented in condition monitoring and fault detection with full dynamic details. Obtaining the required information from the vibration signatures is essential for fault detection in hydraulic systems. Considering the quick progress in signal processing means, Fourier Transform FT, Fast Fourier Transform FFT, and SFFT are very useful for stationary signals [1, 2].

These techniques are very useful for vibration signals with stationery nature, are not suited for transient signal analysis [3]. They fail to disclose the inherent information of the transient signal. They are also limited and cannot be implemented effectively in

© Springer Nature Switzerland AG 2020
A. M. Madureira et al. (Eds.): SoCPaR 2018, AISC 942, pp. 126–136, 2020.
https://doi.org/10.1007/978-3-030-17065-3_13

hydraulic systems machinery diagnostics [4]. To defeat these challenges, Wavelet Analysis (WA) has been introduced. WA is a technique with multi-resolution windows that use the time interval to classify the frequency components (high/low) [5, 6]. The WA breaks down the data into approximation and detail coefficients in a multi-scale, granting a powerful tool for transient signal analysis than the FT. Several studies have introduced the applicability of WA to break down signals for performance enhancement and fault detection in hydraulic systems [7, 8].

In this study, the WA is implemented as a tool for condition monitoring of variable displacement pumps. The signals are obtained experimentally with two different cases: an ideal control system with well-tuned PID controller, and a defective PID controller (improper gains, for the study purpose only). The paper is arranged as follows: Sect. 2 shows the WA method and its mathematical formulation. Section 3 details the components of the variable displacement pumps in general. The problem statement and the represented faults are presented in Sect. 4. Section 5 details the experimental set up along with the recorded vibration signatures, which are presented in Sect. 6. Lastly, Sect. 7 concludes the work.

2 Wavelet Analysis

The type of the signal processing technique is mainly applied according to the nature of the vibration signature and the required information. WA is preferred for all signals with average to high frequencies with discontinuity, which are the transient signals. In WT, the signal is orthogonally hurled on a basis mother wavelet function and the Discrete Wavelet Transform (DWT) coefficients are calculated.

The scale concept is introduced in the DWT to guarantee that the mother wavelet to take the suitable frequency resolution. The DWT is stated as:

$$W(a, b) = \frac{1}{\sqrt{a}} \sum_n \sum f(t) \varphi(\frac{n - b}{a})$$

(1)

where "a" is a scale function, $\varphi((n-b)/a)$ is the mother wavelet or wavelet basis, n is an integer number, and "b" is a dilation number. The parameter $1/\sqrt{a}$ is presented to standardize the mother wavelet and have the same energy level at every scale. The solution of Equation one requires obtaining the value of (a) and (b). For a dyadic discrete sampling, the parameter (a) can be written as

$$a = 2^{\wedge}j$$

(2)

where (j) represents the decompression levels number
(b) can be calculated by

$$b = 2^{\wedge}j.k$$

(3)

(k) is an integer, (t) is the time variable, f(t) is the signal in the time domain, and n is the number of samples. The scale (s) is inversely proportional to the frequency (f). The function (W (a, b)) is represented into scale-time form. Yet, there is no available relationship between the scale and frequency. Mallat introduced the multi-resolution technique to calculate the DWT by employing low/high pass filters and down sampling the signals as illustrated in Fig. 1, [11].

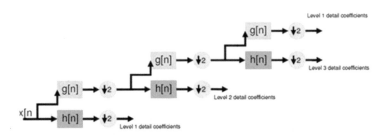

Fig. 1. Filters bank of DWT by multiresolution analysis (Computation of DWT)

Figure 1 demonstrate the DWT calculation procedure. The filter bank employs two sets of functions which are named as scaling and wavelet functions [12, 13]. The bank is associated with high and low pass filter. When the signal passes through a high pass filter (h) and the low pass filter (g). Hence, the first level leads to vectors, which are the approximation coefficient (A1) and the detail coefficient (D1), [14]. At the second level, the approximation (A1) to be decomposed into (A2) and (D2). This process will continue to the required level.

3   Variable Displacement Pumps

Variable displacement pumps are very popular in hydraulic systems with the need for variable pressure/flow requirements. The pump flow rate can be adjusted without changing the prime mover speed by adjusting and controlling the swash plate inclination angle. Theoretically, the pump delivery pressure and the system load pressure should be the same. Many control techniques are applied to enhance the electro-hydraulic pressure control pump systems. For example, Akers and Lin, [15]; Lantto, and Johnson, [16]; Akers and Lin [17] have examined linear control methods in achieving this goal. Manring and et al. [18–21] have implemented proportional solenoid valve to adjust the swash plate inclination angle and output flow accordingly. They have examined the nonlinear control strategy on-line adaptive compensating for the change of the pressure carry over angle. Figure 2 details the pump components, which consists of a limited number of pistons within their cylinders and all are arranged in the common cylindrical block. The pistons are grouped in a circular array within the block at equal intervals about the shaft axis. The cylinder block is held firmly against a port plate using the force of the compressed cylinder-block spring. A ball-and-socket joint connects the base of each piston to a slipper. The slippers themselves are kept in contact with the swash plate and the inclination angle is controlled. For practical applications, the loads on any hydraulic actuator can change from time to time, which in turn requires the hydraulic pump to provide a different operating pressure accordingly.

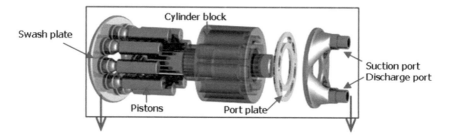

Fig. 2. Main components of a swash plate pump with a cylindrical arrangement

4 Problem Statement

The first generation of the pumps came without a control unit and it was very simple hydraulic mechanical design. A relief valve was implemented to discharge the excessive flow. The valve has solved the extra unwanted flow, however, an overheating for the flow has been witnessed and this has led to a rapid degradation for the hydraulic oil chemical and physical properties and eventually to a rapid pumps' component failure. Hence, there was a need to introduce a better effective design to avoid the pump operation at its maximum output, and the control unit was implemented to generate exactly the right amount of the flow that matches the load requirements. The most recent design comes with a double negative feedback controller to control both the hydraulic spool motion and the swash plate inclination angle. This design is suitable. However, it is costly as it involves extra parts. In [22], a simple design was proposed by using a single PD controller. In that study, the authors compared the proposed control strategy with other control strategies. The main goal of the reducing the manufacturing costs were met, however, the pump performance in the connection with the other pumping system was not investigated and the vibration and noise levels were not investigated as they were not the scope of the author study. In this study, the high level of vibration and noise are considered as the design problem and the pump with the high vibration level is considered as a faulty pump. More work has been conducted in investigating the applicability of wavelet transform in several mechanical engineering applications in the past two years [23–25].

The coming sub-items will explain the control system components, the pump equivalent mathematical model, and the single feedback control strategy.

4.1 Control Unit Structure

The control unit has a secondary power pump, a proportional hydraulic valve, and the electronic control system. In the secondary hydraulic unit, a pressure transducer detects the load pressure and sends it to the control unit thematic unit. The arithmetic unit determines the necessary swash plate inclination angle that meets load needs without surpassing the pump manufacturer limitations such as power, maximum pressure, and flow. The control system structure is represented in Fig. 3.

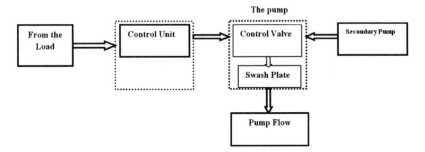

Fig. 3. The pump's control, hydraulic, and electrical unit's interconnection

4.2   Variable Displacement Pump Mathematical Model

The swash plate angular motion about the y-axis is modeled as SDOF with an angular variable and can be represented as φ. The swash plate is subject to the following moments which are the mass moment of inertia and can be expressed as ⟦ J⟧ _eq, the hydraulic piston restoring moment and its damping moment and their coefficients are Keq, and Ceq, respectively.

The equation of motion can be written as:

$$J_{eq}\ddot{\varphi} + C_{eq}\dot{\varphi} + K_{eq}\varphi = M(t) \tag{4}$$

Where Jeq = total equivalent inertial moment for the swash plate [kg.m2]

Ceq = total equivalent angular damping coefficient for the hydraulic piston [N.m. s/rad]

$\ddot{\varphi}$ = inclination angle second derivative (angular acceleration) [rad/s2]

$\dot{\varphi}$ = derivative of the inclination angles (angular velocity) [rad/s]

φ = swash plate inclination angles (angular displacement) [rad]

M = swash plate applied moment [N.m]

Keq = equivalent restoring moment of the hydraulic piston [N.m/rad]

The swash plate transfer function can be written in Laplace form as

$$T = \frac{\varphi(s)}{M(s)} = \frac{\omega_n^2}{s^2 + 2\varsigma\omega_n s + \omega_n^2} \tag{5}$$

Where ωn = Natural frequency rad/s

ζ = Damping ratio

S = Laplace variable

To reach the optimum performance of the variable displacement pump, there is a need to select the appropriate controller kind and its gains, which will lead to developing the required a pump overall performance in terms of pump load sensitivity and flow, which are leading in superb pump operation.

The single feedback controller will be explained in details in the next items with two conditions, which are:

- Faulty pump with single PD controller (Kp = 1, Kd = 0.02, Ki = 0). The inadequacy of the Integral gain leads to amplifying the noise that common in hydraulic pumps
- Healthy pump with a single PID controller (KP = 1.9, KD = 0.1 Ki = 8.4). The gains are obtained by using Nicholas Ziegler method and them tuned by observing the pump-pipe performance.

4.3   Faulty Pump (PD Controller): Faulty Pump

In [22], the authors have proposed a single PD control strategy to control the pump flow rate by controlling the piston stroke via the swashplate inclination angle. They have eliminated the spool movement transducer electrical feedback line and controls only swash plate inclination angle only. The control strategy is illustrated in (Fig. 4).

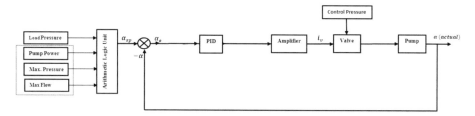

Fig. 4. Swashplate pump controlled with a single control strategy (PID controller)

The PD controller was parameterized by employing the Ultimate Sensitivity technique. The PID gains were KP = 1, Ki = 0 (no integral influence), and KD = 0.02. The proposed strategy was supported experimentally; however, the system vibration and noise levels were not tested.

In this study, the authors proposed a single PID control strategy (KP = 1.9, KD = 0.1 Ki = 8.4), which is meant to be the Healthy Pump, which needs to have a rapid response and low levels of noise and vibrations. The authors have used the WA to evaluate the vibration levels.

5   Experimental Setup

A piezoelectric accelerometer is positioned in the middle of the delivery pipe (the maximum levels of the vibration can be witnessed in the simply supported structures). The accelerometers- pipe arrangement is shown in Fig. 5. The recorded signals are sent to a signal amplifier and conditioner to improve the quality of the recorded signals. The signals have an analog nature and LABVIEW is used to convert the signals into digital signals. A data acquisition card is used to collect the data (DAQ 6062 E). The test set up is shown in Fig. 6. The sampling frequency is selected to be 2000 and hence, the recorded points are enough to form the signal in more than two operating cycles. The signals then are displayed on an oscilloscope (Agilent – 54624 A), and then recorded for further analysis. MATLAB wavelet toolbox is used to analyze the signals in time domain. The pump flow rate is 1 L/s, and the pipe flow speed is 2.94 m/s (which keeps the pipe stable at this speed, and no flutter can be faced). The pipe flutter speed is calculated to be 94 m/s.

For the experimental validation purpose, two arrangements of the experiments are done:

- The variable displacement pump is provided with a single PD controller (proportional-derivative control system) to simulate the inappropriate control system i.e. defective pump. The actual electrical controller is substituted with real-time controller software to model the control system.
- The variable displacement pump is provided with a single PID controller with suitable tuned Proportional, Integral, and Derivative gains. This is considered as the healthy variable displacement pump that can be authenticated experimentally pipe in reference to the vibration levels and the variable displacement pump characteristics.

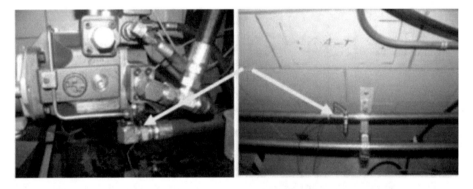

Fig. 5. The positions of the accelerometer on the pump delivery line and pipe

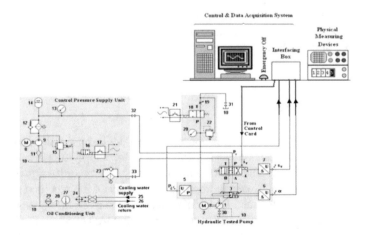

Fig. 6. Hydraulic and control parts/data acquisition unit (Experiment setup)

6   Results

Figure 7 shows the accelerometer readings on the pipe which represent the two studied conditions (the healthy variable displacement pump and the defective pump with the PD controller). With the healthy variable displacement pump (PID controller), the vibration signature exhibits two characteristics: the regularity, and the little vibration amplitude (±0.05 mm), which is very important to prolong the hydraulic systems' components. On the other hand, the faulty variable displacement pump (the single PID controller) has its different characteristics: irregularity and random patterns with high amplitude (±0.5 mm).

The coefficients of DWT for the pipe vibration signature of the faulty pump with a single PD controller are illustrated in Fig. 8. The amplitude of the fifth approximate coefficient equals 0.08, and it has a minimum value (equals zero) at t = 900 ms.

The DWT coefficients for the pipe signature under the healthy pump with a single PID controller are shown in Fig. 9. By comparing the A5, it can be noticed that its maximum amplitude is 0.006, and its minimum value (equals zero) takes place at t = 250 ms (Table 1).

The summarized characteristics of Healthy/Faulty pump conditions are:

Table 1. Comparison between the healthy and faulty pump vibration characteristics

The healthy variable displacement pump provided with a single PID controller	The faulty variable displacement pump provided with a single PD controller
The vibration signal is periodic and small amplitude (0.05 mm)	The vibration signal is random and high amplitude (0.5 mm)
In DWT, A5 is 0.06 and its minimum value takes place at t = 0.25 s	In DWT, A5 is 0.08 and its minimum value takes place at t = 0.9 s

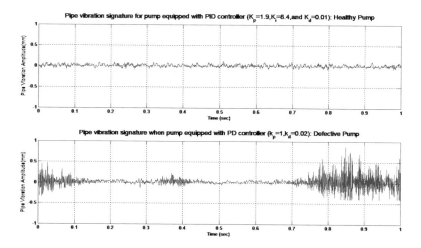

Fig. 7. Pipe vibration signal for the variable displacement pump provided with a single PID controller (Kp = 1.9, Ki = 8.4, and Kd = 0.01) and Pipe vibration signal for the variable displacement pump provided with a single PD controller (kp = 1, kd = 0.02)

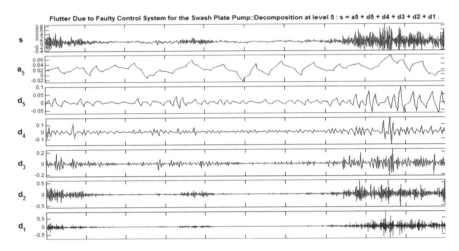

Fig. 8. Time-scale representation of the faulty pump with a single PD controller

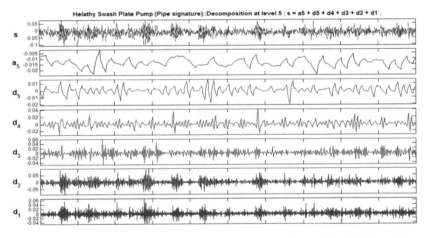

Fig. 9. Time-scale representation of the healthy pump with a single PID controller

7   Conclusions

A set of experiments was carried out to investigate the applicability of the WT in detecting the change in the technical condition of the pump and identify the healthy and faulty pump. For a comparison purpose, the faulty pump was provided with a single PD controller, which produced dynamic instability in all hydraulic components (the variable displacement pump and pipes). The faulty and normal pump recorded signals were analyzed by using discrete wavelet transform with Debauche's mother wavelet, which is preferred for signals with impulsive vibration signals (the variable displacement pump pressure overshooting causes the impulse). The discrete wavelet analysis of the two cases had two different shapes. For the healthy variable displacement pump, a small amplitude for A5, however, for the faulty variable displacement pump; A5 had a higher amplitude (0.08). The findings show the relevance of using wavelet analysis in detecting the defective control unit and can be extended to include other defects.

References

1. Shibata, K., Takahashi, A., Shirai, T.: Fault diagnosis of rotating machinery through visualisation of sound signals. Mech. Syst. Signal Process. **14**, 229–241 (2000)
2. Strang, G.: Wavelet transforms versus Fourier transforms. Bull. Am. Math. Soc. **28**, 288–306 (1993)
3. Barbaroux, P.: Modularité de lorganisation et design des organisations adaptatives: une analyse de la transformation des organisations de défense americaines. Innovations **31**, 33 (2010)
4. Wu, J.-D., Liu, C.-H.: Investigation of engine fault diagnosis using discrete wavelet transform and neural network. Expert Syst. Appl. **35**, 1200–1213 (2008)
5. Forecast of solar irradiance using recurrent neural networks combined with wavelet analysis. Fuel Energy Abstracts **47**, 202 (2006). 06/01359
6. Daubechles, I.: Orthonormal bases of compactly supported wavelets. Fundamental Papers in Wavelet Theory. https://doi.org/10.1515/9781400827268.564
7. Peng, Z., Chu, F.: Application of the wavelet transform in machine condition monitoring and fault diagnostics: a review with bibliography. Mech. Syst. Signal Process. **18**, 199–221 (2004)
8. Djebala, A., Ouelaa, N., Hamzaoui, N.: Detection of rolling bearing defects using discrete wavelet analysis. Meccanica **43**, 339–348 (2007)
9. Daubechies, I.: Ten lectures on wavelets. (1992). https://doi.org/10.1137/1.9781611970104
10. Yang, J., Park, S.-T.: An anti-aliasing algorithm for discrete wavelet transform. Mech. Syst. Signal Process. **17**, 945–954 (2003)
11. Mallat, S.: A Wavelet Tour of Signal Processing: The Sparse Way. Academic Press, Cambridge (2009)
12. Lu, N., Wang, F., Gao, F.: Combination method of principal component and wavelet analysis for multivariate process monitoring and fault diagnosis. Ind. Eng. Chem. Res. **42**, 4198–4207 (2003)
13. Mallat, S.G.: A theory for multiresolution signal decomposition: the wavelet representation. Fundam. Pap. Wavelet Theory **11**(7), 674–693 (1989)
14. Newland, D.E.: Wavelet analysis of vibration: part 2—wavelet maps. J. Vib. Acoust. **116**, 417–425 (1994)
15. Akers, A., Lin, S.: Iterative learning control of double servo valve controlled electro hydraulic servo system. In: 2011 Seventh International Conference on Computational Intelligence and Security (2011)
16. Lin, S.J., Akers, A.: Optimal control theory applied to pressure-controlled axial piston pump design. J. Dyn. Syst. Meas. Control **112**, 475 (1990)
17. Lamento, B., Palmberg, J.-O.: Modeling and simulation of emulsion pump station pressure control system based on electro-hydraulic proportional relief valve. Appl. Mech. Mater. **190–191**, 860–864 (2012)
18. Manring, N.D., Johnson, R.E.: Modeling and designing a variable-displacement open-loop pump. J. Dyn. Syst. Meas. Control **118**, 267 (2016)

19. Manring, N.D., Luecke, G.R.: Modeling and designing a hydrostatic transmission with a fixed-displacement motor. J. Dyn. Syst. Meas. Control **120**, 45 (2011)
20. Manring, N.D., Du, H.: Adaptive robust control of variable displacement pumps. In: 2013 American Control Conference (2013)
21. Manring, N.: The control and containment forces on the swash plate of an axial piston pump utilizing a secondary swash-plate angle. In: Proceedings of the 2002 American Control Conference (IEEE Cat. No. CH37301) (2002)
22. Khalil, M.K.B., Yurkevich, V.D., Svoboda, J., Bhat, R.B.: Implementation of single feedback control loop for constant power regulated swash plate axial piston pumps. Int. J. Fluid Power **3**, 27–36 (2002)
23. He, Z.: Wavelet analysis and signal singularity. In: Wavelet Analysis and Transient Signal Processing Applications for Power Systems, pp. 45–52 (2016)
24. Rusli, M.: Application of short time Fourier transform and wavelet transform for sound source localization using single moving microphone in machine condition monitoring. KnE Engineering (2016)
25. Loutas, T., Kostopoulos, V.: Utilising the wavelet transform in condition-based maintenance: a review with applications. In: Advances in Wavelet Theory and their Applications in Engineering, Physics and Technology (2018)

Characterizing Parkinson's Disease from Speech Samples Using Deep Structured Learning

Lígia Sousa[1]([✉]), Diogo Braga[2,4], Ana Madureira[2,4], Luis Coelho[2,3], and Francesco Renna[5]

[1] Faculdade de Medicina da Universidade do Porto, Porto, Portugal
up201808521@med.up.pt
[2] ISEP/IPP, Porto, Portugal
{1140499,amd,lfc}@isep.ipp.pt
[3] CIETI - Centro de Inovação em Engenharia e Tecnologia Industrial, Porto, Portugal
[4] ISRC - Interdisciplinary Studies Research Center, Porto, Portugal
[5] Instituto de Telecomunicações, Faculdade de Ciências da Universidade do Porto, Porto, Portugal
frarenna@dcc.fc.up.pt

Abstract. An early detection of neurodegenerative diseases, such as Parkinson's disease, can improve therapy effectiveness and, by consequence, the patient's quality of life. This paper proposes a new methodology for automatic classification of voice samples regarding the presence of acoustic patterns of Parkinson's disease, using a deep structured neural network. This is a low cost non-invasive approach that can raise alerts in a pre-clinical stage. Aiming to a higher diagnostic detail, it is also an objective to accurately estimate the stage of evolution of the disease allowing to understand in what extent the symptoms have developed. Therefore, two types of classification problems are explored: binary classification and multiclass classification. For binary classification, a deep structured neural network was developed, capable of correctly diagnosing 93.4% of cases. For the multiclass classification scenario, in addition to the deep neural network, a K-nearest neighbour algorithm was also used to establish a reference for comparison purposes, while using a common database. In both cases the original feature set was optimized using principal component analysis and the results showed that the proposed deep structure neural network was able to provide more accurate estimations about the disease's stage, reaching a score of 84.7%. The obtained results are promising and create the motivation to further explore the model's flexibility and to pursue better results.

Keywords: Voice analysis · Parkinson's disease · Deep neural networks

© Springer Nature Switzerland AG 2020
A. M. Madureira et al. (Eds.): SoCPaR 2018, AISC 942, pp. 137–146, 2020.
https://doi.org/10.1007/978-3-030-17065-3_14

1 Introduction

Parkinson's disease (PD) is a progressively disabling neurodegenerative disease that may result in partial or total loss of movement capacity as well as cognitive abilities [1, 2]. It is the second most common neurodegenerative disease in the world, only surpassed by Alzheimer's [3]. Parkinson's disease affects 7 to 10 million people worldwide, with an incidence of 1 to 2% in individuals over 60 years [4, 5]. It is caused by the loss of certain groups of brain cells responsible for the production of neurotransmitters, such as dopamine, acetylcholine, serotonin and norepinephrine. These chemical molecules produced and released by neurons, which act as messengers in signal transmission in the nervous system, are essential to regulate the motor and non-motor activity of the human body. Their loss causes severe effects, such as impairment of movement control, resulting in disturbances in locomotion, postural instability, impaired vision, and speech abnormalities. Speech disorders are a common symptom of motor disorders in patients with Parkinson's disease, with conditions such as dysarthria (difficulty in articulating sounds), hypophony (low tone of voice), and monotonic tone (change in the fundamental frequency of the voice). Dysarthria is the most common condition and can be identified in about 90% of cases [6, 7]. There is currently no cure for PD, but an early diagnosis and appropriate therapy may be crucial on slowing the symptoms progression. It has been reported that, in about 90% of the diagnosed patients, voice problems can be detected in a pre-clinical stage [1, 8] and that therapy effectiveness can be monitored from speech tasks [9]. Thus, the analysis of the voice can, not only, allow the diagnosis of PD, simply and economically [10, 11], but also, monitor the evolution stage of the disease. Advances in the field of neural networks have led to the creation of new architectures and related algorithms that make them attractive for complex classification problems [12]. In this paper we propose a new methodology for the automatic classification of voice samples regarding the presence of acoustic patterns of PD based on a deep neural network (DNN) approach. It is also our objective to estimate the disease's evolution and current stage.

The rest of this paper is organized as follows: The next section is a brief state-of-the-art that covers important aspects and results reported by other authors while pursuing similar objectives. Then we present in detail the methodology that was implemented for the proposed approach and proceed with an evaluation following widely used metrics. Each result is analyzed and discussed. Finally, we present the main conclusions and some ideas for future developments.

2 Evaluation of Parkinson's Disease Based on Voice Analysis

In [3], the detection of the presence of PD clues is based on the binary classification of voice samples. Using an artificial neural network classifier two distinct sets of parameters are explored: firstly, using input data with 22 features and subsequently, decreasing this number to 16, with results of 80.8% and 83.3% of accuracy, respectively, for the validation dataset. This experience has highlighted the importance of using the right features and how they can affect the performance of an architecture.

Other authors have focused their attention on the data model. In [13] we can find a classification approach based on Artificial Neural Networks while in [14] Support Vector Machines (SVM) are used. The former, using a set of 33 features, has achieved an accuracy of 93.3%. The last, considering a very similar feature set, have obtained an average accuracy of 94% using cross-validation for the prediction models. More recently, in [11], the detection of PD through voice has been performed using SVMs and Random Forests (RF). This project obtained results with an accuracy of 92.38% (SVM) and 99.94% (RF) for the mentioned algorithms, after optimization of the feature set and detailed tuning of the classification models.

Other authors have attributed great importance to the feature selection process of the detection of PD. On Table 1 we present commonly used features for PD classification, where jitter and shimmer have a main role and act as markers of biomechanical instability. Typical variations are the difference of difference of periods (DDP), Relative Average Perturbation (RAP), Period Perturbation Quotient considering n-points (PPQn), and Amplitude Perturbation Quotient for k-points (APQk). For example, in [15], the authors proposed a novel speech algorithms for the binary classification of PD with high accuracy. In their approach, Least Absolute Shrinkage and Selection Operator (LASSO), Minimum Redundancy Maximum Relevance (mRMR), Relief and Local Learning-Based Feature Selection (LLBFS) were used to determine the best speech measures out of 132 features. With these features, the authors were able to achieve an accuracy of up to 98.6% using RF and SVM. In [16], correlation rates, Fisher's Discriminant Ratio (FDR) and ROC curves were used as feature selection techniques to obtain an optimal set of 7 features for the classification task, from 22 possible features. The best features found are variants of shimmer, jitter, pitch and HNR. The K-Nearest Neighbor classifier provided the best accuracy result of 93.82%.

Table 1. Composition of feature vector based on the acoustic analysis of speech samples

Base feature	Derived features
Jitter	Local, local/absolute, DDP, RAP, PPQ5
Shimmer	Local, local/db, DDP, APQ3, AP5, AP11
Pitch	Mean, minimum, maximum, standard deviation
Harmonicity	Auto-correlation (AC), HNR, NHR

3   Methodology

3.1   Materials and Tools

For the binary classification problem we have used the database described in [17], composed of 1040 voice recordings from 20 PD patients and 20 healthy subjects (HS). The average age of the patients was 64.86 years with a standard deviation of 8.97. The recordings were sampled at 96 kHz. For the multiclassification problem we have used the database developed by [16]. This database is composed of 22 speakers with PD (12 females and 10 males), from ages 44 to 79 (mean 67.18, standard deviation 9.4), containing a total of 1002 recording. Each patient is assigned with one of four levels, depending on the disease's progression. To balance the number of elements in each class we have used some additional recordings of healthy individuals [11]. After combining these resources, we have considered a total of 122 utterances, 61 PD and 61 for HS.

For developing the proposed algorithms, we have used Python version 3.5 running on a Windows i7 computer with 12 GB of memory. The DNN implementation was based on the TensorFlow library [18].

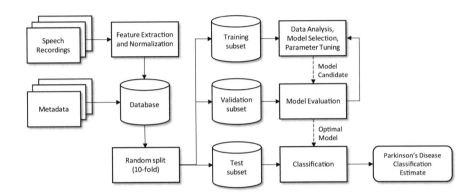

Fig. 1. Development pipeline for the proposed algorithms.

3.2   Development

The development of the proposed approaches has been made according with the pipeline presented in Fig. 1. The first step consisted in extracting features from the recordings and then performing data normalization using a z-score methodology. By combining the resulting information with the related metadata, we have obtained the supporting database. To facilitate the model evaluation, three subsets are randomly created for each session, a training subset, which is used to tune the model parameters, a validation subset, which is used to determine the optimal model, and a test subset, which is used to assess the performance of the optimal model.

To tackle the binary classification problem, we have developed a DNN with a specific structure. This approach conducts to a reduction of dimension of the features due to autoencoders [19]. The first layer, or input layer, uses a number of neurons equal to the number of features presented at the input of the network. Thus, if the number of features vary, for example to optimize results by feature selection, the proposed network can easily adapt to new data. The hidden layer of the model has a number of neurons identical to the input layer. Both layers have a Rectified Linear Unit (ReLU) function as activation function. It exhibits a non-linear behavior that is characterized by the presence of activation of the output only for positive valued inputs. It is a popular option in modern machine learning algorithms since it can eliminate the vanishing gradient problem, in models with several layers, and thus contributes to optimize the training algorithms [20] representing an advantage over older neural networks architectures. Finally, the output layer has only one neuron, because it is only possible to obtain an output for each input, that is activated by a logistic function (sigmoid) followed by a rounding operation to obtain the binary classification result. It is also part of the network architecture the adoption of dropout after the input layer, which is used to avoid overfitting by randomly deactivating half of the neurons of this layer to each epoch during training [21]. Regarding the parameterization of the model we have used 600 epochs, a learning rate of 0.01, a decay rate of 0.01, a momentum of 0.01 and the Adam optimizer using binary cross-entropy as loss function.

For the multiclass problem we have used a slightly different network. The input layer was not changed, maintaining the number of neurons adjusted to the number of features and the activation based on the ReLU function. It was also maintained a dropout in the architecture after the first hidden layer. Three hidden layers with a number of units identical to that of the input layer were also used, also activated by the ReLU function. The number of hidden layers results from the evaluation performed where it was observed that there was no benefit in adding more layers. For the output layer we have used four neurons, according to the four levels of development of the disease, and the softmax function was used as activation function. This layer returns the probability that a set of features belong to each of the 4 classes considered, and the class assigned to the feature set will be the one with the highest probability. For parameters we have considered 600 epochs, a learning rate of 0.01, a decay rate equal to the ratio between learning rate and number of epochs, a momentum of 0.9 and the stochastic gradient descent optimizer using the categorical cross-entropy loss function.

In addition to the deep structured network approach, we have also developed a k-Nearest Neighbors (KNN) architecture for comparison purposes. The KNN uses the entire training dataset, in the sense that the algorithm searches the data set for k most similar instances. To calculate the similarity between instances, the Euclidean distance or Hamming distance is calculated when the data is of the categorical (identification of more than two labels) or binary type. Thus, when it is necessary to make a prediction for new data, the model searches in the training data the k most similar instances returning as prediction the label that is most represented among these k instances. The KNN is considered a non-linear classifier, since it does not assume a form of operation, adapting to different types of problems. It is often used as a baseline for multiclassification problems, mainly due to its low complexity and good performance on clearly clustered data.

Learning models and algorithms can benefit from a prior analysis of the data, depending on its underlying structure, in order to optimize its representation and to decrease the existence of dependencies. To investigate this possibility, we have used principal component analysis (PCA). The main purpose is to obtain linearly uncorrelated variables from the initial set of possibly correlated features and to understand which feature contribute with higher discriminative information to distinguish between classes. For each case we have chosen a set of principal components that was able to cover 80% of the total variance. This means that we were able to better handle the variation within the data set while decreasing the number of features and model complexity.

With the proposed algorithm, we intended, based on the acoustic analysis of voice samples, to, in a first stage, detect the presence of indicators of Parkinson's disease and, in a second stage, to characterize the degree of development of the disease. The second task clearly represents a more challenging problem when compared with the binary classification tasks involved in PD detection. For this reason, a more complex network is adopted to perform the multiclass classification problem, thus enabling a finer data discrimination.

4 Model Evaluation

4.1 Methodology

For evaluating the performance of the proposed approaches, we have used *k-fold* cross-validation during the train/validate/test sessions. This algorithm separates the training dataset in *k* sub-sets of equal dimensions so that one of the subsets is used for validation and the remaining *k-1* for training. This training/testing process is repeated for each subset which represents a total of *k* distinct runs [22]. The main advantages are: a) During training, to decrease the chances of overfitting; b) During testing, to observe the model robustness (to variations in the parameters hyperspace and to weight initialization) and to detect possible local minima. For this project we have used *k* = 10 as a compromise between quality of statistical analysis and training/testing time. Additionally, while trying to detect model overfitting situations, we have used the *k-fold* approach to both the test data (*cross_val_score* function) and validation data (*cross_val_predict* function) and have compared the obtained results and their related statistics (mean and standard deviation).

As a figure of merit for comparing the obtained results we have calculated the accuracy, or score, defined as the percentage of times the model is successful in assigning a label to a set of data, as given in Eq. 1.

$$accuracy = \frac{\sum(true\ positives) + \sum(true\ negative)}{total\ population} \quad (1)$$

To better understand the behaviour of the model and to observe the quality of the classification we have also used confusion matrices that reports true positives (TP), true negatives (TN), false positives (FP) and false negatives (FN).

4.2 Results and Discussion

To observe the learning ability of the model (and effectiveness of the related parameters) we have begun by representing the accuracy and loss curves along the training epochs in order to confirm the correct evolution of the process. In Fig. 2 we can observe the evolution of the accuracy metric, representing the frequency with which the predicted values match the existing labels. In Fig. 3 we have depicted the cross-entropy loss function, showing how accurate are the predictions. From the observation of Figs. 2 and 3 it is also possible to note that, during the training phase, the network weights are iteratively adjusted in order to minimize the value of the loss function. Such evolution of the weights is also shown to lead to increased accuracy values. The network performance is shown to reach a steady state after approximately 45 training epochs. Moreover, the comparison between the accuracy and loss values computed over training and validation data sets testify that the trained model was not affected by overfitting problems.

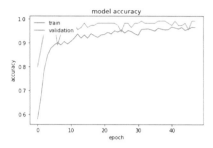

Fig. 2. Evolution of model accuracy along the first training epochs. Accuracy values are normalized on a 0 to 1 range.

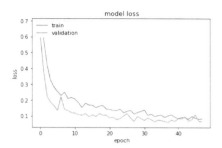

Fig. 3. Evolution of model loss, considering the binary cross-entropy function, along each training epoch.

Binary Classification

The results obtained for this task, using the proposed model and parameters, are presented in Table 2, where we can find the values corresponding to, from top to bottom, the best training accuracy, to the cross-validation using the test subset and to the cross-validation using the validation subset.

Table 2. Accuracy values (percentage) for the binary classification task

	Without PCA	With PCA
Train/Validation	98.361	99.180
Test	92.623	94.262
Predictions	91.803	93.443

By observing Table 2, we can see that the results obtained for the test and prediction stages are good and close which points to the DNN's ability to diagnose PD. We can also observe that the use of PCA resulted in some benefit, but not significant, probably due to the model's generalization ability, given the complexity and size of the available data. The accuracy values, in the "Predictions" row, corresponding to a performance estimation for a production stage data model, are slightly higher than the ones reported in [14], however only a part of the dataset is shared and so the numbers are not directly comparable. Given the importance of this value, since it effectively represents the capacity of the network to correctly provide diagnosis, we have observed some further details. Predictions were made in relation of the output (y_pred) from a dataset (X_test), and the results were compared with the true labels (y_test) of the same set (X_test), giving rise to the confusion matrix presented in Table 3. Again the use of

PCA had just a small influence in the results. This may be due to the differentiating characteristics of each class that might not be represented in the features' variance.

Table 3. Confusion matrices for the classification task using the raw feature set (left) and using principal components analysis (right)

	Without PCA			With PCA		
	True	False	Total	True	False	Total
Pathologic	**60**	**1**	61	**61**	**0**	61
Heathy	**52**	**9**	61	**53**	**8**	61
Total	112	10	122	114	8	122

By observing Table 2, we can see that the use of PCA has helped to eliminate the number of false positives which is especially important in a healthcare related algorithm. The number of false negatives was also reduced but the obtained result can still be subject to improvements. The use of some additional data in future developments could help to improve the obtained values and confirm the observed trends.

Multiclass Classification

For the multiclass classification problem, we have tested the two proposed approaches with and without feature set optimization, i.e., with and without applying PCA. The obtained accuracy values are presented in Table 4, covering each development stage.

Table 4. Accuracy values (in percentage) for the multiclass classification task using KNN and DNN, considering a raw feature set and after a PCA optimization.

	KNN		DNN	
	Without PCA	With PCA	Without PCA	With PCA
Train/Validation	50.522	69.747	74.645	89.915
Test	42.786	61.319	72.403	88.125
Prediction	52.106	59.642	69.936	84.679

When analysing the results of the proposed KNN, we can verify that the obtained results in the test do not present an accuracy as high as the one obtained during training/validation. It is not unusual to find cases where the accuracy for training are higher than testing, however, as they differ by about 8%, the results may suggest that overfitting occurred during the test. Additionally, when observing the detailed statistics, the results may point that the algorithm failed to be able to define neighbourhoods that characterize each class. Overall the obtained results for the KNN approach are not very promising despite our efforts which may indicate that this model is not compatible with the data's nature even when feature optimization is considered.

On the other hand, the DNN model, due to the large number of parameters that can be adjusted, presents a high tuning flexibility which is then reflected in the obtained results. The accuracy variations in the three different stages are acceptable when considering the amount of available data and point to a good generalization ability of the model. The maximum average accuracy values were 69.9% and 84.7% when using the raw feature set and the optimized feature subset, respectively. The results are promising and seem to indicate that the classification model based on neural networks is adequate. However, it is seen as beneficial to increase the size of the database in order to increase the quality of the calculated statistics and to strengthen the proposed model. In this case, proper feature optimization techniques, as PCA, showed to be effective on reducing the complexity of data space, thus leading to a better classification performance

5   Conclusion and Future Work

In this paper we propose a system for characterizing Parkinson's disease from speech samples using deep structured learning. We begin by reviewing some works with similar objectives. For each case, we have identified the aspects that characterize the different approaches as well as the influence they had on the reported results. A particular emphasis has been given to the databases used, to the set of features that was considered and the respective selection process, when applicable, and finally, to the estimation model and its parameters. After showing an overview of the processing pipeline, we have detailed the functions of each stage. We provided a brief presentation of the databases, which were specifically developed for the study of PD. Then, we have thoroughly explained the procedure that was followed to implement the proposed method. This new approach allowed to achieve a final accuracy of 93.4% for the binary classification task, which is higher than previously reported results using a neural networks approach, but slightly lower when compared with studies that have relied on different classification models. For the multiclass classification task, the accuracy value was 84.7%, which is a very promising result. The achievement in the multiclass classification problem motivate the authors to further improve the system. The improvement of the classification model and the use of additional recordings, that can bring improved robustness to the obtained statistical results, are also envisioned.

References

1. Ho, A.K., Iansek, R., Marigliani, C., Bradshaw, J.L., Gates, S.: Speech impairment in a large sample of patients with Parkinson's disease. Behav. Neurol. **11**, 131–137 (1998)
2. Ramig, L.O., Fox, C., Sapir, S.: Speech treatment for Parkinson's disease. Expert Rev. Neurother. **8**, 297–309 (2008)
3. Khemphila, A., Boonjing, V.: Parkinsons disease classification using neural network and feature selection. Int. J. Math. Phys. Electr. Comput. Eng. **6**, 377–380 (2012)
4. Hirsch, L., Jette, N., Frolkis, A., Steeves, T., Pringsheim, T.: The incidence of Parkinson's disease: a systematic review and meta-analysis. Neuroepidemiology **46**, 292–300 (2016)
5. Neurological disorders: public health challenges. World Health Organization (2006)

6. Müller, J., Wenning, G.K., Verny, M., McKee, A., Chaudhuri, K.R., Jellinger, K., Poewe, W., Litvan, I.: Progression of dysarthria and dysphagia in postmortem-confirmed parkinsonian disorders. Arch. Neurol. **58**, 259–264 (2001)
7. Liu, L., Luo, X.-G., Dy, C.-L., Ren, Y., Feng, Y., Yu, H.-M., Shang, H., He, Z.-Y.: Characteristics of language impairment in Parkinson's disease and its influencing factors. Transl. Neurodegener. **4**, 2 (2015)
8. Teixeira, J., Soares, L., Martins, P., Coelho, L., Lopes, C.: Towards an objective criteria for the diagnosis of Parkinson disease based on speech assessment. Presented at the XXXV Congresso Anual de la Sociedad Espanola de Ingeniería Biomedica, Bilbao (2017)
9. Norel, R., Agurto, C., Rice, J.J., Ho, B.K., Cecchi, G.A.: Speech-based identification of L-DOPA ON/OFF state in Parkinson's Disease subjects. BioRxiv Prepr. 420422 (2018)
10. Zhang, Y.N.: Can a smartphone diagnose Parkinson disease? A deep neural network method and telediagnosis system implementation. Park. Dis. (2017)
11. Braga, D., Madureira, A.M., Coelho, L., Abraham, A.: Neurodegenerative diseases detection through voice analysis. In: Abraham, A., Muhuri, P.K., Muda, A.K., Gandhi, N. (eds.) Hybrid Intelligent Systems, pp. 213–223. Springer, Heidelberg (2018)
12. LeCun, Y., Bengio, Y., Hinton, G.: Deep learning. Nature **521**, 436–444 (2015)
13. Gil, D., Magnus, J.: Diagnosing Parkinson by using artificial neural networks and support vector machines. Glob. J. Comput. Sci. Technol. **9**, 63–71 (2009)
14. Saloni, R.K., Gupta, A.K.: Detection of Parkinson disease using clinical voice data mining. Int. J. Circuits Syst. Signal Process. **9** (2015)
15. Tsanas, A., Little, M.A., McSharry, P.E., Spielman, J., Ramig, L.O.: Novel speech signal processing algorithms for high-accuracy classification of Parkinson's disease. IEEE Trans. Biomed. Eng. **59**, 1264–1271 (2012)
16. Proença, J., Veiga, A., Candeias, S., Lemos, J., Januário, C., Perdigão, F.: Characterizing Parkinson's disease speech by acoustic and phonetic features. In: Baptista, J., Mamede, N., Candeias, S., Paraboni, I., Pardo, T.A.S., Volpe Nunes, M.d.G. (eds.) Computational Processing of the Portuguese Language, pp. 24–35. Springer, Heidelberg (2014)
17. Sakar, B.E., Isenkul, M.E., Sakar, C.O., Sertbas, A., Gurgen, F., Delil, S., Apaydin, H., Kursun, O.: Collection and analysis of a Parkinson speech dataset with multiple types of sound recordings. IEEE J. Biomed. Health Inform. **17**, 828–834 (2013)
18. Abadi, M., Agarwal, A., Barham, P., Brevdo, E., Chen, Z., Citro, C., Corrado, G.S., Davis, A., Dean, J., Devin, M., Ghemawat, S., Goodfellow, I., Harp, A., Irving, G., Isard, M., Jia, Y., Jozefowicz, R., Kaiser, L., Kudlur, M., Vasudevan, V., Viegas, F., Vinyals, O., Warden, P., Wattenberg, M., Wicke, M., Yu, Y., Zheng, X.: TensorFlow: Large-Scale Machine Learning on Heterogeneous Distributed Systems. ArXiv160304467 Cs (2016)
19. Ranzato, M., Poultney, C., Chopra, S., LeCun, Y.: Efficient learning of sparse representations with an energy-based model. In: Proceedings of the 19th International Conference on Neural Information Processing Systems. pp. 1137–1144. MIT Press, Cambridge (2006)
20. Glorot, X., Bordes, A., Bengio, Y.: Deep sparse rectifier neural networks. In: Proceedings of the 14th International Conference on Artificial Intelligence and Statistics, pp. 315–323. Fort Lauderdale, Florida (2011)
21. Srivastava, N., Hinton, G., Krizhevsky, A., Sutskever, I., Salakhutdinov, R.: Dropout: a simple way to prevent neural networks from overfitting. J. Mach. Learn. Res. **15**, 1929–1958 (2014)
22. Arlot, S., Celisse, A.: A survey of cross-validation procedures for model selection. Stat. Surv. **4**, 40–79 (2010)

Combinatorial Optimization Method Considering Distance in Scheduling Problem

Yuta Obinata[✉], Kenichi Tamura, Junichi Tsuchiya, and Keiichiro Yasuda

Department of Electrical and Electronic Engineering, Tokyo Metropolitan University,
1-1, Minamiosawa, Hachioji-shi, Tokyo 192-0397, Japan
obinata-yuta@ed.tmu.ac.jp

Abstract. In this paper, we focus on the idea of integral designing of problems/methods/distances in metaheuristics for combinatorial optimization. The above idea is important in combinatorial optimization, where it is necessary to consider the distance according to each problem. Furthermore, the idea is particularly important for methods that use distance for the movement strategy, which was proposed the authors. Therefore, as a practical example of the above idea, and we proposed a method which is introduced a search strategy with more consideration of distance. We report that, when considering the distance, the proposed method has better search performance than the previous method in the flow shop scheduling problem.

Keywords: Combinatorial optimization · Metaheuristics · Distance · Scheduling problem

1 Introduction

Recently, systems have become larger and more complex, and their computational power has increased, while there have been advances in the development of peripheral technologies such as simulation and modeling tools. Based on this background, it is important to develop optimization methods with high versatility and high performance. Many problems that appear in reality, such as facility placement problems and scheduling problems, can be formulated as combinatorial optimization problems. Many of these problems are known to require significant amounts of time to obtain a strict optimum solution; these are called $\mathcal{NP}$-hard problems. In order to solve these problems, metaheuristics which are a framework of the heuristic approximate method, have been attracting much attention.

Most metaheuristics are constructed based on analogies, such as various natural and physical phenomena [1,2]. In contrast, we have constructed and improved combinatorial optimization methods so far, focusing on the characteristics of target problems and employing search strategies that are common to excellent

© Springer Nature Switzerland AG 2020
A. M. Madureira et al. (Eds.): SoCPaR 2018, AISC 942, pp. 147–157, 2020.
https://doi.org/10.1007/978-3-030-17065-3_15

optimization methods [3,4]. Specifically, we introduced "the basin of attraction," which is a set of solutions that arrive at the same local optimal solution using best-improvement Local Search, in the solution space; by doing this, we can interpret the solution space not only as a set of individual solutions, but also as a set of basins of attraction. With the former as a lower structure and the latter as a higher structure, we proposed the combinatorial optimization method based on the hierarchical structure of solution space (HM) [3].

Recently, in metaheuristics for combinatorial optimization problems, research has been conducted to improve the search performance by using distances not only for neighborhood generation but also for selection/movement operations. For these methods, the distance plays a very important role. Therefore, it is important to integrally design problems/methods/distances in combinatorial optimization.

HM shows superior performance in the traveling salesman problem and the 0-1 knapsack problem, which are typical combinatorial optimization problems; however, in the flow shop scheduling problem (FSP), the search performance of Tabu Search (TS) [5] outperforms that of HM; if this issue is solved, HM becomes a more versatile method. In this paper, we aim to improve the search performance of HM for the FSP and we propose a method which is introduced a search strategy with more consideration of distance. We show its usefulness by performing numerical experiments, and we show the importance of design problems/methods/distances integrally in combinatorial optimization.

2   Distance in Metaheuristics

Most metaheuristics can be regarded as performing searches by repeating to generate a neighborhood based on information obtained during the search process, and by selecting the solution from within the generated neighborhood. Thus, the basic structure of metaheuristics is neighborhood generation and selection/movement. We describe that the distance used for its operation.

2.1   Distance in Continuous Optimization Method

Metaheuristics, which solve continuous optimization problems with real-valued vectors as a solution, perform these operations using the Euclidean distance. The following example can be cited as a method that involves the use of the Euclidean distance.

In Particle Swarm Optimization (PSO) [1], when each particle (the search point) moves from the current position to the next position, a new movement vector is generated from a vector toward the best solution from among the solutions that are searched by each search point, a vector toward the best solution among all the searched points, and the previous movement vector.

In Firefly Algorithm (FA) [2], each search point moves to approach other search points. Here, FA determines the degree of movement based on the light intensity (evaluated value of the search point) and attractiveness (coefficient for determining the movement amount). This attractiveness decreases with increasing distance between search points.

2.2   Distance in Combinatorial Optimization Method

We have introduced the concept of distance in the solution space for combinatorial optimization problems.

Mathematically, the distance is defined as a mapping $d : \boldsymbol{X} \times \boldsymbol{X} \rightarrow \mathbb{R}$ that satisfies the axioms of distance in space $\boldsymbol{X}$, and it is a measure of the similarity between solutions. The concept of distance is a flexible framework that can be introduced if it meets axioms, and distances can be defined in the solution space of most combinatorial optimization problems.

Therefore, the concept of distance makes it easy to grasp the search situation and design the search operation. However there are several distances such as following, and we need to select one of them for each problem.

- The distance known as the distance between pairs of any arbitrary permutations, such as the Hamming distance, Spearman's footrule distance, Cayley's distance, and Ulam's distance.
- The minimum value of the neighborhood generation required to move from one solution to another one.

There are the following examples as studies related to problems/distances.

Reeves discussed basic mathematical theory and methods associated with the concept of a fitness landscape [6]. The landscape is represented as $\Lambda = (\boldsymbol{X}, f, d)$, and $\boldsymbol{X}$ denotes the search space, f denotes the objective function, and d denotes the distance defined based on the search space. Bożejko et al. focused on big valleys in the landscape, and proposed an algorithm to verify the differences in the structure of problems due to the difference in distance [7].

There are the following examples as studies related to methods/distances.

In a previous study, the authors proposed methods that use the distance in solution selection/movement. In Ref. [8], the multi-point combinatorial optimization method was proposed, where diversification and intensification are monitored and controlled by the mean distance of search points. In Ref. [9], the multi-point combinatorial optimization method was proposed, and quantitatively estimates the complexity of the problem structure based on the distance obtained during the search process, and it adjusts the balance of the diversification and intensification in the search.

From the above, it is very important to integrally design problems/methods /distances in combinatorial optimization.

3   Combinatorial Optimization Method Based on Hierarchical Structure in Solution Space

The combinatorial optimization method based on hierarchical structures in solution space (HM) is a new combinatorial optimization method proposed by the authors [3]. HM is a method that uses the distance for selection movement.

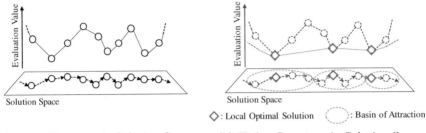

(a) Lower Structure in Solution Space   (b) Higher Structure in Solution Space

Fig. 1. Hierarchical solution space structure

3.1   Higher Concept in Solution Space

We introduce a higher concept in the solution space of combinatorial optimization problems, and it is called "the basins of attraction," which encloses individual solutions. By doing so, HM aims to search for unsearched areas globally. The basin of attraction is a set of solutions that arrive at the same local optimal solution using best-improvement Local Search. The solutions x and y in the same basin of attraction satisfy the following equivalence relation.

$$x \sim y : LS(x) = LS(y)$$

Here, $LS(x)$ represents the execution result of best-improvement Local Search for x. Consequently, this equivalence relation classifies the solution space. Therefore, the solution space can be interpreted not only as a set of solutions, but also as a set of basins of attraction.

In this paper, the solution space that comprises the basins of attraction, which are defined as a set of solutions with common characteristics, is interpreted as a higher structure. In addition, the solution space comprising the individual solutions is interpreted as a lower structure. Figure 1 shows a conceptual diagram of the hierarchical solution space structure for a minimization problem.

3.2   Distance in Combinatorial Optimization Method Based on Hierarchical Structures in Solution Space

HM uses the concept of distance not only in neighbor generation but also in selection/movement. As a result, the distance significantly affects the search of HM.

Specifically, we direct the search of HM far away from the representative in the current basin. In addition, following operation is performed in order to prevent the search point from returning to the searched basins. Here, we define "the diameter-pair set" as pairs of the solutions that are farthest from each other in the set of stored local optimal solutions. That is, other stored local optimal solutions exist between diameter-pair sets. Thus, we limit the neighborhood to depart from the diameter-pair set, and the search point can be prevented from returning to basins of attraction that have been searched.

Actually, we limit the neighborhood to increase the sum of the distance between each element of the diameter-pair set and the search point. As an analogy from the Euclid space, this means going to the outside of the ellipse focus (elements of diameter-pair).

4   Comparison of Search Performance

We compared search performance for difference of a kind of distance by numerical experiment.

In this paper, candidates for the distance are the distances that represent the amount of time required to shift job position as well as the amount of time required to exchange the job position in FSP. The candidates of distance and neighborhood are as follows. Here, n denotes the number of jobs, and $\boldsymbol{F}$ denotes the search space.

The first candidate is the Spearman's footrule distance (Footrule distance) [10] used in previous research.

Spearman's footrule distance: d_{Footrule}
When ranking the position of the target element, it is expressed by the sum of differences in the rank order between orders $\boldsymbol{x}$ and $\boldsymbol{y}$.

$$d_{\text{Footrule}}(\boldsymbol{x}, \boldsymbol{y}) = \sum_{i=1}^{n} |x_i - y_i|$$

Neighborhood based on Footrule distance: $\boldsymbol{N}_{\text{Footrule}}$
$\boldsymbol{N}_{\text{Footrule}}$ was defined as the solutions of $d_{\text{Footrule}} \leq 4$. In FSP, this means an operation of swapping the processing position of a certain job and the job of an adjacent position.

$$\boldsymbol{N}_{\text{Footrule}}(\boldsymbol{x}) = \{\boldsymbol{y} \in \boldsymbol{F} \mid d_{\text{Footrule}}(\boldsymbol{x}, \boldsymbol{y}) \leq 4\}$$

The second candidate is the Ulam's distance [10].

Ulam's distance: d_{Ulam}
When performing an operation to insert arbitrary elements of the order at arbitrary positions, it is represented by the minimum number of operations necessary to change between orders $\boldsymbol{x}$ and $\boldsymbol{y}$.

$$d_{\text{Ulam}}(\boldsymbol{x}, \boldsymbol{y}) = n - (\text{length of longest increasing subsequence})$$

Neighborhood based on Ulam's distance: $\boldsymbol{N}_{\text{Ulam}}$
$\boldsymbol{N}_{\text{Ulam}}$ was defined as the solutions of $d_{\text{Ulam}} = 1$. In FSP, this indicates a shifting operation to move a certain job from its processing position to another position.

$$\boldsymbol{N}_{\text{Ulam}}(\boldsymbol{x}) = \{\boldsymbol{y} \in \boldsymbol{F} \mid d_{\text{Ulam}}(\boldsymbol{x}, \boldsymbol{y}) = 1\}$$

Table 1. Experimental conditions

Problem	FC_{max}	TL		s	
		TS-FD	TS-UD	HM-FD	HM-UD
ta031 (50 jobs 5 machines)	5.0×10^5	118	238	2	5
ta041 (50 jobs 10 machines)	5.0×10^5	47	125	12	12

Table 2. Experimental results of TS and HM

Problem	Indexes	TS-FD	TS-UD	HM-FD	HM-UD
ta031(50 jobs 5 machines)	mean	0.67%	0.51%	1.43%	0.007%*
	best	0%*	0%*	0.18%	0%*
	worst	1.14 %	1.03%	4.22%	0.18%*
	S.D.	0.31%	0.30%	0.87%	0.036%
ta041 (50 jobs 10 machines)	mean	5.15%	3.71%	8.38%	2.72%*
	best	3.61%	2.01%	5.65%	1.54%*
	worst	7.29%	5.22%	13.20%	4.08%*
	S.D.	0.87%	0.83%	1.62%	0.58%

The experiment was carried out using the benchmark problem of FSP [11]. We compared the following:

- TS-FD (TS constructed with Footrule distance),
- TS-UD (TS constructed with Ulam's distance),
- HM-FD (HM constructed with Footrule distance),
- HM-UD (HM constructed with Ulam's distance).

Table 1 shows the experimental conditions. FC_{max}, which is the maximum number of evaluations, represents the termination condition of the algorithm. Each method was executed 50 times until the termination condition. At this time, the initial point had not changed for any of the methods. The unification of FC_{max} makes it possible to compare how good a solution could be obtained by a limited number of evaluations. After preliminary experiments, each parameter showed good results. Here, TL denotes the Tabu list length. The Tabu of TS was the operation memory. The pair of the job and processing number is memorized, and the combination of job and process number is prevented from being repeated.

For each case, the mean, best value, worst value, and standard deviation (S.D.) of the results about the objective function value of 50 times are shown in Table 2. It shows the relative error from the global optimum value for the mean, best, and worst values. In addition, the variation coefficient of the S.D. is shown. Here, "*" indicates the best result.

From the results of Table 2, the proposed method achieves the best result for almost all indicators of all the problems used in this paper. In addition, the improvement rate of results from HM-FD to HM-UD is larger than that of results

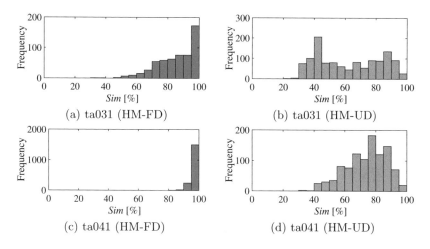

Fig. 2. Histogram of the distribution of Sim

from TS-FD to TS-UD based on all indicators of all problems. In other words, it was shown that HM was larger for the influence of varying the distance.

Here, the similarity rate Sim, which is an index based on distance, is defined in Eq. (1) as follows in order to clarify the proximity between local optimal solutions. This can be represented by 0%–100% in terms of the proximity of solutions, regardless of the kind of distance.

$$Sim(\boldsymbol{x}, \boldsymbol{y}) = \frac{d_{\max} - d(\boldsymbol{x}, \boldsymbol{y})}{d_{\max}} \tag{1}$$

Sim is an index that is normalized by using the maximum distance $d_{\max}$ of each distance. The maximum distance of Footrule distance depends on whether the problem size n is odd or even. In particular, $d_{\max} = (n-1)^2/2$ if n is odd, $d_{\max} = n^2/2$ if n is even. On the other hand, the maximum distance of Ulam's distance is $d_{\max} = n - 1$.

In this paper, we analyzed Sim between $\boldsymbol{x}^{(t,*)}$, which the local optimal solution at a certain point, and $\boldsymbol{x}^{(t-1,*)}$, which the local optimal solution obtained by the previous iteration. That is, we show in the distance between local optimal solutions before and after in the above experiment of HM-FD and HM-UD.

Figure 2(a) and (c) shows the histogram of HM-FD, and Fig. 2(b) and (d) shows the histogram of HM-UD. In Fig. 2(a) and (c), it can be seen that local optimal solutions before and after repetition are close to each other, and this is because the distribution is primarily on the side of 100%. In contrast, in Fig. 2(b) and (d), it can be seen that local optimal solutions before and after repetition exist not only near to each other, but also far from each other, and this is because the distribution covers a wide range. Therefore, it is necessary to consider the difference in structure of solution space determined by the distance.

5   Proposal of Method

Since it is clarified from the above comparison that the behavior of the search point differs depending on the structure of solution space, we propose "a method which is introduced a search strategy with more consideration of distance".

Specifically, the lower limit of movement ($w \cdot d_{\max}$) is determined by using the maximum distance and introduced for escape determination of the higher structure. Here, $w \in [0, 1)$ is a parameter. By this introduction, we aim to improve the flexibility of movement of search points regardless of the kind of distance.

Algorithm 1 Proposed Method

1: **procedure** PM(w,s,$T_{\max}$)
    Step 1: Initialization
2:    Generate initial solution $\boldsymbol{x}^{(0,0)}$
3:    $\boldsymbol{H} = \emptyset$, $t = 0$, $k = 0$
    Step 2: Execution of Search in Lower Structure
4:    **while** $f(\boldsymbol{x}^{(t,k)}) > \min\{f(\boldsymbol{y}) \mid \boldsymbol{y} \in \boldsymbol{N}(\boldsymbol{x}^{(t,k)})\}$ **do**
5:        $\boldsymbol{x}^{(t,k+1)} := \arg\min\{f(\boldsymbol{y}) \mid \boldsymbol{y} \in \boldsymbol{N}(\boldsymbol{x}^{(t,k)})\}, k := k + 1$
6:    $\boldsymbol{x}^{(t,*)} := \boldsymbol{x}^{(t,k)}$
7:    **if** $\mid \boldsymbol{H} \mid \geq s$ **then**
8:        Delete the oldest solution in $\boldsymbol{H}$
9:    $\boldsymbol{H} := \boldsymbol{H} \cup \boldsymbol{x}^{(t,*)}$
    Step 3: Termination Judgment
10:    **if** $t = T_{\max}$ **then**
11:        **return** The best solution in search process
12:    **else**
13:        $\boldsymbol{x}^{(t+1,0)} := \boldsymbol{x}^{(t,*)}$, $k = 0$, $t := t + 1$
    Step 4: Selection of Usage Information
14:    **if** $t \leq 2$ **then**
15:        $\boldsymbol{x}_\alpha = \boldsymbol{x}_\beta = \boldsymbol{x}^{(0,*)}$
16:    **else**
17:        $\{\boldsymbol{x}_\alpha, \boldsymbol{x}_\beta\} = \arg\max\{d(\boldsymbol{x}, \boldsymbol{y}) \mid \boldsymbol{x}, \boldsymbol{y} \in \{\boldsymbol{H} \setminus \boldsymbol{x}^{(t-1,*)}\}\}$
    Step 5: Execution of Search in Higher Structure
18:    $\boldsymbol{N}_1 = \{\boldsymbol{y} \in \boldsymbol{N}(\boldsymbol{x}^{(t,k)}) \mid d(\boldsymbol{x}^{(t,k)}, \boldsymbol{x}^{(t,0)}) < d(\boldsymbol{y}, \boldsymbol{x}^{(t,0)})\}$
19:    $\boldsymbol{N}_2 = \{\boldsymbol{y} \in \boldsymbol{N}(\boldsymbol{x}^{(t,k)}) \mid d(\boldsymbol{x}^{(t,k)}, \boldsymbol{x}_\alpha) + d(\boldsymbol{x}^{(t,k)}, \boldsymbol{x}_\beta) < d(\boldsymbol{y}, \boldsymbol{x}_\alpha) + d(\boldsymbol{y}, \boldsymbol{x}_\beta)\}$
20:    **if** $\boldsymbol{N}_1 \cap \boldsymbol{N}_2 = \emptyset$ **then**
21:        Generate with priority in the order $\boldsymbol{N}_1, \boldsymbol{N}_2$
22:    **else if** $\boldsymbol{N}_1 = \emptyset$ **then**
23:        Return to **Step 2**
24:    $\boldsymbol{x}^{(t,k+1)} = \arg\min\{f(\boldsymbol{y}) \mid \boldsymbol{y} \in \boldsymbol{N}_1 \cap \boldsymbol{N}_2\}, k := k + 1$
    Step 6: Domain Escape Judgment
25:    **if** $\{d(\boldsymbol{x}^{(t,k)}, \boldsymbol{x}^{(t,0)}) > w \cdot d_{\max}\}$ and $\{f(\boldsymbol{x}^{(t,k)}) < f(\boldsymbol{x}^{(t,k-1)})\}$ **then**
26:        Return to **Step 2**
27:    **else**
28:        Return to **Step 5**
29: **end procedure**

Algorithm 1 shows Proposed Method (PM). Here, $\boldsymbol{H}$ denotes the set of searched local optimal solutions, s denotes the size of $\boldsymbol{H}$, $T_{\max}$ denotes the maximum iteration, t denotes the iteration counter, k denotes the number of movements in each iteration, and $\boldsymbol{x}_\alpha, \boldsymbol{x}_\beta$ denotes the diameter-pair set. The method is called PM-FD when the method is constructed with Footrule distance, and the method is called PM-UD when it is constructed with Ulam's distance.

Table 3. Experimental results of proposed method

(a) Results of the method constructing Footrule distance

Problem	Indexes		HM-FD	PM-FD (w is the following fixed value)					
				0.1	0.2	0.3	0.5	0.7	0.9
ta031	$f(\boldsymbol{x})$	mean	1.43%	1.10%*	1.17%	1.47%	1.97%	2.17%	2.20%
		best	0.18%	0.22%	0%*	0%*	0.55%	0.59%	0.59%
		worst	4.22 %	3.23%*	3.41%	3.56%	4.99%	5.32%	5.32%
		S.D.	0.87%	0.57%	0.67%	0.67%	0.93%	0.92%	0.99%
	Sim	mean	85.4	79.1	74.7	67.1	50.4	47.8	47.7
		S.D.	12.3	8.86	5.80	3.64	3.52	7.26	7.34
ta041	$f(\boldsymbol{x})$	mean	8.38%	7.45%*	8.78%	9.73%	10.65%	11.34%	11.33%
		best	5.65%	5.05%	4.51%*	6.95%	8.43%	8.73%	8.49%
		worst	13.20%	10.20%*	11.17%	12.04 %	13.44%	13.84%	13.81%
		S.D.	1.62%	1.14%	1.22%	1.20%	1.17%	1.10%	1.07%
	Sim	mean	96.9	88.8	79.0	69.3	50.1	39.2	38.4
		S.D.	2.37	1.56	1.19	1.33	2.17	7.44	8.82

(b) Results of the method constructing Ulam's distance

Problem	Indexes		HM-UD	PM-UD (w is the following fixed value)					
				0.1	0.2	0.3	0.5	0.7	0.9
ta031	$f(\boldsymbol{x})$	mean	0.007%	0.022%	0.004%*	0.011%	0.007%	0.015%	0.015%
		best	0%*	0%*	0%*	0%*	0%*	0%*	0%*
		worst	0.18%*	0.18%*	0.18%*	0.18%*	0.18%*	0.18%*	0.18%*
		S.D.	0.036%	0.060%	0.026%	0.044%	0.036%	0.050%	0.050%
	Sim	mean	62.5	58.8	50.6	42.1	37.8	36.7	36.7
		S.D.	20.4	19.3	15.6	10.4	5.09	4.50	4.50
ta041	$f(\boldsymbol{x})$	mean	2.72%*	2.80%	3.01%	3.01%	3.12%	3.36%	3.37%
		best	1.54%*	1.54%*	1.81%	1.84%	1.64%	2.24%	2.24%
		worst	4.08%	3.91%*	4.21%	4.01%	4.31%	4.35%	4.55%
		S.D.	0.58%	0.49%	0.50%	0.48%	0.54%	0.50%	0.54%
	Sim	mean	73.9	71.5	66.5	60.4	42.9	35.3	35.4
		S.D.	13.3	11.4	8.84	7.59	6.38	4.81	5.11

6 Numerical Experiment

We performed numerical experiments to evaluate the performance of PM and we compared with HM. We conducted an experiment with varying w under the conditions in Table 1. For each case, the mean, best value, worst value, and standard deviation (S.D.) of the results about the objective function value $f(x)$ of 50 trials are shown in Table 3. Here, "*" indicates the best result. This table also shows the mean and standard deviation (S.D.) about sim.

From the results of Table 3, as w increases Sim is small, so we could change the behavior of the search point. In addition, the result of PM exceeds HM at a specific w value, PM has usefulness at both Footrule distance and Ulam's distance. The result tends to be excellent when w is relatively small. This is considered to be caused by lowering the ability of intensification when w is large. Therefor, for example, there is a possibility that it can be further improved by adding w adaptive adjustment function.

7 Conclusion

In this paper, we showed the importance idea of integral designing of problems/ methods/distances in metaheuristics for combinatorial optimization. The above idea is important in combinatorial optimization. In addition, it is particularly important for methods using distance for the movement strategy, which was proposed the authors. As a practical example of the above idea, we proposed a method which is introduced a search strategy with more consideration of distance. Numerical experiments showed the proposed method has a better search performance than the previous method involving flow shop scheduling problem.

In future, it is desirable to develop adaptive adjustment function in proposed method, or it is desirable to develop of this idea to a multi-point search method.

References

1. Kennedy, J., Eberhart, R.C.: Swarm Intelligence. Morgan Kaufmann, San Francisco (2001)
2. Yang, X.-S.: Firefly algorithms for multimodal optimization. In: Watanabe, O., Zeugmann, T. (eds.) Stochastic Algorithms: Foundations and Applications. SAGA 2009. Lecture Notes in Computer Science, vol. 5792, pp. 169–178. Springer, Heidelberg (2009)
3. Ochiai, H., Tamura, K., Yasuda, K.: Combinatorial optimization method based on hierarchical structure in solution space. In: 2013 IEEE International Conference on Systems. Man, and Cybernetics, pp. 3543–3548. IEEE Press, New York (2013)
4. Yaguchi, K., Tamura, K., Yasuda, K., Ishigame, A.: Basic study of proximate optimality principle based combinatorial optimization method. In: 2011 IEEE International Conference on Systems. Man, and Cybernetics, pp. 1753–1758. IEEE Press, New York (2011)
5. Glover, F., Laguna, M.: Tabu Search. Kluwer Academic, Massachusetts (2002)

6. Reeves, C.R.: Fitness landscapes. In: Burke, E.K., Kendall, G. (eds.) Search Methodologies. Springer, Massachusetts (2005)
7. Bożejko, W., Smutnicki, C., Uchronski, M., Wodecki, M.: Big valley in scheduling problems landscape –Metaheuristics with reduced searching area. In: 2017 22nd International Conference on Methods and Models in Automation and Robotics (MMAR), pp. 458–462. IEEE Press, New York (2017)
8. Yasuda, K., Jinnai, H., Ishigame, A.: Multi-point combinatorial optimization method with distance based interaction. IEEJ Trans. EIS **130**, 6–13 (2010). (in Japanese)
9. Morita, M., Tamura, K., Yasuda, K.: Multi-point search combinatorial optimization method based on neighborhood search using evaluation of big valley structure. In: 2015 IEEE International Conference on Systems, Man, and Cybernetics, pp. 2835–2840. IEEE Press, New York (2015)
10. Diaconis, P.: Group Representations in Probability and Statistics. Lecture Notes-Monograph Series, vol. 11, pp. I–192 (1988)
11. Scheduling instances. http://mistic.heig-vd.ch/taillard/problemes.dir/ordonnance ment.dir/ordonnancement.html. Accessed 13 Mar 2018

An Improved Gas Classification Technique Using New Features and Support Vector Machines

Se-Jong Kang[1], Jae-Young Kim[1], In-Kyu Jeong[1],
M. M. Manjurul Islam[1], Kichang Im[2], and Jong-Myon Kim[1(✉)]

[1] Department of Electrical and Computer Engineering,
University of Ulsan, Ulsan, South Korea
sejong055@gmail.com, kjy7097@gmail.com,
jeonginkeyu@gmail.com, m.m.manjurul@gmail.com,
jmkim07@ulsan.ac.kr
[2] ICT Safety Convergence Center, University of Ulsan, Ulsan, South Korea
kichang@ulsan.ac.kr

Abstract. In this paper, we propose a gas classification technique based on extracting new features and support vector machines (SVM) in a chemical plant. First, various gases are collected using semiconductor gas seniors, and then we calculate the composition ratio of these gasses, which are defined as features. These extracted features are highly discriminative and quantify the presence of gas. Moreover, these features are used as the SVM input for classifying gas types. In addition, we apply a grid search technique in SVM for tuning hyperparameters such as misclassification rate, C, and kernel bandwidth, σ, to improve the classification performance. To verify the proposed technique, we collect various gases composition using a cost-effective self-designed test rig. The experimental results indicate that the proposed method is highly capable of classifying various hazardous gases with good accuracy.

Keywords: Feature extraction · Gas sensor array · Gas classification ·
Support vector machine · Chemical plants

1 Introduction

Chemical plants produce a variety of harmful gases as byproducts in the manufacturing process. According to the Fire Fighting Headquarters (UFFH) in South Korea, there were 79 hazardous material accidents last year, 46 of which (58%) occurred in the two mega-cities, namely Ulsan Mipo National Industrial Complex and Onsan National Industrial Complex, that contain many petrochemical companies [1]. The most common type of accident was gas leakage (26 cases, 33%). Therefore, the use of gas detectors is essential to prevent accidents and human casualties caused by chemical leaks in these industrial fields.

In previous gas classification studies, one of the critical tasks required is to collect hazardous gas samples from the chemical plants. However, a system using an electrochemical gas sensor has the problem of a high construction cost, therefore studies on

© Springer Nature Switzerland AG 2020
A. M. Madureira et al. (Eds.): SoCPaR 2018, AISC 942, pp. 158–166, 2020.
https://doi.org/10.1007/978-3-030-17065-3_16

gas sensors utilizing semiconductor sensors of low gas selectivity and low cost have been actively conducted [1–4]. The general approach of the gas classification algorithm is comprised of data acquisition, feature extraction, and gas classification stages [10]. However, there is a possibility that classification performance can be degraded when the acquired original data is used as an input to the classifier; this is because the acquisition results of the semiconductor sensor tend to be different depending on the surrounding environment. Therefore, we apply a preprocessing step to extract informative features about gases. We extract features by evaluating the composition ratio among collected gases samples, which is a highly effective method to quantify the presence of gases. Once we have a robust features vector, it is applied using a support vector machine classifier (SVM) [11, 12] to classify gas types. SVM is a state-of-the-art classifier that outperforms contemporary classifiers such as k-neatest neighborhood (k-NN), artificial neural network (ANN), and naïve Bayesian [11]. To verify our proposed algorithm, we test on datasets that are recorded using a semiconductor gas sensor from the self-designed gas plant.

The rest of this paper is organized as follows. Section 2 provides an overview of the experiment setup and gas sample collection method. Section 3 provides a detailed methodology including data preprocessing, feature extraction models, and fault classification. Section 4 summarizes results, and finally, a conclusion is given in Sect. 5.

2 Experiment Setup and Gas Samples Collection

As explained in Sect. 1, existing gas sample collection techniques are highly expansive, so we develop a cost-effective system based on an array of semiconductor gas sensors for data collection.

2.1 Experiment Environment

The gas sample collection system utilized for this experiment is shown in Fig. 1. According to Fig. 1, the gas sensors are placed in an array, which contains a combination of six different semiconductor gas sensors. The air is pulled through the developed gas collecting device, and the sensed data from each sensor is transmitted to the server. In the server, the collected data are converted into composition ratios.

Semiconductor Gas Sensor. In this paper, we try to classify various harmful gases by utilizing low-cost sensors in an array format. In general, gas sensors used for gas detection include electrochemical, semiconductor, contact-type, and optical sensors. However, semiconductor-type gas sensors were utilized in this study, because electric-chemical sensors are expensive, contact-fired sensors offer limited response to flammable gases, and optical sensors are too large [2, 3]. Six different semiconductor gas sensors were used in this paper and were purchased from the Winsen Company, as shown in Table 1. The application of each sensor is to measure the concentration of a primary detection gas, but the disadvantage reacting with alternate gases must be considered. In this paper, we attempt to classify six or more gases using six gas sensors by managing these drawbacks.

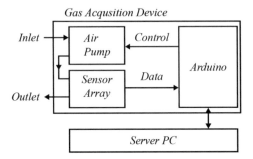

Fig. 1. Gas sample collection system based on an array of sensors

Table 1. Cross-sensitivity of MQ sensor

Types	Main detection gas	Secondary detection
MQ-4	CH_4	C_3H_8, C_2H_6S
MQ-5	C_3H_8	CH_4, C_2H_5OH
MQ-7	CO	CH_4
MQ-135	H_2	NH_3, C_7H_8
MQ-136	H_2S	CO
MQ-137	NH_3	

Sensors Array. Sensors were arrayed as shown in Fig. 2. When sensors are placed in a straight line, the gas sensor cannot operate correctly, because outside air cannot spread effectively across each array [2]. Therefore, a circulation path was made, so that the air pulled from the outside can be appropriately distributed throughout the sensor array.

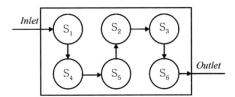

Fig. 2. Sensors array structure

Types of Hazardous Gas used in Our Experiment. Table 2 shows the five standard gases (NH_3, CO, CH_4, C_2H_6S, and C_3H_8) in a single concentration, which are the target gasses used in this paper. In particular, C_2H_6S does not have specific detection sensors, so we want to classify its presence by relying only on data patterns that are output by the six sensors. The collection environment for the experiment was performed by installing collection devices in the experimental chamber of a closed space, as shown in Fig. 3.

Table 2. Five hazardous gases

Types of gases	Concentration	Standard
NH_3	25 ppm	8AL (58L)
CO	50 ppm	8AL (58L)
CH_4	10%LEL	6D (105L)
C_2H_6S	5 ppm	2AL (58L)
C_3H_8	25%LEL	8AL (58L)

Fig. 3. Experimental setup and hazardous gas chamber

3 The Proposed Methodology for Gas Classification

Figure 4 presents the proposed methodology of dangerous gas classification, including data collection based on preprocessing gas samples, feature extraction, and harmful gas classification using the SVM classifier.

3.1 Data Collection

For data collection, a total of six gas datasets are used for experiments, including five hazardous gases and normal atmospheric conditions. Data collected from the gas chamber was sampled at intervals of 1 Hz, with one dataset containing six sensing values. Each gas data was obtained with 50 units per day for two days, and Table 3 shows the composition of all samples. Twenty-five randomly extracted datasets from the acquired data on different days were used for testing using SVM classifier.

Table 3. Number of data collected for each gas sample

Types of gases	Number of samples	Types of gases	Number of samples
Normal	100	CH_4	100
NH_3	100	C_2H_6S	100
CO	100	C_3H_8	100

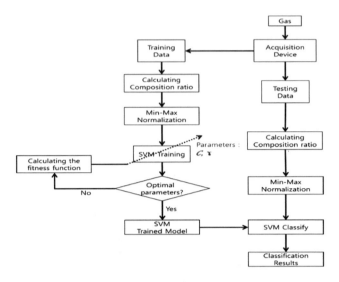

Fig. 4. The proposed gas classification algorithm

3.2   Feature Extraction

The composition rate data was generated with six different sensors that make up one complete dataset, and maximum-minimum normalization was applied for use as an input to the SVM. The composition ratio formula proposed in this paper is as follows:

$$r_{i,j} = \frac{S_j}{S_i}, (i = 0,\ 1,\ \ldots, 6, j = 0,\ 1,\ \ldots,\ i \neq j), \tag{1}$$

where s_j is the sensing value from the sensor and $r_{i,j}$ represents the rate of formation from the i sensor and j sensor. In this paper, the number of composition rates that can be obtained using six sensor data is 30.

3.3   Discriminate Feature Selection Using a Genetic Algorithm (GA)

As our original features vector is high-dimensional, we therefore, apply GA to select the most discriminate feature subset, which is mainly low-dimensional. GA consists of four stages: coding, parent selection, cross section, and displacement. In the genetic algorithm, the fitness function is an essential factor for assessing the quality of each chromosome and is an indicator of the variability between different classes, as well as the density within the class, which can be defined as follows:

$$C = \frac{1}{N_c \times N_s^2} \sum_{i=1}^{N_c} \sum_{j=1}^{N_s} \sum_{k=1}^{N_s} \|S_i(j) - S_i(k)\| \tag{2}$$

$$S = \frac{1}{N_c \times (N_c - 1) \times N_s^2} \sum_{\substack{i=1 \\ }}^{N_c} \sum_{\substack{j=1 \\ j \neq i}} \sum_{k=1}^{N_s} \sum_{l=1}^{N_s} \left\| S_i(k) - S_j(l) \right\| \qquad (3)$$

$$F = \frac{C}{S}. \qquad (4)$$

Equation (2) represents the density within the class, and Eq. (3) represents the distinction between classes. Since the density within the class must be high and the distinction between classes must be low, the objective function consists of Eq. (4) [9].

3.4   Gas Classification Using SVM

Once we obtained a robust features vector for each gas, then further verified the feature vectors using the SVM classifier. To improve classifier performance, SVMs tend to ignore small-sized features when the size of input features is not normalized; a normalization process that unifies the minimum and maximum sizes of features is required. The maximum-minimum normalization of the input characteristics used is shown in Eq. (5) [5, 6].

$$X_{new} = \frac{X - \min(X)}{\max(X) - \min(X)} \qquad (5)$$

Here, X denotes the features vector. To classify nonlinear feature vectors using SVM, we used the radial basis function (RBF) kernel function during kernel ticks [11]. The classification performance of an SVM using an RBF kernel is affected by the argument C, which regulates the penalty for the error, and the argument σ, which regulates the nonlinearity of the kernel function [7]. A grid search method was used to find the optimal combination of C and γ. Multiple combinations of C and σ were created, and a cross-random analysis was performed to select the values of C and σ with the highest accuracy [8]. C is the range from 2^{-5} to 2^{16} at intervals of 2^2 and σ is found as the best value in the range from 2^{-2} to 2^8 at intervals of 2^1.

4   Experiments and Results

To compare the performance of the proposed algorithm, the following four datasets were made as inputs to the SVM, and their accuracies were calculated. Table 4 provides the four dataset types that are used for performance measurement.

Table 5 shows the classification accuracy of each algorithm. The proposed algorithm A.2 with GA-based feature selection showed the highest classification accuracy with several 100% values for the composite rate. On the other hand, original datasets such as A.1 and A.3 for GA and non-GA, respectively, showed lower detection accuracies. According to the results in Table 5, it is evident that the composite ratio-based feature extraction process is highly effective as an improved gases classification technique.

Table 4. Number of datasets used in this experiment

Dataset	Definition
A.1	Original data acquired from each sensor with GA
A.2	Characteristic data obtained by applying the composition ratio with GA
A.3	Original data acquired from each sensor
A.4	Characteristic data obtained by applying the composition ratio

Table 5. Classification accuracy of the four algorithms

	GA		Non-GA	
	Original	Composition rate	Original	Composition rate
Algorithm	**A.1**	**A.2**	**A.3**	**A.4**
Number of features	10	3	30	15
Accuracy [%]	91.7	100.0	83.4	99.7

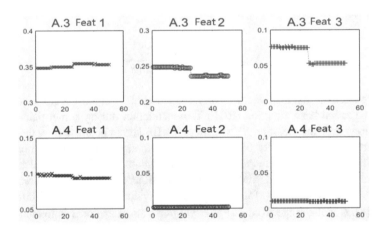

Fig. 5. NH_3 Data from A.3 and A.4

Figure 5 shows the data in A.3 and A.4. According to the results shown in Figure A.3, the data is inconsistent depending on the acquisition environment due to the low precision of the semiconductor sensor. The effects of the unstable data are indicated by the red NH_3 feature vectors and green C_2H_6S feature vectors.

In addition, GA facilities the most discriminate feature subset and to visualize their distribution in a lower dimension. Figure 6 displays a 3-D visualization of the GA-based selected features for A.1 in which gas classes are overlapped. Eventually, the two classes of feature vectors overlapped at the feature space level, so we identified that the classification accuracies for these classes are degraded. For the proposed algorithm, it can be seen in Fig. 7 that the feature vectors in A.2 were highly distributed by class and did not overlap in the feature space, thereby improving the SVM classification performance of the algorithm.

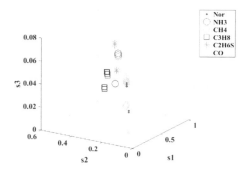

Fig. 6. 3D visualization of A.1 in a feature space

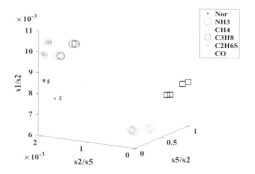

Fig. 7. 3D visualization of A.2 in a feature space

5 Conclusion

This paper proposed a gas classification algorithm using the new feature extraction process and SVM. In addition, we developed a low-cost semiconductor sensor system for gas sample collection. Semiconductor-type sensors are able to create patterns for multiple gases through a sensor array by reversely exploiting disadvantages of responding to multiple gases. By classifying gas through SVM in the generated pattern, the main detection gas of the sensors used in this paper, as well as the secondary detection gas, C_2H_6S, was identified as possible. To improve the performance of the SVM classification, we applied data composition rates and four experiments demonstrated that the apparent performances are improved for experiments using composition rates.

Acknowledgment. This research was supported by the Ministry of Science and Technology, Ministry of Information and Communication, and the Korea Information and Telecommunication Industry Promotion Agency (No. S0702-18-1045).

References

1. Lee, I.S., Shim, C.H.: Gas classification and fault diagnosis of the semiconductor type gas sensor system. J. Korea Inf. Sci. Soc. **7**, 48–57 (2009)
2. Lee, J.H., Cho, J.H., Jeon, G.J.: Concentration estimation of gas mixtures using a tin oxide gas sensor and fuzzy ART. J. Korea Electron. Eng. **43**, 21–29 (2006)
3. Lee, D.S., Jung, H.Y., Ban, S.W., Lee, M.H., Huh, J.S., Lee, D.D.: Fabrication of semiconductor gas sensor array and explosive gas sensing characteristics. J. Korea Electron. Eng. **37**, 9–17 (2000)
4. Lee, K.C., Rye, K.R., Hur, C.W.: Fabrication and yield improvement of oxide semiconductor thin film gas sensor array. J. Korea Inf. Commun. Soc. **6**, 315–322 (2002)
5. Heo, J.Y., Yang, J.Y.: SVM based stock price forecasting using financial statements. J. Korea Inf. Sci. Soc. **21**, 167–172 (2015)
6. Park, J.H., Hwang, C.S., Bae, K.B.: Analysis of target classification performances of active sonar returns depending on parameter values of SVM kernel functions. J. Korea Inf. Commun. Soc. **17**, 1083–1088 (2013)
7. Min, J.H., Lee, Y.C.: Support vector bankruptcy prediction model with optimal choice of RBF kernel parameter values sing grid search. J. Korea Manag. Sci. **30**, 55–74 (2005)
8. Kim, Y.H., Kim, J.Y., Jeong, I.K., Kim, Y.H., Kim, J.-M.: A method of detecting boiler tube leakage using a genetic algorithm and support vector machines. Korea Comput. Inf. Soc. **26**, 55–56 (2018)
9. Kim, J.Y., Kim, J.-M., Choi, B.K.: Bearing fault diagnosis using adaptive self-tuning support vector machine. Korea Comput. Inf. Soc. **24**, 19–20 (2016)
10. Zhai, X., Ali, A.A.S., Amira, A., Bensaali, F.: MLP neural network based gas classification system on Zynq SoC. IEEE Access **4**, 8138–8146 (2016)
11. Islam, M.M.M., Kim, J.-M.: Reliable multiple combined fault diagnosis of bearings using heterogeneous feature models and multiclass support vector Machines. Reliab. Eng. Syst. Saf. **184**, 55–66 (2018)
12. Islam, M.M.M., Islam, M.R., Kim, J.-M.: A hybrid feature selection scheme based on local compactness and global separability for improving roller bearing diagnostic performance. In: Artificial Life and Computational Intelligence, pp. 180–192. Springer (2017)

Superior Relation Based Firefly Algorithm in Superior Solution Set Search

Hongran Wang$^{(\boxtimes)}$, Kenichi Tamura, Junichi Tsuchiya, and Keiichiro Yasuda

Tokyo Metropolitan University, 1-1, Minami-Osawa, Hachioji-shi, Tokyo, Japan
wang-hongran@ed.tmu.ac.jp

Abstract. For many single objective optimization methods, they have only one global optimal solution or suboptimal solution. In this paper, we propose a superior solution set search problem as an optimization problem to simultaneously find multiple excellent solutions in multimodal functions. In addition, we analyzed the search characteristics of Firefly Algorithm (FA), which has a fundamental nature of a Superior Solution Set Search Problem, previously defined in our previous study for single-objective optimization problems. In this paper, we proposed a new FA method based on the former problem. This method, which employs cluster information by K-means clustering, is tested for performance by fundamental numerical experiments.

Keywords: Metaheuristics · Single-objective optimization · Superior Solution Set Search Problem · Cluster · K-means clustering · Firefly Algorithm

1 Introduction

Recently, owing to large-scale and complex actual systems, the demand for obtaining a solution with sufficient optimality for practical use has increased. Meanwhile, with the tremendous development in computer technology and improvement of optimization algorithms and modeling/simulation technologies, the demand for not only practical but also new optimization methods is increasing in the field of optimization. The simulation of optimization problem modeling in the real world has become more closely possible, from the development of computer computing capacity, the development of modeling technology and optimization algorithms related technology. Thus, many of the optimization problems in the real world are often expressed as multimodal problems with multiple local optimal solutions [1, 2].

For many single objective optimization methods, they have only one global optimal solution or suboptimal solution. To select a solution to be used in practice, the development of a method to simultaneously find multiple solutions in a multimodal function is an important research direction. As a result, we proposed a Superior Solution Set Search Problem that searches for various solution sets

© Springer Nature Switzerland AG 2020
A. M. Madureira et al. (Eds.): SoCPaR 2018, AISC 942, pp. 167–176, 2020.
https://doi.org/10.1007/978-3-030-17065-3_17

with similar valuation values and solutions with distance apart. It is expected to present alternative proposals to solve the problems when accidents or technical problems occur.

In addition, the Firefly Algorithm (FA) [3,4] has the property to be divided into multiple groups in the search process of the superior solution set, and multiple promising areas can be searched in parallel. Therefore, unlike many single-objective optimization methods, FA is expected to be able to search efficiently for Superior Solution Set by exploiting the properties it possessed. However, the application of FA for Superior Solution Set Search is improvable. For instance, further improvement of performance can be achieved by adding a mechanism for dividing into more distinct groups.

This paper is organized as follows. Section 2 describes the Superior Solution Set Search Problem. Section 3 outlines the FA and analyzes its search characteristics. Section 4 proposes an adjustment rule of parameter using cluster information. In Sect. 5, the performance of the numerical experiments is carried out. Finally, the conclusion is presented in Sect. 6.

2   Superior Solution Set Search Problem

We have explained a Superior Solution Set Search Problem proposed in our previous study, which searches for various solution sets whose evaluation values are about the same, and the solution distance is far away [5]. Obtaining Superior Solution Set makes it possible to present alternatives for accidents and technical problems, which is considered to be of high engineering value.

In this paper, dealing with the minimization problem of the objective function $f(\boldsymbol{x})$, where decision variables are decided by $\boldsymbol{x} \in \boldsymbol{X}$, where the feasible area is $\boldsymbol{X}$ ($\boldsymbol{X} \subseteq \mathbb{R}^n$). First, we formulate the sublevel set $\mathcal{L}(\delta) \subset \boldsymbol{X}$ of solutions that satisfies the constraints of the evaluation value $\delta \geq 0$ based on the evaluation value of the global optima $f(\boldsymbol{x}^{\star})$.

$$\mathcal{L}(\delta) = \{\boldsymbol{x} \in \boldsymbol{X} \mid f(\boldsymbol{x}) \leq f(\boldsymbol{x}^{\star}) + \delta\} \tag{1}$$

Furthermore, we formulate the ε-neighborhood $\mathcal{N}(\hat{\boldsymbol{x}}; \varepsilon)$, which is an open sphere with a radius of $\varepsilon > 0$ based on $\hat{\boldsymbol{x}} \in \mathbb{R}^n$.

$$\mathcal{N}(\hat{\boldsymbol{x}}; \varepsilon) = \{\boldsymbol{x} \in \mathbb{R}^n \mid \|\boldsymbol{x} - \hat{\boldsymbol{x}}\| < \varepsilon\} \tag{2}$$

Finally, we formulate the Superior Solution Set $\mathcal{Q}(\delta, \varepsilon)$, where the Superior Solution $\hat{\boldsymbol{x}} \in \mathcal{L}(\delta)$ satisfies $f(\hat{\boldsymbol{x}}) \leq f(\boldsymbol{x})$ $(\forall \boldsymbol{x} \in \mathcal{L}(\delta) \cap \mathcal{N}(\hat{\boldsymbol{x}}; \varepsilon))$.

$$\mathcal{Q}(\delta, \varepsilon) = \{\boldsymbol{x}^* \in \mathcal{L}(\delta) | f(\boldsymbol{x}^*) \leq f(\boldsymbol{x})\ (\forall \boldsymbol{x} \in \mathcal{L}(\delta) \cap \mathcal{N}(\boldsymbol{x}^*; \varepsilon))\} \tag{3}$$

$\mathcal{Q}(\delta, \varepsilon)$ is a set of local optimal solution in which the evaluation values are as good as each other and the distance between solutions is far away. We define the problem to aim at finding Superior Solution Set $\mathcal{Q}(\delta, \varepsilon)$.

3   Firefly Algorithm

3.1   Firefly Algorithm

FA is an optimization method based on the analogy of firefly activity. In actual firefly activity, each individual communicates with other fireflies by emitting light, feeding, and applying courtship behavior. Generally, bright firefly is attractive and each firefly is attracted to fireflies that emit bright light around themselves. Xin-She Yang proposed FA in 2007, which is a multi-point search type metaheuristic, and is abstracted as above phenomenon. Yang idealized the courtship of fireflies in developing FA by the following three rules [6].

- All Fireflies are unisex so that any individual firefly will be attracted to all other fireflies.
- The attractiveness is proportional to their brightness, and for any two fireflies, the less bright one will be attracted to the brighter one. However, the intensity decreases as their mutual distance increases.
- If there are no fireflies brighter than a given firefly, it will move randomly.

Each operation of FA will be described as follows. To grasp the light intensity and attractiveness quantitatively, we define the light intensity I_i of FA and attractiveness $\beta_{ij}(d_{ij})$ by Eqs. (4), (5).

$$I_i = \left(\left|f^t_{\min} - f(\boldsymbol{x}^t_i)\right| + 1\right)^{-1} \tag{4}$$

$f^t_{\min}$ is the most excellent evaluation value among the evaluation values in the search point group.

$$\begin{cases} \beta_{ij}(d_{ij}) = \beta_0 e^{-\gamma d^2_{ij}} \\ d_{ij} = \left\|\boldsymbol{y}^t_j - \boldsymbol{x}^t_i\right\| = \sqrt{\sum_{l=1}^{n}(y^t_{j,l} - x^t_{i,l})^2} \end{cases} \tag{5}$$

where β_0 is the maximum value of attraction, and γ is a parameter related to attenuation of light. Each firefly is attracted by all fireflies within stronger light intensity to move. The movement equation is defined by Eq. (6), when firefly i is sucked into firefly j. d_{ij} is the Euclidean distance of the search point $\boldsymbol{x}_i$ with the reference point $\boldsymbol{y}_j$.

$$\boldsymbol{x}^t_i := \boldsymbol{x}^t_i + \beta_{ij}(d_{ij})(\boldsymbol{y}^t_j - \boldsymbol{x}^t_i) + \alpha\boldsymbol{R} \tag{6}$$

where α is a parameter corresponding to the scale of the random number term and $\boldsymbol{R}$ represents a uniform random number vector to be changed in the range of $[-0.5, 0.5]^n$. The brightest firefly $\boldsymbol{x}_{\min}$ moves randomly according to the following Eq. (7).

$$\boldsymbol{x}^t_{\min} := \boldsymbol{x}^t_{\min} + \alpha\boldsymbol{R} \tag{7}$$

FA for the minimization problem of the objective function $f(\boldsymbol{x})$ $(\boldsymbol{x} \in \mathbb{R}^n)$ is shown below.

[Firefly Algorithm]

Step 1 : [Preparation] Set the maximum number of iterations $T_{\max}$, the number of search points m, and each parameter α, β_0, γ. Set the number of iterations $t = 1$.

Step 2 : [Initialization] In the executable area $\boldsymbol{X}$ ($\boldsymbol{X} \subseteq \mathbb{R}^n$), set to generate search point $\boldsymbol{x}_i^1$ ($i = 1, 2, \cdots, m$) randomly. Save the solution as $\boldsymbol{y}_i^1 = \boldsymbol{x}_i^1$ ($i = 1, 2, \cdots, m$) and set $i = 1$.

Step 3 : [Calculation of light intensity] Calculate the light intensity I_i of each search point according to the following formula.

$$f_{\min}^t = \min\{f(\boldsymbol{x}_i^t)|i = 1, 2, \cdots, m\}$$

$$I_i = \left(\left|f_{\min}^t - f(\boldsymbol{x}_i^t)\right| + 1\right)^{-1}$$

Step 4 : [Movement of search points] If $I_i < I_j$, move the search point $\boldsymbol{x}_i^t$ according to the following formula.

$$\boldsymbol{x}_i^t := \boldsymbol{x}_i^t + \beta_0 e^{-\gamma\|\boldsymbol{y}_j^t - \boldsymbol{x}_i^t\|^2}(\boldsymbol{y}_j^t - \boldsymbol{x}_i^t) + \alpha\boldsymbol{R}$$

$\boldsymbol{R} \in [-0.5, 0.5]^n$ represents a uniform random vector. Repeat the above operations and $j := j + 1$, until $j = m$ and move the search point $\boldsymbol{x}_i^t$ according to the following formula.

$$\boldsymbol{x}_i^t := \boldsymbol{x}_i^t + \alpha\boldsymbol{R}$$

Step 5 : [Update search point] If $i = m$, update the search point $\boldsymbol{x}_i^t$ and save the solution $\boldsymbol{y}_i^{t+1}$; otherwise, set $i := i + 1$ and return to Step 3.

$$\boldsymbol{x}_i^{t+1} = \boldsymbol{x}_i^t \ (i = 1, 2, \cdots, m)$$

$$\boldsymbol{y}_i^{t+1} = \boldsymbol{y}_i^t \ (i = 1, 2, \cdots, m)$$

Step 6 : [Termination]   If $t = T_{\max}$, the algorithm is finished; otherwise, return to Step 3, set $t := t + 1$ and $i = 1$.

3.2   Qualitative Analysis of Features of Firefly Algorithm

Metaheuristics have a general search structure that performs a search of "Neighborhood Generation" and "Search Point Update" by repeating [2]. Based on the general structure of metaheuristic, we analyze the search characteristic of FA. FA of "Neighborhood Generation", which takes advantage of the search point $\boldsymbol{x}$ and the reference search point $\boldsymbol{y}$ to generate neighborhood by the Euclidean distance of the search point $\boldsymbol{x}$ and the reference point $\boldsymbol{y}$. FA of "Search Point Update", which is always updated by the generated neighborhood. "Search Point Update" of FA permits movement to become bad, where it is difficult to improve the search point only by random number term. The movement that the update point $\boldsymbol{x}$ is close to the reference point $\boldsymbol{y}$, which greatly contributes to the improvement of the search point of FA.

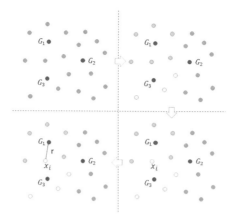

Fig. 1. Image of distance between moving point and cluster center of gravity

However, the parameter α is the step size quantization factor for the random move term in FA, which is to control the range of the exploration areas in FA. Furthermore, the movement amount is adjusted by the parameter β. To solve the search problem efficiently, it is necessary to adjust appropriately the movement amount when approaching an excellent point. However, if the movement amount is relatively small, it is not possible to improve efficiently. On the other hand, if the movement approaching an excellent point is relatively large, all search points converge around the superior solution rather than splitting into multiple groups [7]. For this reason, it is desirable that β is adjusted to an appropriate value according to the distance to the searched reference point. The distribution of β is varied greatly depending on parameter γ, which is a very important parameter for Superior Solution Set Search Problem.

4   Parameter Adjustment Rule Using Cluster Information

Our analysis, shows that the parameter γ has the ability to improve, and the parameter α has the ability to search for exploration areas. To be more scientific and more effective in adjusting the parameters for Superior Solution Set, in this paper, we use the K-means clustering to divide the exploration points into multiple clusters, and use the information of each cluster to adjust parameters γ and α. Simultaneously, we defined diversification and intensifcation of clusters [8], as (a) intensifcation of the search points in the cluster, and (b) diversification of the search points between the clusters. Based on the above definitions, we consider the adjustment of the search capability showing that the ability to improve is stronger than search for exploration areas in the cluster and the ability relationship is inversed between the clusters.

The parameter γ, which is the foundation of FA, has a great influence on the performance of splitting into a plurality of clusters. Specifically, parameter γ of FA has a great influence on the performance of searching for Superior

Solution Set. Furthermore, FA is automatically divided into multiple clusters in the process of searching for Superior Solution Set. Therefore, we propose an adjustment rule of parameter γ considering the property that FA splits into multiple clusters in searching for Superior Solution Set. Various methods can be used to calculate cluster center of gravity, but this time we use K-means clustering which is easy to implement for basic tests. We show the proposed γ adjustment rule algorithm below.

[γ Adjustment Rule Algorithm]

Step 1 : [**Preparation**] Set cluster number K and parameter $\beta_{\min}$.

Step 2 : [**Cluster allocation**] Assign each search point $\boldsymbol{x}_i^t$ ($i = 1, 2, \cdots, m$) to K clusters. Furthermore calculate the centroid $\boldsymbol{G}_k^t$ ($k = 1 \cdots, K$) of each cluster. (We use the K-means clustering this time. Please see Appendix for K-means clustering algorithm.)

Step 3 : [**Determination of** γ] Define γ according to the following equation.

$$\gamma_i = -ln(\beta_{\min}/\beta_0)/||\boldsymbol{x}_i^t - \boldsymbol{G}_k^t||^2$$

Where k is the number of the cluster where the search point $\boldsymbol{x}_i^t$ is assigned.

We use the distance r, as shown in Fig. 1, between the moving point and the cluster center of gravity, and the minimum travel distance $\beta_{\min}$ to solve the corresponding the parameter γ according to Eq. (8). Where $\beta_{\min} = 10^{-6}$ in the cluster and $\beta_{\min} = 10^{-8}$ between the clusters. Furthermore, Fig. 2 shows the image distribution of β and γ adjustment rule.

$$\gamma = -ln(\beta_{\min}/\beta_0)/r^2 \tag{8}$$

The parameter α is the step size quantization factor for the random move term in FA, which is to control the range of the exploration areas in FA. And the parameter α with a method of linearly decreasing. We set the parameter α

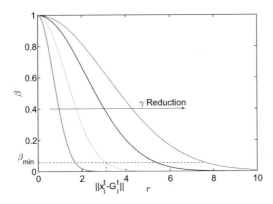

Fig. 2. Image distribution of β and γ adjustment rule

according to Eq. (9), where α is stronger between the clusters and α is weaker in the cluster.

$$\alpha = \alpha_{strat} - \frac{t}{T_{\max}}(\alpha_{strat} - \alpha_{end}) \tag{9}$$

where $\alpha_{strat} = 0.1$, $\alpha_{end} = 0.01$ between the clusters, $\alpha_{strat} = 0.05$, $\alpha_{end} = 0.005$ in the cluster and $T_{\max}$ is the maximum number of iterations.

5   Numerical Experiment

5.1   Condition of Numerical Experiment

In numerical experiments, we compared the fixed parameter γ of FA and the proposed γ adjustment rule of FA which obtains more superior solutions $\boldsymbol{x}^* \in \mathcal{Q}(\delta, \varepsilon)$ to evaluate the performance. However, it is impossible to obtain a strict superior solutions $\boldsymbol{x}^*$, which $\boldsymbol{x}_i^{T_{\max}}$ satisfies $\|\boldsymbol{x}_i^{T_{\max}} - \boldsymbol{x}^*\| \leq 0.5$, and we determine that the $\boldsymbol{x}^*$ has been found.

In this paper, we set up the problem so that the superior solution set $\mathcal{Q}(\delta, \varepsilon)$ is determined regardless of δ and ε for basic study. The benchmark functions have multiple global optimal solutions $\boldsymbol{x}^\star$ while distancing from each other. In this case, we have $\boldsymbol{x}^* = \boldsymbol{x}^\star$ ($\forall \delta \geq 0 \wedge \forall \varepsilon > 0$). From the above, the benchmark functions are defined as follows Table 1.

As a share experiment condition, the number of trials was set at 50. The maximum number of iterations was 1000. In addition, the recommended values setting for other parameters were $\beta_0 = 1.0$ and α is set as Eq. (9). For FA of the parameter γ fixed, we changed γ in 20 types as shown in Table 2 as 0.05, 0.1, $\cdots$, 0.95, 1. The parameters of the proposed method are set as $\beta_{\min}$ referenced by Sect. 4 and cluster number $K = 4$. We determine the evaluation index by

Table 1. Benchmark functions

Functions	Definitions
Function 1	$\min\left(f(\boldsymbol{x} + \boldsymbol{Y}),\ f(\boldsymbol{x} - \boldsymbol{Y}),\ f(\boldsymbol{x} + \boldsymbol{Z}),\ f(\boldsymbol{x} - \boldsymbol{Z})\right)$
Function 2	$\min\left(f(\boldsymbol{x} + \boldsymbol{Y}),\ g(\boldsymbol{x} + \boldsymbol{Z}),\ f(\boldsymbol{x} - \boldsymbol{Y}),\ g(\boldsymbol{x} - \boldsymbol{Z})\right)$
Function 3	$\min\left(f(\boldsymbol{x} + \boldsymbol{Y}) + \mathrm{E},\ g(\boldsymbol{x} + \boldsymbol{Z}) - \mathrm{E},\ f(\boldsymbol{x} - \boldsymbol{Y}) + \mathrm{F},\ g(\boldsymbol{x} - \boldsymbol{Z}) - \mathrm{F}\right)$
$f(\boldsymbol{x})$	Sphere Function : $\sum_{i=1}^n (x_i)^2$
$g(\boldsymbol{x})$	Schwefel's Function : $\sum_{i=1}^n \left(\sum_{j=1}^i x_j\right)^2$
Vectors	Definitions
$\boldsymbol{Y}$	$[2.5, 2.5, \cdots, 2.5, 2.5]$
$\boldsymbol{Z}$	$[2.5, -2.5, \cdots, 2.5, -2.5]$
Constants	Definitions
E	2.5
F	5

Table 2. The results of numerical experiment

Original Firefly Algorithm (γ is fixed and varies between 0.05 and 1.0.)

Functions	Dim.		Proposal	0.05	0.1	0.15	0.2	0.25	0.3	0.35	0.4	0.45	0.5	0.55	0.6	0.65	0.7	0.75	0.8	0.85	0.9	0.95	1.0
Function 1	$n=5$	Mean	3.92 (16th)	1	2.02	4	4	4	4	4	4	4	4	4	4	4	4	4	4	4	3.98	3.96	3.90
		Best	4 (1st)	1	3	4	4	4	4	4	4	4	4	4	4	4	4	4	4	4	4	4	4
		Worst	3 (14th)	1	1	4	4	4	4	4	4	4	4	4	4	4	4	4	4	4	3	3	3
		S.D.	0.27	0	0.14	0	0	0	0	0	0	0	0	0	0	0	0	0	0	0	0.14	0.19	0.30
	$n=10$	Mean	3.56 (1st)	2.26	3.36	3.02	1.86	1.4	0.88	0.96	0.5	0.36	0.22	0.16	0.08	0.08	0.04	0.04	0.02	0.04	0.02	0	0
		Best	4 (1st)	4	4	4	4	3	3	2	2	1	1	1	1	1	1	1	1	1	1	0	0
		Worst	3 (1st)	1	1	1	1	0	0	0	0	0	0	0	0	0	0	0	0	0	0	0	0
		S.D.	0.54	0.69	0.60	0.60	0.85	0.76	0.88	0.70	0.74	0.60	0.42	0.27	0.27	0.19	0.19	0.20	0.14	0.20	0.14	0	0
	$n=20$	Mean	2.82 (1st)	1.82	0	0	0	0	0	0	0	0	0	0	0	0	0	0	0	0	0	0	0
		Best	4 (1st)	3	0	0	0	0	0	0	0	0	0	0	0	0	0	0	0	0	0	0	0
		Worst	2 (1st)	1	0	0	0	0	0	0	0	0	0	0	0	0	0	0	0	0	0	0	0
		S.D.	0.37	0.52	0	0	0	0	0	0	0	0	0	0	0	0	0	0	0	0	0	0	0
Function 2	$n=5$	Mean	3.84 (1st)	1.20	1.32	1.66	2.04	1.92	1.98	1.72	1.58	1.56	1.40	1.16	1.16	1.12	1.06	1	1	1	1	1	1.02
		Best	4 (1st)	1	2	3	4	3	3	3	2	2	2	2	2	2	1	1	1	1	1	1	2
		Worst	3 (1st)	1	1	1	1	1	1	1	1	1	1	1	1	1	1	1	1	1	1	1	1
		S.D.	0.37	0.40	0.47	0.59	0.53	0.53	0.53	0.43	0.50	0.50	0.50	0.49	0.49	0.37	0.33	0.24	0	0	0	0	0.14
	$n=10$	Mean	3.48 (1st)	0.26	0.76	0.7	0.34	0.28	0.12	0	0.02	0	0	0	0	0	0	0	0	0	0	0	0
		Best	4 (1st)	1	1	1	1	1	1	0	1	0	0	0	0	0	0	0	0	0	0	0	0
		Worst	3 (1st)	0	0	0	0	0	0	0	0	0	0	0	0	0	0	0	0	0	0	0	0
		S.D.	0.61	0.44	0.46	0.46	0.48	0.45	0.33	0	0.14	0	0	0	0	0	0	0	0	0	0	0	0
	$n=20$	Mean	1.32 (1st)	0	0	0	0	0	0	0	0	0	0	0	0	0	0	0	0	0	0	0	0
		Best	2 (1st)	0	0	0	0	0	0	0	0	0	0	0	0	0	0	0	0	0	0	0	0
		Worst	0 (1st)	0	0	0	0	0	0	0	0	0	0	0	0	0	0	0	0	0	0	0	0
		S.D.	0.84	0	0	0	0	0	0	0	0	0	0	0	0	0	0	0	0	0	0	0	0
Function 3	$n=5$	Mean	3.80 (1st)	1.00	1.28	1.86	1.96	1.94	1.92	1.92	1.84	1.74	1.56	1.40	1.42	1.38	1.24	1.18	1.14	1.14	1.10	1.16	1.10
		Best	4 (1st)	1	2	3	3	3	3	3	2	2	2	2	2	2	1	2	2	2	2	2	2
		Worst	2 (1st)	1	1	1	0	1	1	1	1	1	1	1	1	1	1	1	1	1	1	1	1
		S.D.	0.40	0.49	0.50	0.57	0.47	0.55	0.49	0.49	0.37	0.44	0.50	0.49	0.50	0.49	0.43	0.39	0.35	0.30	0.30	0.37	0.30
	$n=10$	Mean	3.44 (1st)	0.38	0.28	0.50	0.62	0.42	0.22	0.10	0.04	0.02	0	0	0	0	0	0	0	0	0	0	0
		Best	4 (1st)	1	1	1	1	1	1	1	1	1	0	0	0	0	0	0	0	0	0	0	0
		Worst	2 (1st)	0	0	0	0	0	0	0	0	0	0	0	0	0	0	0	0	0	0	0	0
		S.D.	0.54	0.49	0.45	0.51	0.49	0.50	0.42	0.30	0.20	0.14	0	0	0	0	0	0	0	0	0	0	0
	$n=20$	Mean	1.78 (1st)	0	0	0	0	0	0	0	0	0	0	0	0	0	0	0	0	0	0	0	0
		Best	2 (1st)	0	0	0	0	0	0	0	0	0	0	0	0	0	0	0	0	0	0	0	0
		Worst	0 (1st)	0	0	0	0	0	0	0	0	0	0	0	0	0	0	0	0	0	0	0	0
		S.D.	0.46	0	0	0	0	0	0	0	0	0	0	0	0	0	0	0	0	0	0	0	0

the number of Superior Solutions acquired in one trial. "Best" shows the best evaluation value of 50 runs, "Worst" is the worst value of 50 runs, "Mean" shows the average value of 50 runs and "S.D." shows the standard deviation of 50 runs.

5.2   Results of Numerical Experiment

Table 1 shows the results of the numerical experiment. In the case of 5 dimensions, the proposed method has almost the same performance as results ranked 8th out of 17 types when the parameter is fixed. In the case of 10 dimensions and 20 dimensions, the performance of the proposed method is superior to results of the 1st compared with 17 types in the case of parameter fixed. From the above, the proposed method can testify that it has excellent performance, at the same time, omitting the labor of parameter setting. In particular, it can be concluded that the performance of the proposed method is excellent as the number of dimensions increases. In the proposed method, the adjustment of the parameter γ using the information of the cluster makes it possible to concentrate gradually the search points compared with the case of fixed parameters. Therefore, the performance where the search point group is divided into multiple clusters, in which case, the result has improved.

6   Conclusion

In this paper, we propose an adjustment rule of parameters γ and α using cluster information for Superior Solution Set Search Problem. The parameter γ and the parameter α are important parameters in FA, and have great influence on searching for Superior Solution Set. Moreover, we confirmed the performance improvement of the Superior Solution Set Search by the proposed method in the numerical experiment using the QSphere function. In addition, we compare the case of fixed γ and the proposed method by numerical experiment results. In the proposed γ and α adjustment rule, the search point group is divided into a plurality of clusters relatively smoothly, so that the performance of the search for Superior Solution Set is improved.

References

1. Aiyoshi, E., Yasuda, K.: Metaheuristics and Their Applications. The Institute of Electrical Engineers of Japan, Ohmsha, Tokyo (2007)
2. Yasuda, K.: The present and the future of metaheuristics. SICE J. Control, Meas. Syst. Integr. **47**(6), 453–458 (2008)
3. Yang, X.S.: Firefly algorithms for multimodal optimization. In: Proceedings of the 5th Symposium on Stochastic Algorithms, Foundations and Applications. Lecture Notes in Computer Science, pp. 169–178 (2009)
4. Yang, X.S.: Nature-Inspired Metaheuristic Algorithms, 2nd edn. Luniver Press (2010)

5. Oosumi, R., Tamura, K., Yasuda, K.: Novel single-objective optimization problem and firefly algorithm-based optimization method. In: 2016 IEEE International Conference on Systems, Man, and Cybernetics, pp. 1011–1015 (2016)
6. Yang, X.S.: Engineering Optimization: An Introduction with Metaheuristic Applications. Wiley, Hoboken (2010)
7. Senthilnath, J., Omkar, S.N., Mani, V.: Clustering using firefly algorithm: performance study. Swarm Evol. Comput. **1**, 164–171 (2011)
8. Yasuda, K., et al.: Particle swarm optimization: a numerical stability analysis and parameter adjustment based on swarm activity. IEEJ Trans. Electr. Electron. Eng. **3**(3), 642–659 (2008)
9. Carroll, J.D., Chaturvedi, A.: K-midranges clustering. In: Advances in Data Science and Classification. Springer, Heidelberg (1998)
10. Kanungo, T., et al.: An efficient k-means clustering algorithm: analysis and implementation. IEEE Trans. Pattern Anal. Mach. Intell. 881–892 (2002)

Learning in Twitter Streams
with 280 Character Tweets

Joana Costa[1,2(✉)], Catarina Silva[1,2], and Bernardete Ribeiro[2]

[1] School of Technology and Management, Polytechnic Institute of Leiria,
Leiria, Portugal
[2] Department of Informatics Engineering,
Center for Informatics and Systems of the University of Coimbra (CISUC),
Coimbra, Portugal
{joanamc,catarina,bribeiro}@dei.uc.pt

Abstract. Social networks are a thriving source of information and applications are pervasive. Twitter has recently experienced a significant change in its essence with the doubling of the number of maximum allowed characters from 140 to 280. In this work we study the changes that come from such modification when learning systems are in place. Results on real datasets of both settings show that transferring models between both scenarios may need special treatment, as bigger tweets are harder to classify, making such dynamic environments even more challenging.

Keywords: Social networks · Twitter · Learning algorithms

1 Introduction

Nowadays social networks have become ubiquitous. A vast majority of the western world relies on social media to communicate and to access information, but also for marketing and political intents.

Since its introduction in 2006, messages posted to Twitter have provided a rich source of data for researchers, leading to the publication of over a thousand academic papers [14]. In fact, Twitter has risen to prominence in communities all across the networked world, not only in the research community [2].

The recent doubling of the character count from 140 to 280 is likely to have an effect on the way users express themselves [10]. A possible outcome can be the reduction of acronyms and abbreviations used and also an improvement on the grammar. Nevertheless, other more interesting and not so obvious consequences are likely to occur when learning scenarios are in place.

In this work we study this interesting area for further investigation, analyzing the changes to the way learning can be carried out in Twitter streams after the move from 140 to 280 characters. We focus on pattern recognition and learning in Twitter streams. We learn and evaluate several models with different maximum tweet's length and discuss the changes that occur in different settings.

© Springer Nature Switzerland AG 2020
A. M. Madureira et al. (Eds.): SoCPaR 2018, AISC 942, pp. 177–184, 2020.
https://doi.org/10.1007/978-3-030-17065-3_18

The rest of the paper is organized as follows. In Sect. 2 we introduce the background on the Twitter social network. In Sect. 3 we describe the proposed experimental approach for learning in Twitter streams with different tweets' sizes. Section 4 presents the experimental setup and Sect. 5 includes the results and analysis. Finally, Sect. 6 shows the conclusions and possible lines of future research.

2   Twitter

Twitter[1], created in 2006, rapidly gained popularity with more than 328 million monthly active users in August 2017. According to the company, its mission is to give everyone the power to create and share ideas and information instantly, without barriers. Nevertheless, the concept can be easily described as a network where a user can share a simple message, which is immediately made public.

Twitter took advantage of the worldwide implemented Short Message Service (SMS), and promoted the idea of sharing simple day life events in order to stay connected with friends and family. Although this was the initial concept, Twitter has changed and is now allowing multimedia content, like photos and video.

One of the most significant differences of Twitter as a social network is that the relation between users in not necessarily reciprocal, i.e., in Twitter a user can follow another user without being followed back. Another distinctive characteristic of Twitter is that, by default, all messages are public.

2.1   Tweet

A *tweet* is any message posted to Twitter which may contain photos, videos, links and up to 140 characters of text. There is also the concept *retweet*, described as a tweet that you forward to your followers. Tweets can also include *mentions*, a user name started with the symbol "@", and *hashtags*, detailed in the following. An example of a tweet is shown in Fig. 1.

Shivi Goyal @Shivimuskan · Jul 12
#Portugal #Euro2016Final #Cristiano #beatsfrance breathtaking #imlovingit

Fig. 1. Example of a tweet

2.2   Hashtag

Twitter provides the possibility of including a *hashtag*. A hashtag is a single word starting with the symbol "#", as represented in Fig. 2. It is used to classify

---

[1] http://www.twitter.com.

the content of a message and improve search capabilities. This can be particularly important considering the amount of data produced in the Twitter social network. Besides improving search capabilities, hashtags have been identified as having multiple and relevant potentialities, like promoting the phenomenon described in [6] as *micro-meme*, i.e. an idea, behavior, or style, that spreads from person to person within a culture [1]. By tagging a message with a trending topic hashtag, a user expands the audience of the message, compelling more users to express their feelings about the subject [16].

Fig. 2. Tweet and hashtag representation

The importance of the hashtag in Twitter is already identified in multiple applications, like bringing a wider audience into discussion [8], spreading an idea [12], get affiliated with a community [15], or bringing together other Internet resources [3].

2.3   Twitter Classification Problem

The classification of Twitter messages can be described as a multi-class problem. Twitter messages, represented as $\mathcal{D} = \{d_1, \ldots, d_t\}$, where d_1 is the first instance and d_t the latest. Each instance is characterized by a set of features, usually words, $\mathcal{W} = \{w_1, w_2, \ldots, w_{|\mathcal{W}|}\}$. Consequently, the instance d_i is represented by the feature vector $\{w_{i1}, w_{i2}, \ldots, w_{i|\mathcal{W}|}\}$.

If d_i is a labelled instance it can be represented by the pair (d_i, y_i), where $y_i \in \mathcal{Y} = \{y_1, y_2, \ldots, y_{|\mathcal{Y}|}\}$ is the class label for instance d_i.

Our classification strategy, presented in [5], will be using the Twitter message hashtag to label the content of the message, which means that y_i represents the hashtag that labels the Twitter message d_i.

Notwithstanding being a multi-class problem in its essence, it can be decomposed in multiple binary tasks in a one-against-all binary classification strategy.

3   Proposed Learning Approach

Figure 3 presents the defined baseline learning approach that is divided into two parts:

1. Baseline 140: both train and test sets include only tweets up to 140 characters.
2. Baseline 280: both train and test sets include only tweets between 140 and 280 characters.

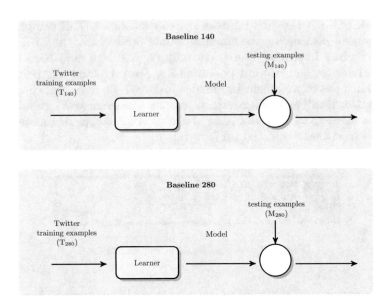

Fig. 3. Baseline approaches for 140 and 280 character tweets

This baseline allows for an independent comparison of learning tweets' classification with different lengths.

Figure 4 presents an overview of the proposed learning approach, representing three different methods. In all three methods, the Twitter stream is used to get a set of Twitter training examples with no filter regarding the tweets' size. Then, a model is created (in this case we have used Support Vector Machines).

The three defined approaches aim to detect differences in the difficulty in classifying tweets depending on their length. Hence, we have defined different test sets:

1. Full Approach: the test set includes tweets of any given dimension.
2. 140 Approach: the test set includes only tweets up to 140 characters.
3. 280 Approach: the test set includes only tweets between 140 and 280 characters.

4   Experimental Setup

4.1   Dataset

To evaluate and validate our strategy, and considering the lack of a labelled dataset with the needed characteristics, we have built a specific dataset. The dataset was constructed by requesting public tweets to the Twitter API[2]. We have collected more than 1,055,000 messages, since 10 March 2018 to 5 April

---

[2] https://developer.twitter.com.

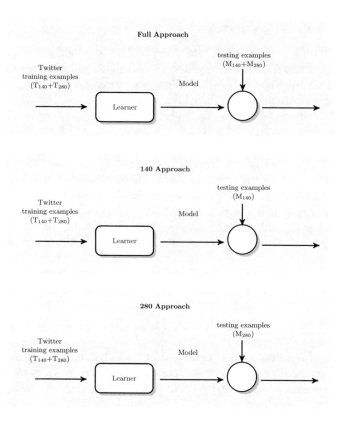

Fig. 4. Full approach, 140 approach and 280 approach

2018, and, considering the worldwide usage of Twitter, tweets were only considered if the user language was defined as English. All the messages that did not have at least one hashtag were discarded, as the hashtags are assumed as the message classification. Finally, tweets containing no message content besides hashtags were also discarded and all the hashtags are removed from remaining tweets. From the 1,055,000 collected messages, we reach 142,000 tweets that have a body part and at least one hashtag.

We have only considered five of the most popular used hashtags, namely #syria, #sex, #trump, #oscars and #ucl.

The tweets were then split into two equal and disjoint sets: training and test. The data from the training set is used to select learning models, and the data from the testing set to evaluate performance. We have used the bag of words strategy to document representation along with tf-idf. Preprocessing methods were applied, namely stopword removal and stemming.

4.2   Learning and Evaluation

The evaluation of our approach was done by the previously described dataset and using the Support Vector Machine (SVM). This machine learning method was introduced by Vapnik [13], based on his Statistical Learning Theory and Structural Risk Minimization Principle. The idea behind the use of SVM for classification consists on finding the optimal separating hyperplane between the positive and negative examples. Once this hyperplane is found, new examples can be classified simply by determining which side of the hyperplane they are on. SVM constitute currently the best of breed kernel-based technique, exhibiting state-of-the-art performance in text classification problems [4,7,11]. SVM were used in our experiments to construct the proposed models.

In order to evaluate a binary decision task we first define a contingency matrix representing the possible outcomes of the classification, as shown in Table 1.

In order to evaluate the binary decision task we defined well-known measures based on the possible outcomes of the classification, such as recall ($R = \frac{a}{a+c}$) and precision ($P = \frac{a}{a+b}$), as well as combined measures, such as, the van Rijsbergen F_β measure [9], which combines recall and precision in a single score:

$$F_\beta = \frac{(\beta^2 + 1)P \times R}{\beta^2 P + R}. \tag{1}$$

F_β is one of the best suited measures for text classification used with $\beta = 1$, i.e. F_1, an harmonic average between precision and recall (1), since it evaluates unbalanced scenarios that usually occur in text classification settings and particularly in text classification in the Twitter environment. Micro-averaged F_1 is computed by summing all values (a, b, c and d), and then use the sum of these values to compute a single micro-averaged performance score that represents the global score.

Table 1. Contingency table for binary classification.

	Class positive	Class negative
Assigned positive	a	b
	(True positives)	(False positives)
Assigned negative	c	d
	(False negatives)	(True negatives)

5   Results and Analysis

Before testing the three proposed approaches, we have defined a baseline to compare learning using 140 characters tweets and 280 characters tweets. Thus, we have defined train and test sets that only have 140 or 280 characters (see 3). Results on Table 2 show that the problem of learning and classifying 280 characters tweets is a harder task than 140 characters tweets.

Table 2. Performance classification results for baseline approaches

	Baseline 140			Baseline 280		
	Precision	Recall	F1	Precision	Recall	F1
#syria	98,54%	95,22%	96,85%	96,10%	91,81%	93,90%
#sex	99,28%	97,33%	98,29%	99,61%	96,68%	98,12%
#trump	97,29%	96,06%	96,67%	93,87%	94,14%	94,00%
#oscars	96,90%	87,60%	92,02%	89,26%	82,74%	85,87%
#ucl	99,93%	98,67%	99,30%	99,75%	97,49%	98,61%
Micro-averaged	**98,34%**	**95,91%**	**97,11%**	**96,18%**	**93,92%**	**95,04%**

Table 3 presents the performance classification results for the three testing scenarios. Analyzing the table, the initial baseline hint that 280 character tweets are harder to classify is somewhat mitigated. There seems to be a clear benefit of constructing models that do not differentiate using the length of the tweets. This heterogeneity seems to benefit the classification of 280 character tweets with little or no loss in the classification performance of 140 character tweets. This result that bigger tweets are harder to classify is somewhat counterintuitive since usually more information is usually positive.

Table 3. Performance classification results for Full, 140 and 280 approaches

	Full approach			140 approach			280 approach		
	Precision	Recall	F1	Precision	Recall	F1	Precision	Recall	F1
#syria	98,68%	93,65%	96,10%	99,06%	94,08%	96,50%	97,07%	92,17%	94,56%
#sex	99,33%	97,18%	98,24%	99,26%	96,51%	97,86%	99,40%	98,71%	99,06%
#trump	97,68%	95,35%	96,50%	98,12%	95,29%	96,68%	95,49%	95,56%	95,52%
#oscars	97,17%	85,98%	91,24%	97,58%	86,56%	91,74%	93,35%	82,39%	87,53%
#ucl	99,93%	98,37%	99,14%	99,93%	98,46%	99,19%	100,00%	93,33%	96,55%
Micro-averaged	**98,54%**	**95,17%**	**96,82%**	**98,76%**	**95,04%**	**96,87%**	**97,56%**	**95,91%**	**96,73%**

6   Conclusions and Future Work

In this paper we have presented learning approaches for Twitter streams with different tweets' sizes, given the recent doubling of the number of maximum allowed characters from 140 to 280 on Twitter social network. Three different settings were proposed to achieve a full comparison. Baseline results showed that using only 280 character tweets can become a harder task than using shorter tweets. However, results of the three proposed settings have mitigated this insight, by presenting a less severe difference when training sets are heterogeneous. Nevertheless, there is a consistency when focusing on 280 character tweets classification performance that indicates results are more challenging.

Future work is foreseen is defining more testing approaches that allow more insight, which will only be possible if 280 character tweets become more common.

References

1. Merriam-Webster's dictionary, October 2012
2. Barnard, S.: Twitter: more than 140 characters. In: Palgrave Macmillan (ed.) Citizens at the Gates, Chap. 2, pp. 13–31. Springer (2018)
3. Chang, H.C.: A new perspective on Twitter hashtag use: diffusion of innovation theory. In: Proceedings of the 73rd Annual Meeting on Navigating Streams in an Information Ecosystem, pp. 85:1–85:4 (2010)
4. Costa, J., Silva, C., Antunes, M., Ribeiro, B.: On using crowdsourcing and active learning to improve classification performance. In: Proceedings of the 11th International Conference on Intelligent Systems Design and Applications, pp. 469–474 (2011)
5. Costa, J., Silva, C., Antunes, M., Ribeiro, B.: Defining semantic meta-hashtags for Twitter classification. In: Proceedings of the 11th International Conference on Adaptive and Natural Computing Algorithms, pp. 226–235 (2013)
6. Huang, J., Thornton, K.M., Efthimiadis, E.N.: Conversational tagging in Twitter. In: Proceedings of the 21st ACM Conference on Hypertext and Hypermedia, pp. 173–178 (2010)
7. Joachims, T.: Learning Text Classifiers with Support Vector Machines (2002)
8. Johnson, S.: How Twitter will change the way we live. Time Mag. **173**, 23–32 (2009)
9. van Rijsbergen, C.: Information Retrieval (1979)
10. Rosell-Aguilar, F.: Twitter as a formal and informal language learning tool: from potential to evidence. In: Rosell-Aguilar, F., Tita Beaven, M.F.G. (eds.) Innovative Language Teaching and Learning at University: Integrating Informal Language into Formal Language Education, Chap. 11, pp. 99–106. Research-publishing.net (2018)
11. Tong, S., Koller, D.: Support vector machine active learning with applications to text classification. J. Mach. Learn. Res. **2**, 45–66 (2002)
12. Tsur, O., Rappoport, A.: What's in a hashtag? Content based prediction of the spread of ideas in microblogging communities. In: Proceedings of the 5th International Conference on Web Search and Data Mining, pp. 643–652 (2012)
13. Vapnik, V.: The Nature of Statistical Learning Theory (1999)
14. Williams, S., Terra, M., Warwick, C.: What do people study when they study Twitter? Classifying Twitter related academic papers. J. Doc. **69**(3), 384–410 (2013)
15. Yang, L., Sun, T., Zhang, M., Mei, Q.: We know what @you #tag: does the dual role affect hashtag adoption? In: Proceedings of the 21st International Conference on World Wide Web, pp. 261–270 (2012)
16. Zappavigna, M.: Ambient affiliation: a linguistic perspective on Twitter. New Media Soc. **13**(5), 788–806 (2011)

Retweet Predictive Model for Predicting the Popularity of Tweets

Nelson Oliveira[2], Joana Costa[1,2(⊠)], Catarina Silva[1,2], and Bernardete Ribeiro[2]

[1] School of Technology and Management, Polytechnic Institute of Leiria,
Leiria, Portugal
[2] Department of Informatics Engineering,
Center for Informatics and Systems of the University of Coimbra (CISUC),
Coimbra, Portugal
njoukov@student.dei.uc.pt, {joanamc,catarina,bribeiro}@dei.uc.pt

Abstract. Nowadays, Twitter is one of the most used social networks with over 1.3 billion users. Twitter allows its users to write messages called tweets that now can contain up to 280 characters, having recently increased from 140 characters. Retweeting is Twitter's key mechanism of information propagation. In this paper, we present a study on the importance of different text features in predicting the popularity of a tweet, e.g., number of retweets, as well as the importance of the user's history of retweets. The resulting Retweet Predictive Model takes into account different types of tweets, e.g, tweets with hashtags and URLs, among the used popularity classes. Results show there is a strong relation between specific features, e.g, user's popularity.

Keywords: Twitter · Popularity prediction · Retweeting

1 Introduction

Twitter is a very popular micro-blogging social network founded in 2006. Twitter plays a role of news media when announcing breaking news [1]. In this social network a user can post a message (tweet), which can be shared (retweet) by other users. These users can define a tweet as favorite and reply to it. Currently, Twitter has over 1.3 billion accounts but it is estimated that only 550 million had ever made a tweet. Twitter is widely used by social media, brands, and celebrities. It can influence a brand's reputation, since 80% of Twitter users mention brands on Twitter. Since Twitter has 500 million tweets daily, this information can be used to marketing purposes, personal interests, like finding the most interesting content. In Twitter the information propagates from the user to his followers, so every time that a user tweets, retweets or favorites a tweet it will appear on the timeline of his followers. The difference here is that the retweet will stick to the user's tweet list. Tweets can also spread based on the hashtag/topic of the tweet, since Twitter allows a user to search tweets according to different criteria, which makes them more visible to the entire Twitter's network.

© Springer Nature Switzerland AG 2020
A. M. Madureira et al. (Eds.): SoCPaR 2018, AISC 942, pp. 185–193, 2020.
https://doi.org/10.1007/978-3-030-17065-3_19

Although Twitter has millions of tweets each day, not all become popular. Predicting this popularity can be very useful to marketing companies to increase the popularity and sales rate of a brand, movies, political campaigns to gather the attentions of the population. Obviously, there are many reasons why certain tweets are more popular than others. These reasons might be, the user who tweeted (celebrity, company, and others), if the topic of the tweet belongs to a trending one, the relation of the user and the follower, if the tweet has a hashtag or not, and many other things for example like the interests of the followers of a user.

Building a Retweet Predictive Model will allow us to identify the popular tweets which can help users to personalize their tweets in order to have higher chances to create a popular message and become popular.

Our main motivation is to find what are the reasons why some tweets are more popular than others, which might help us to recommend the use of hashtags and topics to add in the message, in order to expand our network and provide better content to our followers.

The rest of the paper is organized as follows. Section 2 details the related work on retweet predictive models based on machine learning algorithms. Section 3 presents our proposed approach on building a retweet predictive model, identifying the features and defining the problem. In Sect. 4 we describe the twitter streaming API dataset, the twitter search API dataset, the sampling methods and the features used in our model. The case study and dataset are described, along with the evaluation metrics and experimental results. In Sect. 5 we illustrate the approach with the results of the different experiments. Finally, in the last section the conclusions are presented and some lines of future research addressed.

2   Related Work

Popularity prediction has been studied by multiple authors [2–5] and [6]. Predicting popularity can allow users to better structure their tweets in order to understand how they can have higher possibilities of making a tweet that can achieve a higher popularity, which can be useful, for instance, to companies [7] that want to extend their popularity in order o increase their reputation and profits.

Some studies investigated which are the factors that are most important to popularity prediction. Suh et al. [8] and Janders et al. [3] showed that URLs and hashtags have a positive influence, along Nr. of followers (user's popularity), which is considered one of the most influencing features.

Other studies focused on building retweet prediction models. Petrovic et al. [2] built a Retweet Predictive model that indicates if a tweet is going to be retweeted or not. Janders et al. [3] used Naïve Bayes and Logistic Regression to classify viral tweets and non viral tweets using different thresholds for virality, adding sentiment analysis related features. Zhang et al. [4] used SVM with the Information Gain of the features to predict the popularity of a tweet using multiple classes. A model based on the theory of self-exciting point processes approach was built by Zhao et al. [5] to predict the final number of retweets through time, using information cascades. Wu et al. [6] used a stochastic process-based approach to model user retweeting behavior.

3   Proposed Approach

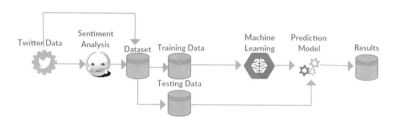

Fig. 1. Retweet predictive model

The main goal of this work is to build a Retweet Predictive Model and show how distinct number of text features (for example tweets with only 1 hashtag, tweets with less than 60 characters) affect its performance regarding different values of popularity. The text features that we tested were the number of hashtags, number of URLs, and length of the text. The tweets that we used here were collected with the Twitter Streaming API [1]. For the secondary goal of this work, we address the problem of how the past activity of a user influences the performance of the model. By that we mean how the usage of the last k number of retweets of a user affects the prediction for the next tweet. The tweets that we used in this paper were collected with the Twitter Search API [2] We used 4 classes as Zhang et al. [4]. Moreover, Class 1 (tweets without popularity) contains tweets that have 0 retweets, class 2 (low popularity tweets) contains tweets that have more than 0 and less than 10 retweets, class 3 (medium popularity tweets) contains tweets that have 10 or more retweets and less than 100, and class 4 (high popularity tweets) contains tweets that have 100 or more retweets. Our model uses tweets that were collected with the Twitter Streaming API and Twitter Search API and performs sentiment analysis of tweets, which were used as training data and testing data. The training data was generated using 2 sampling methods, and fitted to a Random Forest classifier. Finally, we tested the model with different testing sets for each one of the datasets.

4   Experimental Setup

4.1   Twitter Streaming API Dataset Description

In this work we use Twitter Streaming API dataset that contains the tweets gathered using the Twitter Streaming API. The dataset has 12,470,144 (English)

---

[1] https://dev.twitter.com/streaming/overview.
[2] https://dev.twitter.com/rest/public/search.

tweets that go from 1st of July, 2016 until 15th of July. Table 1 shows the distribution of tweets among the 4 different classes. It is clearly visible that the number of tweets of each class is severely skewed.

Table 1. Number of tweets for each class of the Twitter Streaming API dataset

Class	Number of tweets
1	8684496
2	2276806
3	999511
4	509331

4.2 Twitter Search API Dataset Description

To analyze the influence of the users cof retweets and how they affect the predictability for a new tweet that a user posts we used the Twitter Search API. Thus, we have collected from 153k users contained in the previous dataset, their last 200 tweets. From these 200 tweets we have filtered out the data in order to obtain only the pure tweets of each user, we exclude the retweets from tweets that a user has previously retweeted. So far, for each tweet, we also have the number of retweets of each of the user's last 10 tweets.

Table 2. Number of tweets for each class of the Twitter Search API dataset

Class	Number of tweets
1	10088923
2	2803163
3	1118919
4	626580

4.3 Sampling Methods

We used 2 methods to select the training data: **Class balancing** and **Sub class balancing**. We used these two methods to fix the unbalanced classes problem. If we use the random sampling method we have more tweets from the class 1 than the remaining classes, which could compromise the performance of the model.

Class Balancing. Based on the unbalanced distribution of the classes in our dataset, the first sampling method was applied to select randomly 300k tweets from each one of the classes. In this way, we get the same number of samples of each class which will not happen with random sampling.

Sub Class Balancing. In this approach we partitioned the classes into multiple sub classes, in order to avoid the unbalanced distribution inside each class, as tweets with higher number of retweets are scarcer. For example if we look at the class 3, which represents the tweets with 10 or more retweets and less than 100, the first interval for which we count the number of tweets within that range (10–20) has almost 10 times more tweets than the range (90–100). Therefore, when we randomly select 300k tweets from this class most of the tweets are within the first interval. This can influence the prediction for the tweets that are near the edges of each class.

4.4 Features

The features that we have used in our study can be divided in two main groups, user features and tweet features. Most of these features can be extracted from the structure of the data directly when it is gathered from the Twitter Streaming API/Twitter Search API, but we have also added other extra features related to the tweet.

User features are the ones related to the author of the tweet. All of these features already have been used in other studies related to popularity prediction. We used the following features: number of followers, number of statuses (tweets/retweets), number of favorites, number of times the user was listed, number of days of the account. The last feature that we used here is binary, representing if the user is verified (or not).

Tweet features are the ones related to the tweet itself. We used the following features: number of hashtags, URLs, mentions, length of the tweets, number of words, is a tweet a reply, hour of the tweet timestamp. Besides these features we also added the number of videos, number of photos and GIFs, since users tend to engage more with tweets that contain visual content. To our knowledge no previous studies used these features. We have also used features related to Sentiment Analysis, which were positive sentiment score (from 1 to 5) and negative sentiment score (from −1 to −5). These values were obtained using the framework SentiStrength [3], which already was used by Janders et al. [3]. This framework has the particularity that is built for short texts which is the case of tweets. The last feature that we used is binary, representing if a tweet contains (or not) a trending topic.

5   Results and Discussion

In Tables 3 and 4, we present the experimental results for the Twitter Streaming API and Twitter Search API datasets, respectively. Our experimental tests were performed with the Random Forest Algorithm, since it was the one that performed better in a initial phase. The configuration setup for the algorithm includes 300 trees in the forest, and the maximum amount of features to consider when looking for the best split was the square root of the number of used features. The remaining parameters were set to its default values. In our experiments we used the 80:20 ratio for training and testing the model, so in this case

---

[3] http://sentistrength.wlv.ac.uk/.

since we used 300k of tweets for each class, respectively we tested the model
with 300k tweets.

5.1   Twitter Streaming API

Table 3. Results for F1-score of the different experimental tests for the Twitter streaming API dataset

Test	Class			
	1	2	3	4
Randomly selected tweets without balancing	80.80 ± 0.10	39.45 ± 0.15	44.50 ± 0.21	58.60 ± 0.32
Randomly selected tweets	85.19 ± 0.07	45.74 ± 0.26	57.18 ± 0.31	78.18 ± 0.15
Tweets without hashtags	86.36± 0.05	43.22± 0.24	56.41± 0.21	78.93± 0.52
Tweets with 1 hashtag	78.65 ± 0.14	51.55 ± 0.11	58.21 ± 0.12	75.32 ± 0.26
Tweets with 1 or more hashtags	80.23 ± 0.24	**52.12 ± 0.10**	58.61 ± 0.16	75.72 ± 0.11
Tweets without URLs	83.58 ± 0.12	44.46 ± 0.23	56.38 ± 0.12	78.44 ± 0.36
Tweets with 1 or more URLs	**87.72 ± 0.07**	48.50 ± 0.12	**59.38 ± 0.46**	76.57 ± 0.36
Tweets with 0 to 60 characters	87.04 ± 0.08	39.77 ± 0.29	54.51 ± 0.34	**80.34 ± 0.24**
Tweets with 60 to 100 characters	85.36 ± 0.08	44.76 ± 0.16	56.40 ± 0.35	78.01 ± 0.25
Tweets with 100 to 140 characters	82.62 ± 0.09	50.83 ± 0.07	58.81 ± 0.12	77.08 ± 0.22

Comparing the global results (when random tweets were tested) when we used
the sub class balancing, we noticed a major improvement in all of the classes
especially in the class with higher popularity (class 4). One possible reason why
this happened might be because, as indicated, the distribution within each class
is also unbalanced; this means that dividing each class into multiple sub classes
helps in better selection of the training samples which improves the performance
of the model.

Analyzing the performance of the model regarding the different text features that were tested, we showed that some features are more related to certain
classes. As an example, shorter texts are more related to tweets without popularity (class 1) and tweets with more popularity (class 4), while longer tweets
tend to be more related to the classes 2 and 3. In this case, the performance of
the model increases with the length of the tweets.

In the case of hashtags it is possible to conclude that tweets that do not have
any hashtags are more easily identified when they do not get any retweets. These
tweets of the class 1 can also be more related to the tweets that have at least one
URL; this means that regarding other used features, whenever at least one URL
exists, the model performs better with respect to the class 1, which indicates a
correlation between these type of tweets and this class.

In the case when the tweet has one or multiple hashtags the classes 2 and 3
are the ones which have improvements in the performance, while the class 4 gets
worse. This can seem strange since the presence of hashtags is associated with
popularity. However, it is worth to notice that the number of followers (user's
popularity) plays one of the greatest roles when predicting the popularity of a
tweet. So in this case, we can say that popular users do not need hashtags in

their tweets to have more retweets. This might be related to the performance of the model when we did not use any hashtags regarding the class 4 where most of the tweets were made by users that are more popular than the ones contained in the other classes. This performance was better than when testing with tweets that have at least one hashtag. In the classes 2 and 3 we can see that the model has the same behavior for these 2 classes; it means that in testing with different values of the text features, the performance values for both of the classes, either increases or decreases. This can mean that the number of retweets between the both ranges of these classes are equally affected regarding to the different text features and their values.

5.2 Twitter Search API

Table 4. Results for F1-Score of the different experimental tests for the Twitter Search API dataset

Tested tweets	Class			
	1	2	3	4
Random tweets without balancing	84.74 ± 0.05	46.80 ± 0.19	59.93 ± 0.21	72.89 ± 0.40
Random tweets	87.27 ± 0.06	50.42 ± 0.11	59.28 ± 0.19	75.53 ± 0.21
Nr. of retweets of the last tweet	87.27 ± 0.06	50.42 ± 0.11	59.28 ± 0.19	75.46 ± 0.41
Nr. of retweets of the last 2 tweets	87.45 ± 0.08	50.85 ± 0.18	59.57 ± 0.39	75.59 ± 0.4
Nr. of retweets of the last 3 tweets	87.55 ± 0.04	51.02 ± 0.09	59.87 ± 0.18	75.82 ± 0.19
Nr. of retweets of the last 4 tweets	87.64 ± 0.04	51.27 ± 0.09	60.13 ± 0.20	75.99 ± 0.16
Nr. of retweets of the last 5 tweets	87.72 ± 0.04	51.51 ± 0.12	60.33 ± 0.15	76.13 ± 0.12
Nr. of retweets of the last 6 tweets	87.79 ± 0.03	51.70 ± 0.16	60.43 ± 0.23	76.13 ± 0.18
Nr. of retweets of the last 7 tweets	87.84 ± 0.05	51.80 ± 0.22	**60.66 ± 0.20**	**76.25 ± 0.24**
Nr. of retweets of the last 8 tweets	87.89 ± 0.03	51.86 ± 0.22	60.57 ± 0.28	76.18 ± 0.26
Nr. of retweets of the last 9 tweets	87.87 ± 0.03	51.93 ± 0.21	60.51 ± 0.25	75.99 ± 0.35
Nr. of retweets of the last 10 tweets	**87.93 ± 0.06**	**51.96 ± 0.08**	60.45 ± 0.39	75.99 ± 0.31

For this dataset there were also improvements when we used the sub class balancing, but not that much as with the Twitter Streaming API dataset. The main reason is that since we have gathered tweets directly from the users and not from the Twitter Stream, the tweets are not that dispersed in terms of number of retweets. When analyzing the results of the Twitter Search API dataset, we have noticed that, mainly the performance of the model increased for the first 2 classes when using more number of retweets of the last k tweets of a user. The same thing did not happen in the classes 3 and 4 since it mainly starts to decrease after $k = 7$. The reason that we found is that the number of retweets for less popular users tend to be more similar across their last tweets. On the other hand, for some of the more popular tweets, they might have been tweeted by users who generally got less retweets than those obtained for that tweet.

6   Conclusions and Future Work

In this work we have presented a retweet predictive model that is able to predict the popularity of a tweet. An analysis of the importance of different factors such as the user's popularity on a Retweet Predictive Model was performed, using two sampling techniques with Twitter Stream API dataset and Twitter Search API dataset. We showed that dividing each class into multiple samples helps to increase the performance of the model. We also empirically showed that the performance of the model is different regarding each class for tweets with different characteristics (Twitter Stream API dataset), which might suggest that there are some characteristics that are more related to a certain class. For example, the best performance achieved by the model regarding the class 4 was when we used tweets with a short text (0 to 60 characters). Regarding the Twitter Search API dataset, we showed that there is a correlation with the past numbers of retweets of a user, especially for the classes 1 and 2. As for the more popular tweets (classes 3 and 4), that correlation is not so strong, since popular tweets are often unpredictable, e.g., a popular tweet from a user which has few followers.

This work can be further extended. One way is to analyze the performance of the model with different user model profiles (e.g., users with many followers, users with few followers and many favorites). This will help to get personalized profiles for different types of users. It will be then possible to determine the odds to get a given range of retweets, through the values of the features that they introduce in their text message.

References

1. Park, H., Kwak, H., Lee, C., Moon, S.: What is Twitter, a social network or a news media? In: 19th International World Wide Web Conference (WWW 2010) (2010)
2. Petrovic, S., Osborne, M., Lavrenko, V.: Rt to win! Predicting message propagation in Twitter. In: Proceedings of the Fifth International Conference on Weblogs and Social Media, Catalonia, Spain, Barcelona (2011)
3. Kasneci, G., Jenders, M., Naumann, F.: Analyzing and predicting viral tweets. In: Association for Computing Machinery, pp. 657–664 (2013)

4. Xu, Z., Zhang, Y., Yang, Q.: Predicting popularity of messages in Twitter using a feature-weighted model. Int. J. Adv. Intell. (2012)
5. He, H.Y., Rajaraman, A., Zhao, Q., Erdogdu,. M.A., Leskovec, J.: A self-exciting point process model for predicting tweet popularity. In: Proceedings of the 21th ACM SIGKDD International Conference on Knowledge Discovery and Data Mining, KDD 2015, pp. 1513–1522 (2015)
6. Wu, B., Shen, H.: Analyzing and predicting news popularity on Twitter. Int. J. Inf. Manag. **35**, 702–711 (2015)
7. Nitins, T., Burgess, J.: Twitter, brands, and user engagement. Twitter Soc. **89**, 293–304 (2014)
8. Pirolli, P., Suh, B., Hong, L., Chi, E.H.: Want to be retweeted? Large scale analytics on factors impacting retweet in twitter network. In: Proceedings of the 2nd IEEE International Conference on Social Computing, SOCIALCOM, pp. 177–184 (2010)

Handcrafted Descriptors-Based Effective Framework for Off-line Text-Independent Writer Identification

Abderrazak Chahi[1,2]($\boxtimes$), Youssef El Merabet[1], Yassine Ruichek[2], and Raja Touahni[1]

[1] Laboratoire LASTID, Département de Physique, Faculté des Sciences, Université Ibn Tofail, BP 133, 14000 Kenitra, Morocco
abderrazak.chahi@uit.ac.ma,
y.el-merabet@univ-ibntofail.ac.ma,
rtouahni@hotmail.com
[2] Le2i FRE2005, CNRS, Arts et Métiers, Université Bourgogne Franche-Comté, UTBM, 90010 Belfort, France
yassine.ruichek@utbm.fr

Abstract. Feature extraction is a fundamental process in writer identification that requires efficient methods to characterize and handle local variations of the characters shape and writer individuality. This paper presents a reliable and novel learning-based approach for off-line text-independent writer identification of handwriting. Exploiting the local regions in the writing samples (document or set of word/text line images), we propose a simple, yet adaptive and discriminative feature representation based on handcrafted texture descriptors including Local Binary Patterns (LBP), Local Ternary Patterns (LTP) and Local Phase Quantization (LPQ) to describe characteristic features of the writing style variability. The overall system extracts from the input writing a set of connected components, which seen as texture sub-images where each one of them is subjected to LBP, LPQ or LTP to compute the feature matrix. This matrix is then fed to dimensionality reduction process followed by segmentation into a number of non-overlapping zones, which are afterward subjected to sub-histogram sequence concatenation to form the holistic concatenated component feature histogram. In the identification stage, the writing samples are recognized and classified through their feature histograms using the nearest neighbor 1-NN classifier. Experiments conducted on two handwritten representative databases (AHTID/MW and IAM) indicate that our system performs better than the state-of-the-art systems on the Arabic AHTID/MW database, and show competitive performance on the English IAM database.

Keywords: Off-line writer identification · Histogram sequence concatenation · Text-independent · Handwritten connected components · Dissimilarity measure · Texture descriptors

© Springer Nature Switzerland AG 2020
A. M. Madureira et al. (Eds.): SoCPaR 2018, AISC 942, pp. 194–206, 2020.
https://doi.org/10.1007/978-3-030-17065-3_20

1 Introduction

With the technological advances and the frequent use of computers in all aspects of life, the analysis of handwriting is becoming one of the important and challenging problems in the recent decades, which receives growing interest in the field of pattern recognition. Handwriting recognition is an area of research in full expansion. It covers a wide range of applications. One can cite on-line/off-line writer identification [1–3], on-line/off-line verification of handwritten signatures [4, 5], handwritten musical scores [6], classification of ancient documents [7] and forensic document analysis to identify the authenticity of unknown handwritten documents within a large database (e.g. the analysis of ransom notes, fraudulent and threatening letters, etc.) [8, 9]. The writing shape and its characteristics differ from one person to another. Besides, a handwritten text will not be written in exactly the same way twice by the same person. For this reason, writer identification remains a challenging and attractive area of research, since there is no ideal model that can characterize such extreme variations, and hence it is difficult to describe the large variability between handwritten samples. In some cases of specialization, the working concept and value of computerized analysis of handwriting is the same one as fingerprint systems. Indeed, contrary to the electronic and printed documents, handwriting has been shown to be a reliable biometric component, which carries useful and distinguishing information about the person who produced it. Furthermore, handwriting analysis has a great importance, as it presents an active research area for psychologists, graphologists, forensic experts, and historians.

Writer identification is the task of characterizing the writer from his handwritten based on machine-learning and computer image processing approaches. It is an effective biometric process that uses the one-to-many recognition technique by returning a list of writer candidates, which sorted in classes, and whose writing style is similar to the questioned handwritten sample. However, it remains more difficult to obtain promising identification results on some massive databases, even though an only small subset of handwritten samples is used or a reduced number of writers is considered in evaluations. In point of fact, computerized writer identification frameworks assist the forensic experts, by reducing the search space in comparing efficiently and automatically an unknown handwritten text within a huge handwritten database.

Research in writer identification has primarily focused on one of the two following streams: (*i*) on-line approach, which uses a set of dynamic characteristics of the writing such as the order of strokes, writing trajectory, writing time and speed, pen pressure etc. These spatial and temporal writing characteristics are captured through digitizing acquisition devices at the real time of writing; (*ii*) off-line writer identification strategy (delayed or static), where the digitization of the written trace is carried out after its generation and it is available on two-dimensional (2D) image representation (scanned handwritten images). The off-line approach exploits different image processing and feature extraction techniques to enhance the system performance in characterizing, in an efficient way, the writing style of each writer.

Within off-line identification systems, depending on how the writing content is available and presented in both training and testing sets, the writer identification approaches are traditionally classified into two main categories, namely text-dependent and text-independent. The text-dependent methods rely on employing exactly the same fixed handwritten text in the test and training sets, while the text-independent method does not depend on the text content, and hence it uses any given text to characterize the writer with a free choice of the pen.

The presented paper discusses a novel and effective system for offline text-independent writer identification of handwritten samples (document or set of word/text line images). The proposed framework involves three main stages: (*1*) using prepro-cessing and segmentation techniques, the input handwritten samples are segmented into a set of connected components, where each one of them is labeled by its corresponding class (writer). Afterward, a uniform widow size is used for all labeled components (on training and testing) by resizing them into 50×50 pixels; (*2*) feature extraction process, which consists, respectively, in characterizing each resized component using the most commonly used texture descriptors in writer identification [1, 10–13] including Local Phase Quantization (LPQ) [14], Local Binary Patterns (LBP) [15] and Local Ternary Patterns (LTP) [16], and segmenting the obtained feature image to normalized zones where each one of them is represented by its respective histogram, and then concatenating the histograms of all zones to form the final feature histogram; (*3*) writer identification is performed using nearest neighbor classification (1-NN) with Hamming distance measure. The overall system is evaluated on two challenging handwritten databases (AHTID/MW [17] and IAM [18]) comprising handwritten samples in Arabic and English languages respectively, and compared with several recent state-of-the-art writer identification systems.

The outline of the reminder of this paper is organized as follows. Section 2 describes in details the main stages of the proposed framework. Section 3 reports the experimental results and discussion, in comparison with the recent state-of-the-art systems. A short summary of this work and some future research directions are given in Sect. 4.

2   The Proposed Methodology

The proposed system consists of three main steps: (1) pre-processing and segmentation into connected component sub-images, (2) feature extraction and dimensionality reduction, and (3) writer identification (classification). The flowchart of the proposed system is illustrated in Fig. 1. The description of each stage of our methodology is given in the following subsections.

2.1   Pre-processing and Segmentation

The two-dimensional (2D) input handwritten image (document or set of word/text line images) is subjected to a series of basic pre-processing techniques (i.e. binarization, noise reduction, and segmentation of the writing sample) in the objective to get an enhanced and clear image representation to extract the most useful information from it. In the first step, we carried out binarization process to convert each handwritten sample available in grayscale format into a binary image using the global thresholding tech-nique introduced in [19]. The interest of this step is to clearly discern the foreground (writing style) and background, as its strength is to preserve the useful maximum writing pixels and to discard unwanted background traces associated with the scanned input image. A noise reduction filter is then used to remove all accidental writing traces and diacritics from the binary image, which are considered as noise. Thereafter, we proceed to the segmentation phase, in which, the obtained pre-processed image is cropped into a set of connected component sub-images where each one of them is labeled by its corresponding class. Figure 2 shows pre-processing and segmentation of an Arabic handwritten word.

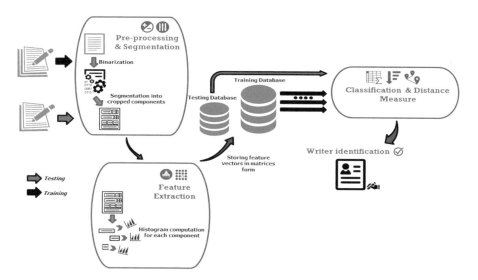

Fig. 1. Structure of the writer identification system

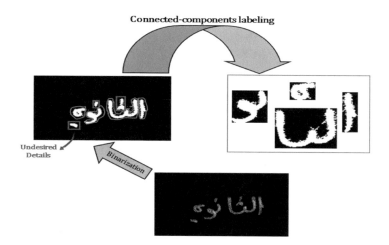

Fig. 2. Extracted connected components of an Arabic word from the AHTID/MW database

2.2   Feature Extraction

Prior to the classification step, handwriting sub-images i.e., extracted connected components, need to be characterized by a set of feature histograms, which are essential to accurately identify the writers. In this context, extracting significant and useful features is conditioned by the discriminative capability of the used feature extraction method, which is a key part to reduce the misclassification and improve the identification performance. In this paper, we propose an effective feature extraction representation based on Local Binary Patterns (LBP), Local Ternary Patterns (LTP) and Local-Phase Quantization (LPQ) texture operators, which are the most frequently used methods in writer identification [1, 10–13].

Traditional Local Binary Patterns

Local Binary Patterns (LBP) model, originally proposed by Ojala et al. in [15], is a fast and powerful gray-scale invariant operator for texture classification. The kernel function of LBP model summarizes the gray-level structure in a local region by computing the local contrast in texture analysis. As illustrated in Fig. 3, the pixels of the gray-scale image I are labelled in a 3×3 local neighborhood by thresholding the eight-neighbor pixels with the center pixel. The LBP code for a referenced pixel x_c is defined as follows:

$$LBP_P(x_c) = \sum_{k=1}^{P} S\{I(x_k) - I(x_c)\} \cdot 2^{k-1} \tag{1}$$

where $I(x_c)$ is the central pixel value, $I(x_k)$ is the value of its neighbors, P is the number of neighbors. The function S is given as:

$$S(z) = \begin{cases} 1, & \text{if } z \geq 0 \\ 0, & \text{if } z < 0 \end{cases} \tag{2}$$

For a given image, the distribution of the texture densities of all central pixels and their neighbors represents the LBP histogram. LBP generates 256 (2^8) possible patterns (cf. Table 1).

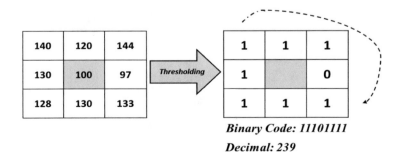

Binary Code: 11101111

Decimal: 239

Fig. 3. An example of LBP encoding scheme

Feature Extraction Procedure

As shown in Fig. 4, the proposed feature extraction method mainly relies on computing, within local regions of the input handwritten image, spatial information captured by texture operators to efficiently characterize the writer individuality and the large variability between handwritten samples. The adopted feature extraction process, illustrated in Fig. 4, involves four main stages: (*1*) extracted component sub-images obtained after pre-processing step (cf. Sect. 2.1) are first resized into the same fixed window size of 50 × 50 pixels. Each resized component is afterward fed to LBP, LTP or LPQ encodings to get a uniform feature image F; (*2*) we perform a simple dimensionality reduction technique, which consists in normalizing the obtained feature image F by a factor R, aiming to reduce the computing time of the classification step. This technique allows eliminating unnecessary features that may not be instructive and informative in the feature image F since we dispose of, in the first stage, connected component images in binary form; (*3*) in order to hold more discriminative measures to characterize, in an efficient way, the writing style variability, the normalized feature matrix $\frac{F}{R}$ is scanned in top-down, left-right and portioned into $N_r \times N_r$ non-overlapping regions, where each one of them is subjected to LBP, LTP or LPQ histogram codes. Formally, each region $RG^i(i = 1, 2\ldots N_r \times N_r)$ is represented by its feature histogram h_i computed through Eq. 3; (*4*) feature sub-histograms $h_i(i = 1, 2\ldots N_r \times N_r)$ of dimensionality D_{im} (cf. Table 1) computed, one by one, within each region, are concatenated using Eq. 5 to constitute the final feature vector H of dimensionality $D_c = (N_r \times N_r) \times \frac{D_{im}}{R}$.

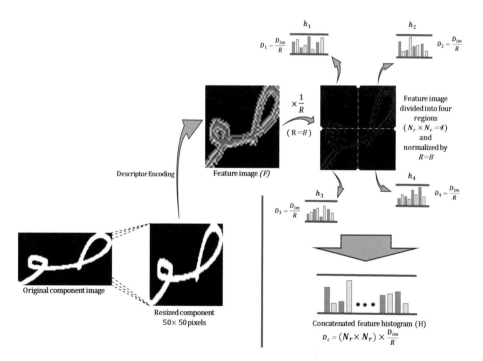

Fig. 4. The proposed feature extraction framework

The writing style of each writer is better characterized, mainly thanks to the connected component-histogram sequence concatenation, as it produces a complete spatial distribution of the writing pixels, yielded promising classification results.

$$h_i(v) = \sum_{x_p \in R_g^i} \delta\left(v, RG^i(x_p)\right) \tag{3}$$

where $RG^i(x_p)$ is the value of pixel x_p in the region $RG^i(i = 1, 2 \ldots N_r \times N_r)$, $v \in [0, N_{bins}]$, $N_{bins} = D_{im} - 1$ is the number of bins of the feature histogram h_i, D_{im} is the descriptor dimension given in Table 1, R is the dimensionality reduction factor, and $\delta(\cdot)$ being the Kronecker delta function given in the following:

$$\delta(a, b) = \begin{cases} 1, & \text{if } a = b \\ 0, & \text{if } a \neq b \end{cases} \tag{4}$$

$$H = \prod_{i=1}^{N_r \times N_r} h_i \tag{5}$$

where Π is the concatenation operator. The dimensionality D_c of the feature histogram H and the computing time of the classification process rise gradually as the number of regions $N_r \times N_r$ increases, and this is naturally due to the comparison of distances between the component histograms. This issue is solved thanks to the dimensionality reduction factor R, which permits reducing computing time, as D_c decreases when R grows according to $D_c = (N_r \times N_r) \times \frac{D_{im}}{R}$. It should be noted that both dimensionality reduction factor R and number of regions $N_r \times N_r$ highly influence the system performance in writer identification. Effectively, different coding of LBP, LPQ and LTP histograms are obtained with different values of N_r and R.

Table 1. The tested feature extraction operators

Feature extraction methods	Dimension (D_{im})
LBP	256
LPQ	256
LTP	512

2.3 Writer Identification

At this stage of our proposed system, the writing samples (document or set of word/text line sub-images) in the testing and training sets, are represented, using LBP, LPQ and LTP texture operators, by their respective set of feature histograms computed from all the connected components extracted in each sample. Note that, for each texture operator, the number of feature histograms in each sample is equal to the number of connected components extracted from it. The next step consists in performing writer identification process using a (dis)similarity measure and Hamming distance metric.

For that, the Hamming distances between each component of the testing sample and all of the other components in the training sample for a given writer is calculated and the training component that reports the minimum Hamming distance, is returned as the most similar component. Subsequently, the (dis)similarity measure between the testing sample and all of the other ones in the training set is computed. The (dis)similarity of a testing sample T_s, which needs to be identified, and a training one R_s is given by:

$$DISM(T_s, R_s) = \frac{1}{card(T_s)} \sum_{i=1}^{card(T_s)} \min\left\{D\left(H_{T_s}^i, H_{R_s}^1\right), D\left(H_{T_s}^i, H_{R_s}^2\right), \ldots, D\left(H_{T_s}^i, H_{R_s}^{card(R_s)}\right)\right\}$$

(6)

with $card(T_s)$ and $card(R_s)$ being the number of connected components in the writing samples T_s and R_s respectively. $H_{T_s}^i$ and $H_{R_s}^j$ are the respective feature histograms of ith and jth components in the samples T_s and R_s. respectively. The Hamming distance function $D(.,.)$ between the two feature histograms $H_{T_s}^i$ and $H_{R_s}^j$ is defined as follows:

$$D\left(H_{T_s}^i, H_{R_s}^j\right) = \sum_{n=1}^{D_c} \left|H_{T_s}^i(n) - H_{R_s}^j(n)\right|$$

(7)

where D_c is the feature histogram dimension. The writer of the testing sample T_s is then recognized as the writer of the writing sample in the training set, which returns the minimum dissimilarity value (cf. Eq. 8):

$$Writer(T_s) = argmin\left\{DISM\left(T_s, R_s^1\right), DISM\left(T_s, R_s^2\right)\ldots, DISM\left(T_s, R_s^{card(Train)}\right)\right\}$$ (8)

with $card(Train)$ is the number of writing samples in the training base.

3   Experimental Results and Discussion

3.1   Databases Setup

The overall writer identification system has been tested on two challenging representative handwritten databases: the Arabic AHTID/MW [17] and English IAM [18] databases. In our experiments, the databases setup is adopted as that given in [20] for AHTID/MW, and in [10] for IAM to have a fair and clear comparison with the writer identification systems reported in the literature.

AHTID/MW [17]: The Arabic Handwritten Text Images Database written by Multiple Writers (AHTID/MW) is a publicly available database for research in the recognition of Arabic handwritten text with open vocabulary, word segmentation and writer identification. It contains 53 writers from different educational levels and various ages with a free choice of the pen. The writers were asked to fill 70 text lines, and hence allow collecting a total of 22.896-word images scanned and stored in PNG format as grayscale images with a resolution of 300 dpi. The database was portioned into 4 sets of word samples, where 3 sets are used in the training and the last one as the testing set.

IAM [18]: The IAM database is the most popular English handwriting database, which was mainly used for writer identification/verification and handwritten text recognition systems. It comprises 657 writers who contributed to produce 1539 forms with 13.353 isolated and labeled handwritten English text lines of variable content (given in resolution of 300 dpi and available as PNG images with 256 gray levels). We evaluated the proposed framework on the complete set of 657 writers by fixing a maximum of 14-text line samples for each writer. In experiments, a percentage of 60% of the text line samples are used as training set while 40% are used as testing set.

3.2   System Evaluation Results

The quality of the recognition task, which aims to identify the persons who wrote the query samples, is greatly conditioned by the capacity of the feature extraction method to characterize the sensitivity of characters shape and between writer variability. In our proposed system, each writing sample is captured by a set of feature histograms computed from all connected components extracted from it. For evaluation, we record the identification rate as the proportion of the testing samples, which are classified and matched correctly through the simple nearest neighbor classifier (1-NN) using Hamming distance metric. In all experiments, the proposed system is assessed on the total set of writers for each handwritten dataset. On AHTID/MW database, we use the full 4-fold cross validation protocol, where four split permutations were generated for each writer. This configuration is adopted in order to investigate all possible scenarios in the testing stage and further evaluate the performance stability of the overall proposed system. Table 2 reports the obtained Top-1, Top-3 and Top-5 writer identification rates of the proposed system on IAM database, and the Top-1 average identification accuracy as well as the identification rates recorded over each split separately on AHTID/MW database are summarized in Table 3.

From these Tables, it is easily observed that all the tested feature extraction methods demonstrate a significant performance stability over the evaluated databases as they allow achieving high identification rates. On AHTID/MW database (cf. Table 3), the LBP, LPQ and LTP operators show consistent identification rates over the four subdivisions, which are more or less close to the overall average identification accuracy. As depicted in Tables 2 and 3, the LPQ operator provides significantly the highest performance on the two tested databases, in comparison with LBP and LTP operators to characterize the writing style of each writer.

Table 2. Writer identification accuracies on IAM (657 writers) database

Feature extraction operators	Writer identification accuracy (%)			Dimension (D_c)
	Top-1	Top-3	Top-5	
LPQ	**91.17**	94.21	96.04	**512**
LTP	90.56	94.21	95.43	1332
LBP	88.73	93.15	93.91	1152

Table 3. Writer identification accuracies on AHTID/MW (53 writers) database

Feature extraction operators	Splits				Average identification accuracy (%)	Dimension (D_c)
	Split.1	*Split.2*	*Split.3*	*Split.4*		
LPQ	**100**	**100**	**100**	98.11	**99.53**	**288**
LTP	98.11	96.22	98.11	90.56	95.75	423
LBP	96.23	96.23	98.11	92.45	95.75	304

It should be noted that the results reported in Tables 2 and 3 are obtained using the best configurations of the parameters R (dimensionality reduction factor) and $N_r \times N_r$ (number of regions in the normalized feature matrix $\frac{F}{R}$). The optimal values of these parameters are determined experimentally for each of the evaluated feature methods and over each handwritten database. Indeed, using LPQ operator as the best performing feature extraction method on IAM database, we achieve the best Top-1 identification rate of 91.17% with the following configuration: number of regions of $N_r \times N_r = 16$ and $R = 8$ for the dimensionality reduction factor, which results a dimension of $D_c = 512$ for the feature histogram H (cf. Eq. 5). On AHTID/MW database, the configuration of $(N_r \times N_r = 16, R = 8)$ proved to be, using LPQ operator, the best setting to extract textural features from the handwritten samples with a dimensionality of $D_c = 288$ that corresponds to the final feature histogram H according to $D_c = (N_r \times N_r) \times \frac{D_{im}}{R}$.

3.3 Performance Evaluation with State-of-the-Art Systems

In the recent decades, extensive researches in the area of text-independent-based offline writer identification have been focusing on the development of efficient and robust techniques that would better characterize the writing variability and reduce the processing time. Comparing the overall system performance with the state-of-the-art writer identification, we considered only the well-known systems, which are evaluated on the complete set of writers of the two tested databases. The results obtained by these approaches along with those achieved by the proposed system on IAM and AHTID/MW databases, are summarized in Table 4. As can be seen from Table 4, the proposed system reaches, on IAM database, an identification accuracy of 91.17% using LPQ descriptor for feature extraction, surpassed only by the system introduced in [20], and outperforms several existing systems reported in [10, 21–23]. The proposed system provides, using all evaluated texture operators for feature extraction, a very accurate and reliable solution for characterizing the large variability between handwritings on AHTID/MW database. Indeed, our system achieves a significant performance (rate of 99.53%), which greatly outperforms all the literature identification rates.

Table 4. Performance comparison with the state-of-the-art writer identification systems

System	Year	Language	Database	Writers	Writer identification rate
Kumar et al. [21]	2014	English	IAM	650	88.43%
Bulacu and Schomaker [22]	2007	English	IAM	650	89%
Hannad et al. [10]	2016	English	IAM	657	89.54%
Siddiqi and Vincent [23]	2010	English	IAM	650	91%
Khan et al. [20]	2017	English	IAM	650	97.2%
Proposed framework + LPQ		**English**	**IAM**	**657**	**91.17%**
Proposed framework + LTP		**English**	**IAM**	**657**	**90.56%**
Proposed framework + LBP		**English**	**IAM**	**657**	**88.73%**
Schomaker and Bulacu (*implemented* in [20])	2017	Arabic	AHTID/MW	53	66.4%
Slimane and Märgner [24]	2014	Arabic	AHTID/MW	53	69.48%
Khan et al. [20]	2017	Arabic	AHTID/MW	53	71.6%
Hannad et al. (*implemented* in [20])	2017	Arabic	AHTID/MW	53	77.3%
Khan et al. [11]	2016	Arabic	AHTID/MW	53	87.5%
Proposed framework + LPQ		**Arabic**	**AHTID/MW**	**53**	**99.53%**
Proposed framework + LTP		**Arabic**	**AHTID/MW**	**53**	**95.75%**
Proposed framework + LBP		**Arabic**	**AHTID/MW**	**53**	**95.75%**

4 Conclusion

This paper introduced an original and simple offline-feature extraction model for text-independent writer identification. The proposed system relies on local analysis of regions of interest in the writing sample referred to as connected component using handcrafted descriptors for feature extraction. Our framework generates for each component sub-image a feature vector computed through histogram sequence concatenation process within writing zones based on effective dimensionality reduction technique. These feature vectors were subjected to 1-NN classifier to recognize the writers of the testing samples. In experiments, we achieved very satisfactory classification results on the Arabic AHTID/MW database and reached competitive performance on the English IAM database.

In future work, we intend to evaluate the system effectiveness on more challenging handwritten databases. Moreover, we plan to apply other sophisticated classifiers such as SVM and artificial neural network to further investigate the system performance in writer identification. Off-line text-independent writer verification task will also be addressed in future research directions.

References

1. Bertolini, D., Oliveira, L., Justino, E., Sabourin, R.: Texture-based descriptors for writer identification and verification. Expert Syst. Appl. **40**(6), 2069–2080 (2013)
2. Abdi, M., Khemakhem, M.: A model-based approach to offline text- independent arabic writer identification and verification. Pattern Recogn. **48**(5), 1890–1903 (2015)
3. Tan, G., Viard-Gaudin, C., Kot, A.C.: Automatic writer identification framework for online handwritten documents using character prototypes. Pattern Recogn. **42**(12), 3313–3323 (2009)
4. Plamondon, R., Lorette, G.: Automatic signature verification and writer identification—the state of the art. Pattern Recogn. **22**(2), 107–131 (1989)
5. Kumar, R., Sharma, J.D., Chanda, B.: Writer-independent off-line signature verification using surroundedness feature. Pattern Recogn. Lett. **33**(3), 301–308 (2012)
6. Fornés, A., Lladós, J., Sánchez, G., Bunke, H.: Writer identification in old handwritten music scores. In: 2008 the Eighth IAPR International Workshop on Document Analysis Systems, pp. 347–353. IEEE (2008)
7. Arabadjis, D., Giannopoulos, F., Papaodysseus, C., Zannos, S., Rousopoulos, P., Panagopoulos, M., et al.: New mathematical and algorithmic schemes for pattern classification with application to the identification of writers of important ancient documents. Pattern Recogn. **46**(8), 2278–2296 (2013)
8. Franke, K., Köppen, M.: A computer based system to support forensic studies on handwritten documents. Int. J. Doc. Anal. Recogn. **3**(4), 218–231 (2001)
9. Said, H., Tan, T., Baker, K.: Personal identification based on handwriting. Pattern Recogn. **33**(1), 149–160 (2000)
10. Hannad, Y., Siddiqi, I., El Kettani, M.: Writer identification using texture descriptors of handwritten fragments. Expert Syst. Appl. **47**, 14–22 (2016)
11. Khan, F., Tahir, M., Khelifi, F., Bouridane, A.: Offline text independent writer identification using ensemble of multi-scale local ternary pattern histograms. In: 6th European Workshop on Visual Information Processing (EUVIP). IEEE, pp. 1–6 (2016)
12. Hannad, Y., Siddiqi, I., El Kettani, M.: Arabic writer identification using local binary patterns (LBP) of handwritten fragments. In: 7th Iberian Conference on Pattern Recognition and Image Analysis, pp. 237–244. Springer (2015)
13. Bertolini, D., Oliveira, L.S., Sabourin, R.: Multi-script writer identification using dissimilarity. In: 2016 23rd International Conference on IEEE Pattern Recognition (ICPR), pp. 3025–3030 (2016)
14. Ojansivu, V., Heikkilä, J.: Blur insensitive texture classification using local phase quantization. In: Image and Signal Processing - 3rd International Conference, ICISP, pp. 236–243. Springer (2008)
15. Ojala, T., Pietikainen, M., Maenpaa, T.: Multiresolution gray-scale and rotation invariant texture classification with local binary patterns. IEEE Trans. Pattern Anal. Mach. Intell. **24**(7), 971–987 (2002)
16. Tan, X., Triggs, B.: Enhanced local texture feature sets for face recognition under difficult lighting conditions. IEEE Trans. Image Process. **19**(6), 1635–1650 (2010)
17. Mezghani, A., Kanoun, S., Khemakhem, M., El Abed, H.: A database for arabic handwritten text image recognition and writer identification. In: 2012 International Conference on Frontiers in Handwriting Recognition, pp. 399–402. IEEE (2012)
18. Marti, U.-V., Bunke, H.: The IAM-database: an English sentence database for offline handwriting recognition. Int. J. Doc. Anal. Recogn. **5**(1), 39–46 (2002)

19. Choudhary, A., Ahlawat, S., Rishi, R.: A neural approach to cursive handwritten character recognition using features extracted from binarization technique. In: Complex System Modelling and Control Through Intelligent Soft Computations. Springer International Publishing, pp. 745–771 (2015)

20. Khan, F., Tahir, M., Khelifi, F., Bouridane, A., Almotaeryi, R.: Robust off-line text independent writer identification using bagged discrete cosine transform features. Expert Syst. Appl. **71**, 404–415 (2017)

21. Kumar, R., Chanda, B., Sharma, J.D.: A novel sparse model based forensic writer identification. Pattern Recogn. Lett. **35**, 105–112 (2014)

22. Bulacu, M., Schomaker, L.: Text-independent writer identification and verification using textural and allographic features. IEEE Trans. Pattern Anal. Mach. Intell. **29**(4), 701–717 (2007)

23. Siddiqi, I., Vincent, N.: Text independent writer recognition using redundant writing patterns with contour-based orientation and curvature features. Pattern Recogn. **43**(11), 3853–3865 (2010)

24. Slimane, F., Märgner, V.: A new text-independent GMM writer identification system applied to Arabic handwriting. In: 14th International Conference on Frontiers in Handwriting Recognition, pp. 708–713. IEEE (2014)

Server Load Prediction on Wikipedia Traffic: Influence of Granularity and Time Window

Cláudio A. D. Silva[1(✉)], Carlos Grilo[1,2(✉)], and Catarina Silva[1,3(✉)]

[1] School of Technology and Management, Polytechnic Institute of Leiria,
Leiria, Portugal
klaudio.ads@gmail.com,
{carlos.grilo,catarina}@ipleiria.pt
[2] CIIC, Polytechnic Institute of Leiria, Leiria, Portugal
[3] Center for Informatics and Systems of the University of Coimbra,
Coimbra, Portugal
catarina@dei.uc.pt

Abstract. Server load prediction has different approaches and applications, with the general goal of predicting future load for a period of time ahead on a given system. Depending on the specific goal, different methodologies can be defined. In this paper, we follow a pre-processing approach based on defining and testing time-windows and granularity using linear regression, ANN and SVM learning models. Results on real data from Wikipedia servers show that it is possible to tune the size of the time-window and the granularity to improve prediction results.

Keywords: Load forecasting · Linear regression · Artificial Neural Networks ·
Support Vector Machines · Server load prediction · Wikipedia

1 Introduction

Load prediction, also known as load forecasting, emerged as a support to predict and prevent resource imbalance [1], rapidly proving its importance in the industry. Organizations that do not use load forecasting or have incorrect forecasts have a high chance of being negatively affected by this. The inability to identify load problems may trigger a snowball effect on the long run. Overestimation may lead to overstock, which can be impossible to resell with profit. On the other hand, underestimation may result in revenue loss or in the incapability to fulfil the organization's goals. When forecasting load, it is of the utmost importance to keep a balance between resources and demand, avoiding stock outs or over stocking.

Load forecasting has become one of the major research fields in various areas, e.g., electric load forecasting, sales forecasting, dam water levels forecasting, traffic forecasting and load forecasting in computer servers, commonly known as server load prediction [2–4], which is the area of study of this paper. Server load prediction is a fairly new research area when compared to other areas such as energy load forecasting. New areas of applicability of server load prediction appear at an increasingly high rate, such as, cloud computing, grid computing, utility computing, virtualization or automated service provisioning.

© Springer Nature Switzerland AG 2020
A. M. Madureira et al. (Eds.): SoCPaR 2018, AISC 942, pp. 207–216, 2020.
https://doi.org/10.1007/978-3-030-17065-3_21

In server load prediction, one of the first actions taken by researchers is the identification of the load forecasting type they are working with. These types are categorized according to the forecast time horizon, which is usually problem dependent. It can be set by the specific goal of the prediction to achieve, or it can be constrained by the availability and accuracy of data. Therefore, each forecast horizon is used for different purposes and/or restrictions. According to Gross [5], these horizons can be divided as: Very-Short Term Load Forecasting (VSTLF), Short-Term Load Forecasting (STLF), Medium-Term Load Forecasting (MTLF) and Long-Term Load Forecasting (LTLF). VSTLF has a forecast horizon lower than an hour, while STLF has a forecast horizon of up to 30 days, MTLF has a forecast horizon of up to one year and LTLF has a forecast horizon higher than a year.

Along with the distinct types of load forecasting, a large variety of methods exist that can be used. Some methods evolved with the goal of achieving better results in a certain type of load forecasting. The case study presented in this paper is of hourly load prediction, which categorizes it as STLF.

In server load prediction, the goal is frequently to determine the load that will be put on the servers by users, either in terms of data volume or number of connections. Efficient resource management is essential in the viability of the datacenter, ensuring that workloads have the necessary resources without breaching service level agreements. To efficiently adjust resource allocation, prediction algorithms are used to forecast incoming load and/or resources needed to guarantee the performance of the datacenter for every kind of specific load [6]. The dynamic nature of load in servers, however, handicaps the performance of common linear models such as linear regression. This is mainly due to the fact of data being nonlinear and non-stationary [7].

Two common alternatives used by researchers in server load prediction are Artificial Neural Networks (ANN) [8–12] and Support Vector Machines (SVM) [13]. ANNs tend to handle better nonlinear and non-stationary data. For instance, the ANN can model any function linear or not and the activation layers can be linear or not. However, ANNs bring some chronic problems with them, e.g., the global minima [14].

SVMs are based in the statistical learning theory and adopt a structural risk minimization principle, which produces a better generalization and is superior to the empirical risk minimization used by ANNs. One benefit that SVMs bring is the guarantee of a global minima. Moreover, SVMs can handle better the input of corrupted data.

In this paper, we propose a novel pre-processing method based on enrichment of data by introducing temporal factors. This method aims to increase the performance of server load prediction models in an initial step before the parametrization step/tuning phase.

The paper is structured as follows: Sect. 2 presents related work in the field of server load prediction, analysing and describing different approaches that share the same or similar learning methods as our paper. Section 3 details the experimental environment presented in this paper. In Sect. 4 the results are presented and discussed. Finally, Sect. 5 presents conclusions and future lines of research.

2   Related Work

The two most commonly used approaches for server load prediction are a time series approach and a regression approach. In the case of the time series approach, the idea is to predict the load using mainly data with a time dimension. Predictions tend to be daily, weekly or seasonal. Auto-regressive moving average (ARMA) models, Kalman filtering, spectral expansion techniques and the Box-Jenkins methods are commonly used in time series approaches [15].

For the regression problem approach, the methodologies used are the definition of basic functional elements, identification of the coefficients needed in the linear combination of the functional elements previously defined and, lastly, the selection of all the relevant variables [15]. The created models along with load forecast also identify the relationships between the dependent and independent variables in the data.

In the literature review bellow, our focus relies on the results and models created through Linear Regression, Neural Networks and Support Vector Machines by other researchers.

2.1   Linear Regression

The linear regression field is the one with the least publications available. Most papers found use linear regression mostly as a stepping stone for the final solution.

In [16], a method was presented that predicted CPU usage based on linear regression. The proposed approach approximated the short time future CPU utilization based on the history of usage in each host. The researchers used as evaluation metrics an average SLA violation percentage, which represents the percentage of average CPU performance that was not allocated to an application when requested. Their error in real workload lied between 7.97% and 9.94%.

2.2   Artificial Neural Networks

In the ANN field, a study was conducted on work done by researchers that evaluated the performance of different Neural Network algorithms in the desire to correctly predict the CPU load of a host [17]. They designed a multilayer neural network that takes as input patterns collected from the load traces and that predicts future load statistics. Subsequently, they made tests using Quick Propagation, Back Propagation with and without Momentum and the Resilient Propagation algorithm for the load traces. They proved that Resilient Propagation algorithms had a better prediction accuracy compared to other ANN algorithms. Their results were summarized in a graph without concrete values. From it, it is possible to see that their mean error varies between 2.5% and 40% with the lowest results presenting a standard deviation of more than double of its mean error.

Another group of researchers used a variance of ANNs, dynamic neural networks, for server load prediction [8]. Their approach used neural networks for the time series predictions. The sample dataset used was derived from the request count logs on a AWS EC2 container. The goal of their research was to correctly predict the load in servers and with this information ensure that the node count was close to optimal, preventing over-usage of computing time.

The created model presented average errors between 2.95% and 3.6%. It also had difficulty in predicting peak values in the load. However, the authors justify these peaks as Distributed Denial of Service (DDOS) attacks.

2.3 Support Vector Machines

The SVM field is currently one of the most invested by researchers, being some of the reasons for this, the versatility and performance of this learning methods when compared to others.

In [18], researchers used SVM for their server load prediction problem. Their SVM model predicted the server load. Along with their model, they proposed a novel method for selecting free parameters by which they could increase the precision of their prediction. The SVM prediction process was divided into three different steps. Firstly, they reconstructed the phase space, where the embedded dimension was the key value to be decided. Secondly, they tested different parameter values and tried to achieve the best model possible. Lastly, they defined the predictions steps with whom they wanted to get the prediction values. They concluded that by using SVMs to predict server load in future, in addition with their approach to select free parameters for SVM, they could efficiently predict the time series, including stationary and nonstationary. Unfortunately, their paper does not present values regarding the mean error value of the resulting model.

In [19], the researchers proposed a novel approach for day ahead forecasting using clustering and SVMs. Just like in our approach, the researchers also applied a preprocess step to their data. In their case, this was done through clustering of data. Their SVM model was trained considering two different approaches. Firstly, they considered all the previous patterns found in the data without clustering and secondly by using the daily average loads this time with clustering. Their average error varied between days, but it ranged around the 2%–2.5%.

2.4 Conclusion

All the above studies created solutions for server load prediction, achieving reasonable results. All of them used algorithms based in Linear Regression, Neural Network or Support Vector Machines. We propose an approach on which we use these algorithms and improve their performance by applying specific data pre-processing methodologies.

3 Proposed Approach

The devised approach entails an iterative process with the goal of a constant model improvement. Firstly, the data is retrieved and saved into a dataset, secondly a set of procedures is applied in order to clean and re-format the dataset. In this paper all the attributes with a correlation coefficient higher than 90% were removed. Subsequently, the dataset transformation module is applied, which creates different sets of training and test sets. For each training and test set created, a model based on different learning strategies is develop. The last step involves the evaluation of each model in terms of performance.

3.1   Data Source

The initial data was collected from the Wikimedia foundation (dumps.wikimedia. org/other/pageviews). Whenever a request of a page is made to Wikipedia, whether for reading or editing, the made request always passes through one of Wikipedia's internal hosts. From these requests, the project name, the size of the page requested, and the title of the page requested were collected. The resulting data was written into a page view raw dataset, where one record is the hourly aggregate of the pageviews/requests and size of an individual page. Figure 1 exemplifies five records of data of the initial dataset:

```
en.b 19th_Century_Literature 1 8422
en.b 360_Assembly/360_Instructions/BC 1 12217
en.b 360_Assembly/360_Instructions/LM 2 0
en.b 360_Assembly/Branch_Instructions 1 53068
en.b 360_Assembly/Pseudo_Instructions 1 8853
```

Fig. 1. Record structure example

Each attribute is split by a space, having a record a total of four attributes. The first attribute refers to the language the page is written in, In the example above, all records start with *en* followed by a dot as second parameter, e.g., the first line has *en.b*. The *en* refers to the Wikipedia language, which in this example is English, the second parameter after the dot indicates what type of page was requested. The second attribute is the name of the retrieved page. In the example above, for the first record this would be *19_Century_Literature*. The third attribute indicates the number of times a specific page was requested within that hour, these request numbers are not unique visits. The last attribute is the size of the returned content. In the given example, the first record had just one request, which accounted in total for 8422 response bytes.

From this initial data, a dataset was created containing the generated load and number of requests of English pages only.

3.2   Pre-processing

The pre-processing method developed in this paper consists of a 4-step procedure. The first step consists in applying a granularity value to each record, which defines the level of detail in the data. In this dataset, granularity is always in an hourly scale. Then, the time window definition step is applied, in which the number of hours back in time a record contains is defined. The appliance of the granularity step upfront guarantees that the created history maintains the same level of detail. The last step consists in applying normalization on the resulting dataset. For each created dataset that is normalized, there is another one that is not. This is done in order to evaluate how normalization affects the results (Fig. 2).

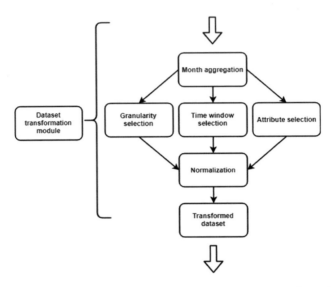

Fig. 2. Diagram depicting the different steps involved in the data transformation module

3.3   Learning Methods Parameterization

The created linear regression model used the Akaike criterion with attribute selection method M5 and ridge parameter with value 1.0E-8.

The ANN model was created with a Multilayer Perceptron architecture and with Backpropagation to classify the instances. The model configuration adopted was: learning rate: 0.3; validation threshold of 20; training time of 500 iterations; number of neurons in hidden layers was defined by the formula *hidden layer neurons = (number of input attributes + number of output neurons)*/2 and momentum value: 0.2.

The SVM model implemented was based on the SVM variation SMOreg. The created model used a *c* value of 1, a Poly Kernel variation with a cache size of 250007 and an *e* value of 1 and the regression optimizer: RegSMOImproved, with an *e* of 1.0E-12 and tolerance of 0.001.

3.4   Evaluation Metrics

The selected evaluation metric was Mean Absolute Percentage Error (MAPE). MAPE is a measure of prediction accuracy of a forecasting method. It expresses accuracy as a percentage. *At* is the actual value and F_t is the forecasted value:

$$MAPE = \frac{100\%}{n} \sum_{t=1}^{n} \left| \frac{A_t - F_t}{A_t} \right|$$

In the formula above, A_t is the actual value and F_t is the forecasted value. When using MAPE, it is possible to compare results between models, since it is scale independent. With this metric, it is possible to compare results of different models, such as the ones presented in the related work section.

4 Results and Discussion

We used Wikipedia's data from January 2016 to August 2016, history values of 1, 2, 4, 6, 8, 24 and 168 were applied and granularity values of 1, 2 and 7 were used. A grid search approach was used and the different possible combinations of these attributes throughout different months were tested with three different learning methods.

From the literature review presented in Sect. 2, three different works were used for comparison. These were reference [16], which achieved errors between 7.97% and 9.94% by using real workload data, reference [17], with errors ranging between 2.5% and 40% and, lastly, reference [19], with errors varying between 2% and 2.5%.

The average MAPE values per learning method across all months are presented in the table below for an initial comparison with the above mentioned works. These values were calculated using all the results from the application of our grid search methodology.

Table 1. Average MAPE values for all months for linear regression, ANNs and SVMs

	Linear regression	ANN	SVM
	Average MAPE	Average MAPE	Average MAPE
January	2.15	2.60	1.97
February	1.68	2.34	1.56
March	2.54	2.98	2.60
April	3.52	3.50	2.30
May	2.65	2.66	2.26
June	1.72	2.20	1.74
July	4.99	6.54	4.97
August	2.18	2.60	2.10

Table 1 shows that most average errors fall between 2% and 3%, which places them in the range of the best results achieved in references [16, 17] and [18]. On average, the best performing months were February and June with the lowest errors being 1.56% and 1.72%, respectively. The months of February and June achieved lower average errors when compared with the above references, where the lowest error was 2%, in [19]. An error lower than 2% was also achieved for January when using SVMs. The worst results were obtained for July, with errors between 4.97% and 6.54%, placing it above 2% but below the 7.97% mark achieved in [16]. In short, most of our runs led to very appealing results, with average errors bellow the state of art.

Table 2 contains all the lowest MAPE values per learning methods with their respective preprocessing configuration. As can be seen, the minimum MAPE values are below 1% for some months. Although all learning methods achieved good results with our approach, SVMs achieved the best results from an average perspective. Linear regression, on the other hand, had the best results from an absolute point of view, achieving the lowest MAPE value of 0.93% with a granularity of 7 and time window of 8 for August.

Table 2. Minimum MAPE values with respective preprocessing configuration

	Linear regression			ANN			SVM		
	Min MAPE	Granu-larity	Time window	Min MAPE	Granu-larity	Time window	Min MAPE	Granu-larity	Time window
Jan	1.33	7	8	1.36	1	8	1.20	7	4
Feb	0.97	2	8	1.16	2	4	0.96	2	8
Mar	1.05	7	8	1.22	1	4	1.16	1	8
Apr	1.37	2	24	1.64	2	8	1.48	1	168
May	1.33	7	1	1.61	1	8	1.32	7	1
Jun	1.02	1	4	1.11	1	4	1.02	1	8
Jul	2.06	2	24	2.01	1	4	2.13	1	24
Aug	0.93	7	8	1.21	2	8	1.04	7	8

From the absolute point of view, the proposed methodology achieved always better results when compared to the state of art in all months with the exception of July, for which the worst error value was 2.13%.

The approach presented also tackled the problem of predicting the peak values mentioned in [8]. The data regarding the month of February, where an error of 0.96% was obtained using SVMs, was used to compare the predicted value to the actual one. Figure 3 depicts the resulting chart. The two depicted red arrows represent the peak values that occurred on February. It is possible to confirm that through the application of our methodology the created solution could successfully predict these peak values.

The results achieved indicate that the approach applied has a high suitability to the data used. The viability of the methodology is also validated due to the use of real life data. Lastly, the use of different training and test sets across multiple months, induces us to believe that our methodology achieves interesting results, without creating an

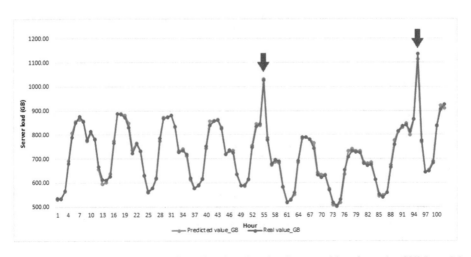

Fig. 3. Graphical representation of predicted and real values resulting from the SVM model regarding the month of February.

overfitted model. Further testing should however be done with data from different years and from different problems in order to verify if our approach presents a risk in creating overfitted models.

5 Conclusions and Future Work

This paper describes an approach to the server load prediction problem on Wikipedia's load. The proposed pre-processing method achieved better results than other related works. Results also show that our methodology achieves satisfactory results when applied with linear regression, ANNs and SVMs, with SVMs standing out.

On average, MAPE values are between 1.55% and 6.5%, with the best created models achieving results under the 1% mark. Since the approach is applied on different training and test sets across multiple months, results also indicate that this methodology increases the generalization capabilities of the models.

We would like to stress that the proposed preprocessing approach is generic and can thus be applied in different scopes that involve time series, not only server load forecast.

Future work is suggested in the use of data from different years, the addition of external data and the application on different load forecasting problems to test the adaptability of the approach.

References

1. Di Persio, L., Cecchin, A., Cordoni, F.: Novel approaches to the energy load unbalance forecasting in the Italian electricity market. J. Math. Ind. **7**, 5 (2017)
2. Park, D.C., El-Sharkawi, M.A., Marks, R.J., Atlas, L.E., Damborg, M.J.: Electric load forecasting using an artificial neural network. IEEE Trans. Power Syst. **6**(2), 442–449 (1991)
3. Dalrymple, D.J.: Sales forecasting practices: results from a united states survey. Int. J. Forecast. **3**(3–4), 379–391 (1987)
4. Hipni, A., El-shafie, A., Najah, A., Karim, O.A., Hussain, A., Mukhlisin, M.: Daily forecasting of dam water levels: comparing a support vector machine (SVM) model with adaptive neuro fuzzy inference system (ANFIS). Water Resour. Manag. **27**(10), 3803–3823 (2013)
5. Gross, G., Galiana, F.D.: Short-term load forecasting. Proc. IEEE **75**(12), 1558–1573 (1987)
6. Lorido-Botran, T., Miguel-Alonso, J., Lozano, J.A.: A review of auto-scaling techniques for elastic applications in cloud environments. J. Grid Comput. **12**(4), 559–592 (2014)
7. Dinda, P.A.: Online prediction of the running time of tasks. In: 10th IEEE International Symposium on High Performance Distributed Computing. IEEE (2001)
8. Pukach, P., Hladun, P.: Using dynamic neural networks for server load prediction. Comput. Linguist. Intell. Syst. **2**, 157–160 (2018)
9. Aljabari, G., Tamimi, H.: Server load prediction based on dynamic neural networks. In: Students Innovation Conference. Palestine Polytechnic University (2012)
10. Ahmed, A., Brown, D.J., Gegov, A.: Dynamic resource allocation through workload prediction for energy efficient computing. In: Advances in Computational Intelligence Systems. Springer, Cham, pp. 35–44 (2017)

11. Herbst, N., Amin, A., Andrzejak, A., Grunske, L., Kounev, S., Mengshoel, O.J., Sundararajan, P.: Online workload forecasting. In: Self-Aware Computing Systems. Springer, Cham, pp. 529–553 (2017)
12. Caballé, S., Xhafa, F.: Distributed-based massive processing of activity logs for efficient user modeling in a Virtual Campus. Clust. Comput. **16**(4), 829–844 (2013)
13. Vapnik, V.: The Nature of Statistical Learning Theory. Springer, New York (1995)
14. Gori, M., Tesi, A.: On the problem of local minima in backpropagation. IEEE Trans. Pattern Anal. Mach. Intell. **1**, 76–86 (1992)
15. Rojas, I., Pomares, H., Valenzuela, O.: Time Series Analysis and Forecasting: Selected Contributions from ITISE 2017. Springer (2017)
16. Farahnakian, F., Liljeberg, P., Plosila, J.: LiRCUP: linear regression-based CPU usage prediction algorithm for live migration of virtual machines in data centers. In: 39th EUROMICRO Conference on Software Engineering and Advanced Applications (SEAA). IEEE (2013)
17. Naseera, S.: A comparative study on CPU load predictions in a computational grid using artificial neural network algorithms. Indian J. Sci. Technol. **8**, 35 (2015)
18. Yu, Y., Zhan, X., Song, J.: Server load prediction based on improved support vector machines. In: 2008 IEEE International Symposium on IT in Medicine and Education (2008)
19. Jain, A., Satish, B.: Clustering based short term load forecasting using support vector machines. In: PowerTech, Bucharest. IEEE (2009)

Evolutionary Genes Algorithm to Path Planning Problems

Paulo Salgado[1](✉) and Paulo Afonso[2]

[1] CITAB, Universidade de Trás-os-Montes e Alto Douro,
5000-801 Vila Real, Portugal
psal@utad.pt
[2] Instituto de Telecomunicações/ESTGA,
Universidade de Aveiro, Águeda, Portugal

Abstract. Genes are fundamental pieces for reproductive processes and one force field creator of the evolutionary mechanisms of the species, whose laws and mechanism are not well known. In this paper a new evolutionary optimization strategy that combines the standard genetics algorithms (GA) with selfish perspective of evolution of genes is presented. Natural selection theory is explained by a mechanism, which is centred in individuals, that are the elements of a population, characterized by their chromosomes. The primary variables are the genes (characters or words), which are non-autonomous entities, grouped in a Chromosome structure (phrases of live). However, genes make their influence felt far beyond the chromosome structure (entity of the individual). Based on this paradigm, we propose the Evolutionary Genes algorithm (EGA) that enriches the GA with a new line field generating of evolutions. Genes-centred evolution (GCE) improve the search engine of Chromosome-centred evolution (CGE) of the GA. Its impact is apparent on the increased algorithm speed, but mainly on the improvement of genetics solutions, which may be useful to solve complex problems. This approach was used to path-planning problems, in a continuous search space, to show its effectiveness in complex and interdependent sub-paths and evolution processes. GCE improved local sub-paths search as sub-processes that catalyse the CCE engine to find an optimal trajectory solution, task that the standard genetic algorithm have no ability to solve.

Keywords: Genes algorithms · Genetic algorithms · Path planning

1 Introduction

Evolutionary algorithm (EA) is a family of stochastic problem solvers based on principles of natural evolution. During the last four decades successive EA and artificial life have been proposed to solve complex engineering problems through computational simulations [1, 2], particularly in optimization tasks. Genetic algorithms (GA) have become one of the most famous EA. The base of GA paradigm is that the solution(s) for a given problem can be seen as a survival task where possible solutions compete with each other for survival and the right to reproduce. This competition is the driving force behind the progress that desirable leads to an optimal solution. GA are now used

© Springer Nature Switzerland AG 2020
A. M. Madureira et al. (Eds.): SoCPaR 2018, AISC 942, pp. 217–225, 2020.
https://doi.org/10.1007/978-3-030-17065-3_22

to solve complex and multidimensional problems where other traditional methods fail or are difficult to be use [3, 4].

For the theory of natural selection [5, 7], reproduction is the leading mechanism of evolution [6], where individuals are its main agents. These base elements are selected to participate on the reproductive process, each individual favoured by its set of characteristics revealed in the performance tests. The characteristics that favour their Fitness Function (FF) are then transferred to the next generation [5] making these traits more common at each generation. Hereditary traits are transferred via DNA [8]. Here, the main processes involved are natural selection and reproduction, implemented by a set of genetics operators. DNA sequences can change through mutations, producing new alleles and affecting the characteristics of the underlying gene, altering the phenotype.

A Gene-centred view of evolution is another point of view on the evolution process, initially proposed by Richard Dawkins in his book "The Selfish Gene" (1976). In this theory, genes are the essential units of selection [9, 10], having the design to replicate themselves in order to secure and perpetuate their own existence. Individual organisms contribute weakly to the evolution process; their main role is the manufacture and hosting of genes. Genes highlight their merits or failures across individuals, defining their characteristics and behaviours.

The performance of a gene is revealed through multiple individuals, generally taking into account the qualities of the best individuals of the latter set. Gene fitness is inevitably associated with these values, recorded in a gene performance table (collective memory of gene qualities), but also by the performance of individuals who have such a gene in their chromosome. On the other hand, each individual is subject during his or her life to a multiplicity of performance evaluation tests, where the individual and combined actions of a subset of their genes (activated genes) is revealed. These values measure the individual's different performance perspectives. Both measures (gene and individual performance) are taken into account in the natural selection of individuals.

This idea is transferred to the proposed Evolutionary Gene Algorithms (EGA). It adopts the principle of the two mechanisms described above as main force fields of the evolutionary processes [12, 13]. These use individuals as the main agents for natural evolution by measuring their phenotype characteristics. Genes with advantageous phenotypic effects benefit the individuals and increase their survival probability at the same time that alleles remain in gene population, promoting their own propagation and so maximizing their representation in future generations. In this perspective fitness function of individuals and of genes are correlated functions, two visible parts of a global fitness function [14]. As in GA algorithm, there is a population of individual, each one with their own chromosome (association of genes) and a fitness function. In parallel structure, all the genes of the population are placed on (virtual) population of genes whose fitness value was inferred from the performance values of individuals of the population that have that gene. Both populations were evaluated and subject to a set of genetic operations.

EGA has been tested in trajectory planning problems within a continuous and non-convex space of solutions. The main task is to find a sequence of continuously linked straight segments that connect a point of departure and a point of arrival, avoiding obstacles scattered in the navigation region with reasonable margin of safety precautions [15]. An evolutionary algorithm based on selfish and altruistic strategies has been used to solve this kind of path-planning problem, with good results [16]. Here, three algorithms were used to solve hard path-planning problems. All of them use the same number of individuals from one population or divided among a set of *nPop* parallel populations (with migration process) and the same genetic operators functions. They are: GA, the standard genetics algorithms; MGA, multi-populations GA (with migration); EGA, the GA with genes evolution strategy, and MEGA, MGA (with migration) with genes evolution strategy.

This paper is organized as follows. The path-planning problem is described in Sect. 2. Next section presents the main structure of the Evolutionary Gene Algorithms (EGA). The main experimental results are presented in Sect. 4. These results are compared with the results of the other algorithms to verify the improvements of the evolutionary gene strategy. Finally, the last section presents the main conclusions of this study and of the proposed algorithm.

2   Path-Planning Problem

In this work, the path-planning problem is formulated as the task of determining a safe and continuous path connecting two points in a workspace, W: the starting and the ending points. It should avoid collision with known obstacles, while, if possible, maximizing the safety clearance distance to them with a minimum distance route.

Without losing the generality, we consider that the trajectory is composed by a set of connected line segments, which join the initial and final node of the trajectory passing through n intermediate ordered nodes, i.e., the set $N = \{P_0, P_1, \cdots, P_n, P_{end}\}$. Obstacles are independents straight segments, where O is the set of their pairs of endpoints. Let $S = \{s_1, \cdots, s_n, s_{end}\}$ be a sequence of linked $(n + 1)$ segments, where $s_i = \overline{P_{i-1}P_i}$ is the segment that connects two consecutive nodes, the $(i - 1)^{\text{th}}$ node with i^{th} node. N are inside of W region, with position values $P_i \in W$. Their optimal values are unknown, with the exception of P_0 and P_{end} points, whose values are given by initial problem conditions. In most cases, these values cannot be obtained by analytical means or their estimation task is very hard to do. Generally, only near optimal values can be computed trough strategies or algorithms based on meta-heuristics or by optimization methods.

The solution of the presented path-planning problem, $N^* = \{P_0, P_1^*, \cdots, P_n^*, P_{end}\}$ is part of all set of solutions in the continuous space $W^n = W \times \cdots \times W$, that maximize a set of criterions and restrictions that could be (or could not) formalized through a function. The resolution of this (not convex) optimization problem is almost of type NP-complex.

The main objective of path-planning algorithms is to find the optimal ordered nodes N of the trajectory that maximize the fitness function:

$$Fitness(N) = Safety(N, O) + Feasibility(N, O)/Dist(N) \tag{1}$$

where $Dist(N)$ is the length of trajectory, the sum of segments length of the path N.

This function is a sum of two parts: the *Feasibility* function, which it is correlated with the number of segments of trajectories that intersect segments of the obstacles set O, and a *safety* function, that measures features such as suavity and safety proximity of trajectory segments to obstacles segments.

The path N must be free of collisions. In this case, the *Feasibility* measure is a higher value, F_{max}. If all segments of the trajectory S collide with obstacles, the *Feasibility* is a value zero. So, the *Feasibility* function measures the free intersections number between the S and O, given by:

$$Feasibility(S, O) = F_{max} \left(1 - \frac{nc}{n+1}\right)^{\alpha} \tag{2}$$

where nc is the number of cross overlapping that happens between S and O segments and $\alpha \geq 1$ a shape factor, usually with value 1.

Straight line segments that link the initial-point and the end-point have the lowest distance value. However, in most cases that situation is not possible and $Dist(N) \geq \overline{P_0 P_{end}}$. Typically, this distance is about 2 to 10 times this value. The second term of Eq. (1) is a ratio between the feasibility and the distance of the trajectories, but where the sensibility to variable nc is higher and lower for the *Dist* variable. For that reason, $F_{max} > \overline{P_0 P_{end}}$.

Safety function values reflects the higher or lower proximity between the straight segments of the path S and straight segments of the set of obstacles O, given by:

$$Safety(N, O) = S_{max} \left(1 - \frac{1}{n} \sum_{s \in S} e^{-D_s/\sigma}\right) \tag{3}$$

where Ds is the minimal distance between the segment $s \subset S$ and the set of segments O, i.e., $D_s = \min(|s - o_j|)$, $\forall o_i \in O$, where $|s - o_j|$ is the minimal distance between s and o_j segments. Its value is near zero if all trajectory segments are very close or intersect obstacle segments and with value S_{max} when $D_s \gg \sigma$, situation where all segments of S are fairly distant from the segments O (i.e., for $D_s \gg \sigma$). Generally, the *Safety* function has lower value when compared with second part of Eq. (1).

Path-planning problem is here taken as an optimization problem that will be solved by GA, MGA, EGA and MEGA algorithms. They will be used to find the best way-points at right sequence to define the optimal path, by maximization the fitness Eq. (1). This task is hard complex because the choice of ith node is dependent of the choice made for the previous one, $i - 1$. The following parameters values were used: $F_{max} = 200$; $S_{max} = 20$, $\alpha = 1$, $\sigma = 5$ and $n = 20$ for an example with 20 randomly placed obstacles.

3 Evolutionary Algorithm

Evolutionary Gene Algorithms (EGA) use the GA structure, but they have a new additional (virtual) population whose elements are genes. These are collected from all chromosomes of the population.

In the context of the path-planning problems, a chromosome of the ith individual represents a potential solution given by a sequence of nodes, $N_i = \{P_0, P_{i,1}, \cdots, P_{i,n}, P_{end}\}$ codified by a set of genes G_i and with a performance value given by the fitness function:

$$F_i = Safety(N_i, O) + Feasibility(N_i, O)/Dist(N_i) \tag{4}$$

The fitness of the ith individual at k^{th} generation, $FI_i^{(k)}$, is given by a randomly weighed sum of $F_i^{(k)}$ with a sub-set of the most relevant genes, $G_i^* \subset G_i$:

$$FI_i^{(k)} = r \cdot F_i^{(k)} + (1 - r) \cdot \frac{1}{ng} \sum_{g \in G_i^* \subset G_i} FG_g^{(k-1)} \tag{5}$$

where $FG_g^{(k)}$ is the fitness value of gene $g \in G_i^* \subset G_i$. ng is the number of elements of the set G_i^*. The random parameters r have values in interval [0, 1] with density of probability of the uniform distribution, i.e., $r \in U([0,1])$.

The gene fitness value measures its performance in context of its historical behaviour as well as of its participation in the behaviours of news chromosomes, given by:

$$FG_g^{(k)} = \max\left(FG_g^{(k-1)}, \max_{i \supset g} F_i^{(k)}\right) \tag{6}$$

If g is a new gene, then $FG_g^{(k)} = 0$.

Moreover, if g mutates then its fitness value is computed based on a similarity factor with other genes. Let $S_{gh} \in [0, 1]$ be a similarity factor between gene g and gene h.

$$\text{If } S_{gh} > S_{thresh} \text{ and } FG_h > FG_g, \text{then } FG_g^{(k)} = FG_g^{(k)}(1 - S_{gh}) + S_{gh}FG_h^{(k)} \tag{7}$$

with S_{thresh} the threshold of similarity.

The processes of the proposed EGA are as follows:

Step 1. Create a random initial population or multi-population of n individuals.

Step 2. Evaluate the population through the fitness function (4).

Step 3. Calculate the fitness of genes with Eq. (6).

Step 4. Select individuals for reproduction based on fitness values given by (5).

Step 5. Crossover parent's chromosomes to produce a child solution. Transferred genes carry their performance values.

Step 6. Mutate some chromosomes of the population. Update the fitness of individuals and genes (7).

Step 7. Renew the population with these offspring individuals.

Step 8. Repeat steps 2–6 until the specified number of generations is reached.

For the selection process are used the *Tournament Selection* and *Roulette Selection* strategies are used. For genetic crossover operator has been used the "Natural" and "Real" crossover have been used. Four mutation operators we used: *Uniform, Border,*

Perturbation and *WeigthedGenesMutation*, the last one being used only by EGA and MEGA algorithms. With the last mutation function, genes with poor fitness have higher probability to mutate. Most of these methods and operators are well described in literature and are part of most practical implementations of evolutionary algorithms [17].

GA, EGA, MGA and MEGA are used to solve the same path-planning problem. Each gene consists of a sequence of pair-wise positive real values that represent a node of the trajectory. A chromosome is a sequence of genes, i.e. a sequence of waypoints of the path. Structures with one or multiple population (with migration facilities) are tested to solve this problem. The EGA and MEGA are the algorithms that incorporate the proposed evolutionary gene strategy.

4 Results

Conventional GA and MGA algorithms are used to solve the same complex path-planning problem. EGA and MEGA algorithms have the same structure of preceding algorithms, but incorporate evolutionary gene strategy. The total number of individuals was 250, divided among 10 populations of the multidimensional structures (MGA and MEGA). These algorithms incorporate migration processes with a probability value of 0.5%. Each chromosome has 22 elements, corresponding to 11 waypoints (nodes) N, all inside the workspace. Each experiment is executed until the 50th generation. The statistical results here presented are for the last generation, namely the mean fitness values, the means of maximum and minimum of fitness values of populations.

The workspace, W, has the square frontier border with length side length of 100 units, and 20 objects randomly placed inside W. This workspace is used in all experiments. Geometrical components are described in two dimensions (2D).

The problem has various feasible solutions, but to find the optimal value is not a trivial task. Moreover, this test example has enough pitfalls to make it difficult to execute most algorithms, including those of the evolutionary type.

GA results are shown in Fig. 1, where the red line represents the best solution. It was not able to find a feasible path, i.e., by avoiding obstacles. Moreover, the population evolutions converge for a restrict zone of W. With multi-population GA structure (MGA) the results are a little better and present a feasible solution, but with the same convergence problem of GA, as shown in Fig. 2. Results are significantly better when the algorithms incorporate the evolutionary gene strategy, as is the case of EGA and MEGA, as shown in Figs. 3 and 4 (right side). There, the best paths solutions found by each population are plotted, where red line is the best solution. All solutions have fitness values around 11.3. Most present solutions circumvent the obstacles safely with a minimum length path. Moreover, the genetic diversity of the population also ensures a more global demand for the solution, with each populations providing a good and un-repeated solution. MEGA needs 25 generations to achieve the average performance of value 10 (rising time, *rt*) whereas the EGA algorithm needs 42 generations. These and other values are presented in Table 1.

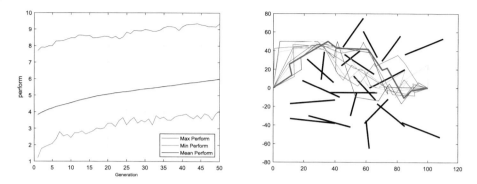

Fig. 1. Results of GA algorithms (1 population with 250 individuals): (Left) fitness value of populations (minimum, maximum and mean values). (Right) trajectories results.

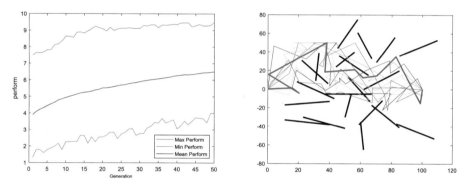

Fig. 2. Results of MGA algorithm (10 populations each with 25 individuals): (Left) fitness value of populations (minimum, maximum and mean values). (Right) trajectories results.

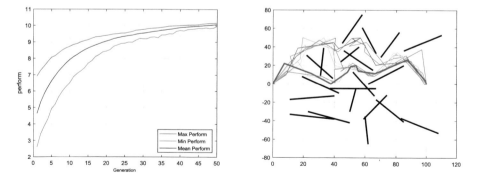

Fig. 3. Results of EGA algorithm (1 population with 250 individuals): (Left) fitness values of populations (minimum, maximum and mean values). (Right) trajectories results

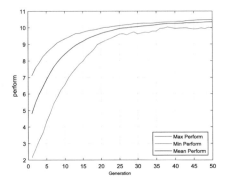
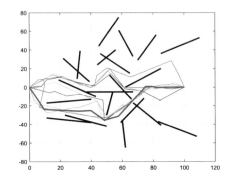

Fig. 4. Results of MEGA algorithm (10 populations each with 25 individuals): (Left) fitness values of populations (minimum, maximum and mean values). (Right) trajectories results.

Table 1. Main results of GA, EGA, MGA and MEGA algorithms.

Algorithm	Performance of BI	Mean performance of BI's	rt (level 10)
GA	9.803	–	>>50
EGA	10.624	–	>>50
MGA	11.368	11.212	42
MEGA	11.315	11.265	25

BI- best individual; *rt*- rising time;

5   Conclusion

In this paper, a new evolutionary gene strategy was proposed for improved evolutionary algorithm, as well-known AG and MGA algorithms. Essentially, this new strategy gives new field forces to the evolution mechanisms. Its impact is apparent on the improvement of algorithm speed, but mainly on the improvement of quality and numbers of genetics solutions, particularly when applied to solve complex problems. So, this approach was used on path-planning problems, in a continuous search space. It has shown to be effective in complex and interdependent sub-paths and evolution processes. GCE improved local sub-paths search as sub-processes that catalyse the CCE engine to find an optimal trajectory solution, a task that the standard genetic algorithm is not able to solve.

References

1. Simon, D.: Evolutionary Optimization Algorithms. Wiley, Hoboken (2013)
2. Holland, J.H.: Adaptation in Natural and Artificial Systems: An Introductory Analysis with Applications to Biology, Control, and Artificial Intelligence. MIT Press, Bradford Books Editions (1975). (Reprint, ISBN 978-0262581110, 1992)

3. Qu, B.Y., Zhu, Y.S., Jiao, Y.C., Wu, M.Y., Suganthan, P.N., Liang, J.J.: A survey on multi-objective evolutionary algorithms for the solution of the environmental/economic dispatch problems. Swarm Evol. Comput. **38**, 1–11 (2018)
4. Zelinka, I.: A survey on evolutionary algorithms dynamics and its complexity – mutual relations, past, present and future. Swarm Evol. Comput. **25**, 2–14 (2015)
5. Gregory, T.R.: Understanding natural selection: essential concepts and common misconceptions. Evol.: Educ. Outreach **2**(2), 156–175 (2009)
6. Godfrey-Smith, P.: Conditions for evolution by natural selection. J. Philos. **104**, 489–516 (2007)
7. Zeigler, D.: Natural selection. In: Zeigler, D. (ed.) Evolution. Academic Press, Chap. 2, pp. 9–22 (2014)
8. Efremov, V.V.: Equilibrium between genetic drift and migration at various mutation rates: simulation analysis. Russ. J. Genet. **41**(9), 1055–1058 (2005)
9. Burt, A., Trivers, R.: Genes in Conflict: The Biology of Selfish Genetic Elements. Belknap Press, Cambridge (2006)
10. Avise, J.C.: 1976 selfish genes. In: Avise, J.C. (eds.) Conceptual Breakthroughs in Evolutionary Genetics. Academic Press, Chap. 48, pp. 101–102 (2014). ISBN 9780124201668
11. Okasha, S.: Population genetics. In: Zalta, E.N. (ed.) The Stanford Encyclopedia of Philosophy (Fall 2015 Edition) (2015)
12. Sibly, R.M., Curnow, R.N.: Evolution of discrimination in populations at equilibrium between selfishness and altruism. J. Theor. Biol. **313**, 162–171 (2012)
13. Demsetz, H.: Seemingly altruistic behavior: selfish genes or cooperative organisms. J. Bioecon. **11**(3), 211–221 (2009)
14. Kleiner, K.: The selfish gene that learned to cooperate. New Sci. **191**(2564), 13 (2006)
15. Salgado, P., Igrejas, G., Afonso, P.: Hybrid PSO-cubic spline for autonomous robots optimal trajectory planning. In: INES 2017 of 21st International Conference on Intelligent Engineering Systems, pp. 131–136, 20–23 October, Larnaca, Cyprus (2017)
16. Salgado, P., Igrejas, G., Afonso, P.: Evolutionary based on selfish and altruism strategies - an approach to path planning problems. In: IEEE International Conference on Intelligent Systems (IS), Madeira, Portugal (2018)
17. Kramer, O.: Genetic Algorithm Essentials. Studies in Computational Intelligence. Springer, Cham (2017)

An Efficient and Secure Forward Error Correcting Scheme for DNA Data Storage

Anouar Yatribi[1(✉)], Mostafa Belkasmi[1], and Fouad Ayoub[2]

[1] ICES Team, ENSIAS, Mohammed V University in Rabat,
Rabat, Morocco
anouar.yat@gmail.com
[2] MTIC Team, LaREAMA Lab, CRMEF, Kenitra, Morocco

Abstract. In this paper, a new efficient error correcting scheme for DNA archival digital data storage is proposed. We devise a double protection scheme for DNA oligos, aiming to ensure the protection of both information and indexing header data from both symbol flipping and erasure-burst errors, using two different cyclic ternary difference-set codes, which are known to be completely orthogonalisable and very easy to decode using a simple majority-logic decoding algorithm. We show that the proposed scheme is efficient and easily scalable, and provides a coding potential of 1.97 bit per nucleotide, and a reasonable net information density of 0.75 bit/nt under the considered experimental conditions, with relatively a lower decoding complexity and costs compared to other DNA data storage approaches.

1 Introduction

Recently, DNA (Deoxyribonucleic Acid) based data storage systems has seen a large interest from research and industrial communities, due to the ever growing data density required by the latest technologies, as big data, IoT and high quality movies and holographic data. Indeed, the current magnetic based data storage mediums, namely optical, magnetic and digital cloud data storage, has proven their physical density limitations, and needs to be maintained regularly. It was reported by [1] that the total amount of data generated in 2016 has reached 16.1 Zeta Bytes, and will exponentially grow to reach 163 Zeta Bytes in the horizon of 2025. Clearly, the classical data storage mediums will fail to handle this challenge and must furthermore be replaced by other storage candidates. Indeed, a natural approach beyond the classical storage mediums, towards storing massive data into the DNA molecule has been developed [2]. This natural approach for data storage does prove the low-density, scalability and long-term stability that can provide DNA-based data storage systems.

A DNA molecule is defined as a nucleic acid that contains the genetic code of a living organism, and is composed from a sequence of nucleotides, each one taking a value from the four bases (A: Adenine, C: Cyanine, G: Guanine, T: Thymine). Nucleotides are organized into single or double stranded chains of

© Springer Nature Switzerland AG 2020
A. M. Madureira et al. (Eds.): SoCPaR 2018, AISC 942, pp. 226–237, 2020.
https://doi.org/10.1007/978-3-030-17065-3_23

DNA chunks, called oligonucleotides (or oligos). Theoretically, 1 gram per DNA can store 455 EB (Exabytes) of information in low-maintenance environments, with a long-term storing longevity [3], which outperforms by far the current classical digital storage mediums.

Many efforts has been devoted recently for storing data into the DNA code. In 2012, Georges Church and his team [4] proved experimentally the concept of storing data in DNA molecules, by storing 22 MB of data in DNA. In 2013, Goldman et al. [3] made the breakthrough by proposing and testing a new efficient approach for storing data of size 739 KB on DNA and retrieving it back. They proposed an efficient DNA encoding scheme with the given sequencing and synthesizing technologies in 2013 for storing archives of several MegaBytes (MB) in a 500–5000 years horizon, in low-maintained environments. Authors have used four folds redundancy to retrieve data from one of the DNA strands, and proposed the use of the ternary Huffman code and differential encoding for avoiding homopolymer runs. However, this approach provides an increase in the DNA length, which limits it for the commercial use due to the cost increase. Later, the synthesizing and reading-access techniques has been improved, giving the possibility to synthesize a larger amount of DNA data. Based on the Goldman's approach, many different approaches were proposed in the literature for error control coding in DNA data storage [5–10].

In this paper, we propose a new efficient encoding scheme for DNA archival data storage systems, by the use of ternary cyclic Difference-Set Codes (DSC) over $GF(3)$ (ie. the Galois Field of order 3). The proposed scheme aims to ensure at the same time the protection of information and indexing headers in DNA oligos for long-term storage in low-maintained environments. Additionally, simple low-cost decoding algorithms can be used for decoding the retrieved data.

The paper is organized as follows. In Sect. 2, we briefly describe the DNA channel model, and the modulation process for this channel, in the case of ternary symbols. Previous works on DNA forward error correction and the Goldman's encoding scheme [3] are also reviewed. Section 3 introduces the proposed scheme, where cyclic ternary DS codes are presented followed by the presentation of the proposed DNA encoding scheme. Section 4 presents performance analysis of the proposed DNA data storage scheme, with a comparison with the relevant previous works. Finally concluding remarks are presented in Sect. 5.

2 DNA Channels and Forward Error Correcting

2.1 Channel Model and Capacity

DNA data storage systems can be considered as a classical digital transmission over a noisy channel. The DNA information is transmitted over the channel by synthesizing DNA oligos. The information is received by sequencing the DNA oligos and decoding the sequenced sequence. The channel noise is caused by various experimental factors, including DNA synthesis imperfections, PCR dropouts, degradation of DNA molecules over time, and sequencing errors. In contrast to other classical theoretic channels, where the noise is identical and independently distributed, the error patterns in DNA depends essentially on the input sequence [10, 11].

It was shown that biological, bio-chemical and bio-physical processes are causing errors, while the physical and chemical effects introduces by itself errors to DNA oligos by the time [7]. Church and his team [4] gathered DNA channels characteristics by conducting several experimental analysis. Technically, 3 types of errors were observed in DNA channels. First, flipping errors (swapping) occurs when a DNA nucleotide symbol is replaced by another one. Additionally, insertion and deletion errors were also detected. Oligos that were not found in the DNA are called missing oligos. The obtained experimental results showed that the swap error rate lies approximately between $6.0.10^{-4}$ and $1.4.10^{-3}$, while insertion and deletion errors are 10^{-3} and 5.10^{-3} respectively [4]. Authors in [7] claimed that the DNA channel is a data memory-less channel. In [11], authors described the DNA channel as a constrained channel, concatenated to an erasure channel. Previous studies has shown that the homopolymer runs (consecution of identical DNA symbols) and GC content are the major constraints that impacts synthesis and sequencing errors.

Figure 1 illustrate the transmission model applied to data storage systems.

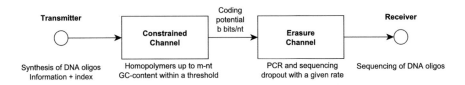

Fig. 1. DNA data storage channel model

The transmitter generate a synthesized DNA oligo of length N, which include information data and an indexing header. As DNA are organized in a mixed pool due to the multiplexing architecture of synthesis reactions and high throughput sequencing, it is important to index each oligo before transmitting it. We will denote the appended index header oligo by h_{index}. The synthesized oligo is transmitted over a constrained channel. Here an oligo is considered a valid sequence if its GC content is within $0.5 \pm c_{gc}$, which means that $50 \pm (100c_{gc})\%$ of GC content is allowed in the DNA sequence, and if its longest homopolymer length is up to m nucleotides, ie. there's at most m identical consecutive DNA symbols. Otherwise, the sequence is considered invalid and cannot be transmitted. Then valid sequences are exposed to an erasure channel with low dropout rate δ_v. Finally, the transmitted DNA oligos are sequenced for reading the received data. Additionally, forward error correction can be introduced, by appropriately choosing a code C for encoding data before the transmission, and decoding the received data after sequencing DNA oligos. Erlich et al. [11] established a theoretical study of the DNA channel capacity, by assigning realistic values to the parameters m, c_{gc}, h_{index} and δ_v into his model. Authors developed the analytic expression of the information capacity per nucleotide, as well as the coding potential, such that the constrained channel is taken into consideration, with the homopolymer and GC content constraints.

2.2   DNA Modulation

From the DNA channel characteristics mentioned above, a suitable modulation must be applied in order to limit the error propagation phenomena. A feasible modulation must take into consideration the following conditions :

- When a nucleotide symbol is erroneous, the error should be propagated to the minimum number of digits after the demodulation.
- In order to avoid homopolymers, the maximal run-length of similar nucleotides should be limited to 3.
- Self-complementary DNA sections has to be avoided because it causes amplification issues of the corresponding oligo, and also a significant information density loss.

Based on these constraints, an efficient modulation scheme was proposed in [7] for handling these limitations. Interested readers are referred to [7]. Also, the differential encoding (base 3 to DNA) technique used by Goldman [3], presented next in Table 1, provides satisfying results. However, this modulation technique is only useful when ternary symbols are considered for modulation.

2.3   DNA Forward Error Correction

Many academic and industrial research communities has proposed error-correcting schemes for DNA data storage, after that the pioneering works of Church's [4] and Goldman's [3] teams made a remarkable advance in long-term data storage. Using the next generation synthesizing and sequencing technology in 2012, Church [4] proposed on an efficient one bit per base encoding algorithm for storing digital information into a fixed length of DNA chunks (99 bases). For the header, flanking primers (headers) were inserted at the beginning and the end of information data in order to indicate the specific DNA segment in which the information data was encoded. A net information density of 0.83 bits per nucleotide was achieved. However, Church's approach suffers from the existence of homo polymers repeated DNA sequences that introduces writing and reading errors. This problem was resolved by Goldman in 2013 [3], by introducing the improved base-3 Huffman code with 3 symbols called trits (0, 1 and 2) in addition to differential encoding for data modulation. Thus, using Goldman's approach, binary data (one byte) was first encoded by the Huffman code, and then converted (modulated) to its corresponding DNA triplet where each trit digit was converted to one of the 3 nucleotides different from the last one, in order to avoid homopolymers. The differential encoding (modulation) technique used by Goldman to avoid homopolymers is presented in Table 1.

In the Goldman's scheme, a four redundancy protection system was used, where for each DNA strand of length 117 nt, comprising 114 nt for information and indexing data, and one parity symbol, plus a beginning and ending

Table 1. Goldman's base-3 to DNA modulation for avoiding repeated nucleotides

Previous nt	Next trit to encode		
	0	1	2
A	C	G	T
C	G	T	A
G	T	A	C
T	A	C	G

symbol. Two similar copies of the DNA strand were added and two odd indexed strands were reverse complemented. Additionally, prep-ending paddings and parity were used to allow error detection. The obtained net information density in this scheme was 0.34 bits per nucleotide.

Clearly, the Goldman's approach uses a large amount of redundancy to information data, providing lower DNA information density and higher costs.

This approach can be efficiently improved by introducing error correcting codes for the protection of oligos. In [7], authors has achieved a net information density of 0.92 bits per nucleotide by encoding information data using the Reed Solomon (RS) code $(n, k, d_{min}) = (255, 223, 31)$ over $GF(2^8)$, and separately encode the header using a strong binary BCH code with parameters $(63, 39, 9)$. In [12], the extended $(8, 4, 4)$ binary Hamming code, or repetition coding were used for encoding data. A sub-code $(11, 256, 5)_3$ of the ternary Golay code $(11, 6, 5)$ was proposed in [5] for encoding information data, and a DNA storage capacity of 115 ExaBytes (EB) was achieved. However, the indexing header remains unprotected, which impact error rates performance of the DNA storage medium. In [6], authors has introduced a double protection scheme, which consists of an inner Reed Solomon (RS) code over $GF(47)$ with parameters $(n, k) = (39, 33)$ for correcting individual errors, and an outer RS code over the extension field $GF(47^{30})$ with parameters $(n, k) = (713, 594)$ for the correction of erasures and errors from the inner decoder. The field $GF(47)$ was used in order to avoid homopolymers of length $m > 3$. Authors achieved a net information density of 1.14 bits per nucleotide, and claimed that their proposed encoding scheme is robustly suitable for digital data storage in DNA for thousands of years. Later, in [10], a Fountain Code technique was proposed to screen potential valid oligos to reach the maximal coding capacity of $b = 1.98$ bit/nt. 2 bytes of RS code redundancy was additionally added to protect both the seed and data payload. A net information density of 1.55 bits per nucleotide was achieved in this scheme.

3   The Proposed DNA Data Storage Scheme

3.1   Cyclic Difference-Set Codes over $GF(3)$

Difference-Set Codes (DSC) represents an infinite class of algebraic majority-logic decodable codes derived from finite geometries, namely from projective geometry (PG). DS codes were discovered in their cyclic form independently by Weldon [13] and Rudolph [14]. Weldon has shown that these codes are approximately powerful as BCH codes, with simpler decoding implementation. Later in [15,16], it was shown that OSMLD codes derived from finite geometries can be considered as finite geometry LDPC codes, and performs better than BCH codes and other linear block codes under the BP iterative decoding algorithm.

DS codes are completely orthogonalizables, as a consequence, the minimal distance is exactly defined by $d_{min} = J+1$, where J is the number of orthogonal parity-check equations on each symbol been decoded, providing that the code is capable to correct each error pattern of weight $t = \frac{J}{2}$ (t erroneous symbols) or less, by a simple majority-logic vote. Due to the orthogonality provided by the dual code, the length n of a cyclic DS code takes the form $n = J(J-1) + 1$.

The expressions of the code length n and the minimum distance d_{min} were given by Weldon [13].

For a prime p, and a positive integer $s > 0$, the code length, dimension, and minimum distance of a cyclic DS code over $GF(p)$ are given by:

$$n = p^{2s} + p^s + 1 \tag{1}$$

$$k = n - \left(\left[\binom{p+1}{2} \right]^s \right] + 1 \right) \tag{2}$$

$$d_{min} = p^s + 2 \tag{3}$$

When $p = 3$, a cyclic difference-set code over $GF(3)$ is defined by the following parameters:

$$n = 3^{2s} + 3^s + 1 \tag{4}$$

$$k = n - 6^s - 1 \tag{5}$$

$$d_{min} = 3^s + 2 \tag{6}$$

The construction of cyclic DS codes is derived from combinatorial objects called Difference-Sets, introduced by Singer [17], which in turn are derived from finite geometry, namely Projective Geometries, well investigated later in [18].

Table 2 displays a set of cyclic Difference-Set codes over $GF(3)$, constructed from $PG(m, q = 3)$. The difference-set D for constructing each code is only displayed for short codes due to spatial constraints.

Erasure-burst-correction capabilities of cyclic LDPC codes derived from two-dimensional finite geometries (EG (Euclidean Geometry) and PG) was investigated in [19]. Let σ be the zero-covering-spam of maximum length contained in the regular parity-check matrix H associated to a DS code C (considered also as an LDPC code). Then any erasure-burst of length $\sigma + 1$ or less is guaranteed to

Table 2. A set of ternary cyclic difference-set codes

s	(n, k, d_{min})	$r = \frac{k}{n}$	D
1	$(13, 6, 5)$	0.46	$\{0, 1, 3, 9\}$
2	$(91, 54, 11)$	0.59	$\{0, 1, 37, 39, 51, 58, 66, 69, 82, 86\}$
3	$(757, 540, 29)$	0.71	-
4	$(6643, 5346, 83)$	0.80	-

be recoverable regardless of its starting position. In fact, the erasure-burst capacity l_b of a DS code is lower-bounded by $l_b \geq \sigma + 1$. The erasure-burst-correction efficiency of the code C is given by:

$$\eta = \frac{l_b}{n - k} \tag{7}$$

When $\eta = 1$, and therefore $l_b = n - k$, then the code is said to be optimal for erasure-burst-correction.

3.2 Encoding Information Data Block Using the Code $(91, 54, 11)_3$

We propose a double protection to data and header chunks using a set of DS codes presented in Table 2. In this work, each byte is mapped to 6 ternary symbols (trits), that will be modulated to a DNA sequence following Table 1. Note that the use of ternary symbols avoids automatically the homopolymer runs constraint. To realize this scheme, a mapping table of ASCII symbols and a set of 256 ternary sequences of length 6 must be fixed from the set of possible sequences (729 sequences). Therefore, each data-block (DB) of length $k_1 = 54$ trits corresponds to 9 bytes of data information. Then each DB is encoded using the DS code C_1 with parameters $(n, k, d_{min}) = (91, 54, 11)$, which adds 37 trits of redundancy for protecting data.

For sake of scalability of the proposed scheme, we introduce a positive integer parameter $\lambda > 0$, which acts as a multiplier for defining the length N of oligos such that the available and current synthesis and sequencing technologies are capable to handle. Thus, the total length of encoded data to modulate into DNA is given by:

$$N = \lambda n \tag{8}$$

Thus, in each data block of length n, the code C_1 is capable to detect 10 error trits, and to correct any error pattern of 5 trits or less. Additionally, the code C_1 has an erasure-burst capacity for correcting configurations of burst erasures in the stored nucleotides. We note that the code C_1 can be replaced by a longer DS code if the available technologies are suitable for it.

3.3   Encoding the Index Header Using the Code $(13, 6, 5)_3$

Information data is lost if the indexing headers are corrupted. Henceforth, a strong protection to headers must be set up. In this work, we propose, for each oligo of length N, a prep-ending index of length $n_H = 13$, which corresponds to an encoded sequence of $k_H = 6$ trits using the DS code C_2 over $GF(3)$ with parameters $(n_H, k_H, d_{min_H}) = (13, 6, 5)$. Thus, with $k_H = 6$ trits, it is possible to index $3^6 = 729$ distinct oligos for each processed file. Note that, similarly to C_1, the code C_2 is also capable of correcting erasure-bursts. The weight enumerator polynomial $A_{C_2}(x)$ of the code C_2 is :

$$A_{C_2}(x) = 1 + 156x^6 + 494x^9 + 78x^{12} \tag{9}$$

With the given proposition, the total length l of protected oligos including its indexing headers is given by:

$$l = N + n_H \tag{10}$$

Consequently, with the proposed protection scheme, robust forward error correction is ensured for both data and indexing headers, while efficiency and scalability are provided.

The proposed encoding scheme for DNA data storage systems is fully described in Fig. 2.

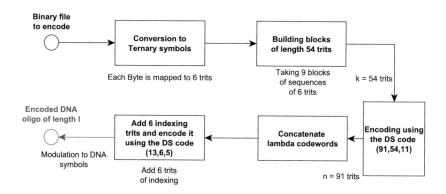

Fig. 2. The proposed DNA data encoding scheme

3.4   Decoding DNA Data

For decoding sequenced DNA oligos, each DNA chunk is demodulated to base-3, and the header is decoded first using C_2. Syndrome calculation can be performed using a simple division of the encoded sequence by the generator polynomial $g_2(x)$ of the code C_2. If errors are detected (non-zero syndrome), a majority-logic decoding is performed for correcting the header, or it can be decoded using the Maximum Likelihood (ML). Then, decoding of data information is performed by using the majority-logic decoding algorithm (OSMLGD), for decoding each of the λ codewords of C_1 contained in the oligo of length N.

4   Performance Analysis

We propose to evaluate the proposed forward error correcting scheme for DNA data storage. Performance analysis of the proposed scheme is performed based on the coding potential and the net information density. Note that, due to the absence of the experimental materials, the obtained results are based on an analytical study as in [11], which tends to approximate the realistic case.

4.1   Coding Potential

Suppose that a DNA sequence is valid only if its GC content is within $0.5 \pm c_{gc}$, and its longest homopolymer length is up to m nucleotides.

The coding potential, b, represents the entropy of each nucleotides in valid sequences. The coding potential per nucleotide is defined based on the GC and homopolymer constraints, and is given by :

$$b = 2 - \frac{3\log_2 e}{4^{m+1}} - \frac{\log_2\left[2\Phi(2\sqrt{l}c_{gc}) - 1\right]}{l} - \frac{\log_2 K}{l} \text{ (bit/nt)} \qquad (11)$$

where l is the oligo length, m is the maximum homopolymer length, K is the index length and $\phi(.)$ is the cumulative function of a standard normal distribution. In this study, a conservative set of constraints has been chosen, as in [11], given by the following parameters : $m = 3$ and $c_{gc} = 0.05$. The oligo length of $l = (3 \times 91) + 13 = 286$ (nts) and $\lambda = 3$ are set to match our proposed scheme.

By the given parameters, the obtained coding potential is $b = 1.97$ bits per nucleotide.

4.2   Net Information Density

In the proposed scheme, each byte is encoded into 6 nucleotides with a coding potential of $b = 1.97$ bit/nt. A set of λ blocks of nine bytes is encoded into $\lambda n + n_H$ nucleotides. Therefore, the net information density D_n of the proposed scheme is calculated by:

$$D_n = \frac{72\lambda}{\lambda n + n_H} \text{(bit/nt)} \qquad (12)$$

In this work, we have $n = 91$, $n_H = 13$ and λ is a parameter to be optimized used for improving the net density, which depends on the current sequencing technologies. When using $\lambda = 2$, a net density of 0.74 bit per nucleotide is achieved. For $\lambda = 3$, the net density is 0.75 bit/nt. Note that the use of longer codes with higher rates can further increase the net information density of the proposed scheme.

4.3   Comparison with Previous Works

Table 3 depicts a comparison between the proposed scheme and various DNA storage encoding schemes proposed in the literature. The comparison is performed based on a set of parameters, such that a subset of these parameters are defined based on realistic experience. Due to the absence of experimentation for the proposed scheme, the fields associated to these parameters in Table 3 are omitted. Table 3 includes the length of input data in the experiment, the coding potential, redundancy, robustness of dropouts, error detection and correction existence in the scheme, and the full recovery parameter which indicates if the DNA oligos were completely recovered in the experiment. Also the net and physical information densities are included, in addition to the realized capacity and the number of oligos used in the experiment. We observe that the proposed scheme is comparable with the other approaches, at the cost of lower complexity. In addition, it is shown that the proposed scheme outperforms the Goldman's approach in performance with a lower required complexity. In contrast to the methods proposed by Church [4] and Goldman [3], our scheme does not need redundant copies of the DNA sequences. Also, as ternary error correcting codes are used, the use of screening techniques for choosing valid sequences, as proposed in [10], is not required.

Table 3. Comparison of DNA storage encoding schemes

Parameter	Church et al. [4]	Goldman et al. [3]	Grass et al. [6]	Bornhort et al. [8]	Blawat et al. [7]	Erlich et al. [10]	This work
Input data (Mbytes)	0.65	0.75	0.08	0.15	22	2.15	-
Coding potential (bits/nt)	1	1.58	1.78	1.58	1.6	1.98	1.96
Redundancy	1	4	1	1.5	1.13	1.07	1.05
Robustness to dropouts	No	Repetition	RS	Repetition	RS	Fountain	DSC
Error detection and correction	No	Yes	Yes	No	Yes	Yes	Yes
Full recovery	No	No	Yes	No	Yes	Yes	-
Net information density (bits/nt)	0.83	0.33	1.14	0.88	0.92	1.57	0.75
Realized capacity	45%	18%	62%	48%	50%	86%	
Number of oligos	54,898	153,335	4,991	151,000	1,000,000	72,000	-
Physical density (Pbytes/g)	1.28	2.25	25	-	-	214	-

5   Concluding Remarks

We have proposed a robust cost-effective DNA encoding scheme with a net information density of 0.75 bit per nucleotide, under the consideration of realistic parameters, allowing to detect and correct errors in DNA oligos, using cyclic ternary difference-set codes. The proposed system aims to ensure the protection of both information and indexing headers from erasures and symbol errors by the use of two different DS codes, with a simple majority-logic decoding process providing a significant reduced data storage and decoding cost, and a high DNA information density. As shown in the paper, the proposed scheme is scalable and may be improved by different ways, based on the optimization of the parameters involved in the designed system. With the current evolution of synthesizing and sequencing technologies, longer codes with higher correction capacities can be considered in the proposed scheme for future DNA data storage mediums.

References

1. Reinsel, D., Gantz, J., Rydning, J.: Data age 2025: the evolution of data to life-critical. Don't Focus on Big Data (2017)
2. Baum, E.B.: Building an associative memory vastly larger than the brain. Science **268**(5210), 583–585 (1995)
3. Goldman, N., Bertone, P., Chen, S., Dessimoz, C., LeProust, E.M., Sipos, B., Birney, E.: Towards practical, high-capacity, low maintenance information storage in synthesized DNA. Nature **494**(7435), 77 (2013)
4. Church, G.M., Gao, Y., Kosuri, S.: Next generation digital information storage in DNA. Science **337**(6102), 1628 (2012)
5. Limbachiya, D., Dhameliya, V., Khakhar, M., Gupta, M.K.: On optimal family of codes for archival DNA storage. arXiv preprint arXiv:1501.07133 (2015)
6. Grass, R.N., Heckel, R., Puddu, M., Paunescu, D., Stark, W.J.: Robust chemical preservation of digital information on DNA in silica with error-correcting codes. Angew. Chem. Int. Ed. **54**(8), 2552–2555 (2015)
7. Blawat, M., Gaedke, K., Hutter, I., Chen, X.-M., Turczyk, B., Inverso, S., Pruitt, B.W., Church, G.M.: Forward error correction for DNA data storage. Procedia Compute. Sci. **80**, 1011–1022 (2016)
8. Bornholt, J., Lopez, R., Carmean, D.M., Ceze, L., Seelig, G., Strauss, K.: A DNA based archival storage system. ACM SIGOPS Oper. Syst. Rev. **50**(2), 637–649 (2016)
9. Jain, S., Hassanzadeh, F.F., Schwartz, M., Bruck, J.: Duplication-correcting codes for data storage in the DNA of living organisms. In: 2016 IEEE International Symposium on Information Theory (ISIT), pp. 1028–1032. IEEE (2016)
10. Erlich, Y., Zielinski, D.: DNA fountain enables a robust and efficient storage architecture. Science **355**(6328), 950–954 (2017)
11. Erlich, Y., Zielinski, D.: Capacity-approaching DNA storage. bioRxiv, pp. 074237 (2016)
12. Heider, D., Barnekow, A.: DNA-based watermarks using the dna-crypt algorithm. BMC Bioinf. **8**(1), 176 (2007)
13. Weldon Jr., E.J.: Difference-set cyclic codes. Bell Syst. Tech. J. **45**(7), 1045–1055 (1966)
14. Rudolph, L.D.: Geometric configurations and majority logic decodable codes. Ph.D. thesis, MEE-University of Oklahoma (1964)
15. Lucas, R., Fossorier, M.P.C., Kou, Y., Lin, S.: Iterative decoding of one-step majority logic decodable codes based on belief propagation. IEEE Trans. Commun. **48**(6), 931–937 (2000)

16. Kou, Y., Lin, S., Fossorier, M.P.C.: Low-density parity-check codes based on finite geometries: a rediscovery and new results. IEEE Trans. Inf. Theor. **47**(7), 2711–2736 (2001)
17. Singer, J.: A theorem in finite projective geometry and some applications to number theory. Trans. Am. Math. Soc. **43**(3), 377–385 (1938)
18. Shu, L., Costello, D.J.: Error Control Coding. The Second International edition, Prentice-Hall, pp. 704–712 (2004)
19. Ryan, W., Lin, S.: Channel Codes: Classicaland Modern. Cambridge University Press, Cambridge (2009)

A Blockchain-Based Scheme for Access Control in e-Health Scenarios

João Pedro Dias$^{(\boxtimes)}$, Hugo Sereno Ferreira, and Ângelo Martins

Faculty of Engineering, INESC TEC and Department of Informatics Engineering,
University of Porto, Rua Dr. Roberto Frias, Porto, Portugal
{jpmdias,hugosf}@fe.up.pt, angelo.martins@inesctec.pt

Abstract. Access control is a crucial part of a system's security, restricting what actions users can perform on resources. Therefore, access control is a core component when dealing with e-Health data and resources, discriminating which is available for a certain party. We consider that current systems that attempt to assure the share of policies between facilities are mostly centralized, being prone to system's and network's faults and do not assure the integrity of policies lifecycle. Using a blockchain as store system for access policies we are able to ensure that the different entities have knowledge about the policies in place while maintaining a record of all permission requests, thus assuring integrity, auditability and authenticity.

Keywords: e-Health · Access control · Blockchain · Distributed Ledger Technology · Security · Distributed systems

1 Introduction

Healthcare is experiencing an explosion of data partially due to the widespread of health data collection systems such as wearables (e.g. fitness trackers) [14], health tracking applications (e.g. diet tracking) [17], ambient assisted living systems such as CAALYX (Complete Ambient Assisted Living Experiment) [2] and other Internet-of-Things based systems. By now, it is estimated that medical data will grow at a rate of 48% per year, reaching 2.3 zettabytes by the year of 2020 [12].

The explosion of data being collected and, *a posteriori*, analyzed by different entities, leads to the debut of data security and privacy issues. By one hand, these issues are taken into account because such smart devices may be connected to the Internet at some point for accessing its collected data anytime and anywhere [18]. The data being collected from those devices may be part of the Personal Health Records (PHR) and this is typically owned by the patients and may be or not, shared with third-entities [19]. On the other hand, Electronic Medical Records (EMR) and Electronic Health Records (EHR) store individual information that is required by the healthcare professionals and may be shared among different institutions and facilities [19].

A. M. Madureira et al. (Eds.): SoCPaR 2018, AISC 942, pp. 238–247, 2020.
https://doi.org/10.1007/978-3-030-17065-3_24

Hence, there is the need to control the accesses to this data resources by third-entities. Access control is concerned about determining the allowed activities of certain users, mediating every attempt by a user to access a resource in the system [10]. Dealing with the user access control to health data, personal or medical, held by different parties, which may be required to be accessed by third-parties with different goals (e.g. insurance companies *versus* doctors), is not an easy task. This is especially problematic when we are still moving towards a unified and interoperable electronic health (e-Health) systems [16].

In this paper, we suggest an approach to the problem of access control in large scale and distributed systems, such as e-Health, where different entities and users should be able to access data with different permission levels and granularities. The `Data Keepers` should be able to manage the accesses to their data by the means of adding, changing or revoking permissions. Such system should be also capable of defining fine-grained permissions both, at the user level and, at the resource level.

This paper is structured as follows: firstly, it is given an overview over the related work in the scope of permission management in e-Health systems, focusing also blockchain approaches for access control. Afterwards is it given a description of the purposed solution architecture. Then we address some core details of the *proof-of-concept* implementation. Finally, some final remarks are presented, summing up the contributions as well as pointing out further developments.

2   Background and State of the Art

2.1   Blockchain

Distributed Ledger Technology (DLT) consists of a consensus of replicated, shared and synchronized digital data distributed along a set of nodes, working as a distributed database, generally geographically dispersed [6]. It is important to note that, despite all blockchains being distributed ledgers, not all distributed ledgers are blockchains.

The blockchain is a specific type of distributed ledger conceptualized by Satoshi Nakamoto and used as a core component of the digital currency Bitcoin [15]. Data in a blockchain should be tamper-proof, specifically accomplished by the use of cryptography, by the means of digital signatures and digital fingerprints (hashing) [15]. Also, consensus must be assured among peers considering scenarios where some of the peers are providing erroneous data, by partially or completed computer/network failures or, even, by malicious intent when some party tries to subvert the ledger [15].

A blockchain consists of a chain of blocks that contains information about transactions. Each one of these transactions, is digitally signed by the entity emitting them. It can be used as a state transaction system (state machine), where there is a state that corresponds to the snapshot of the chain (the result of all transaction until now) and, after adding a new block of transactions to the chain, we got a new snapshot that corresponds to a new state of the system, as result of the new transactions [8].

In order to validate a block, it is necessary a *proof-of-work*, that is used in order to get a consensus in the peer-to-peer network [15]. There are alternatives to *proof-of-work*. In the *proof-of-stake*, as it is being considered to be used in Ethereum [8], the creator of the next block to be pushed in the chain is chosen in a deterministic way based on the wealth of the node [6]. Another one, as used in the Sawtooth Lake [5], uses a *Proof of Elapsed Time* (PoET), which is a lottery-based consensus protocol that takes advantage of the trusted execution environments provided by Intel's Software Guard Extensions.

Blockchains can be considered of three main kinds, as stated by Buterin [3], namely: public, fully-private and consortium. Public blockchains (e.g. Bitcoin), is a type of blockchain in which anyone can read, send transactions to and expect to see them included if they are valid, and, further, anyone in the world can participate in the consensus process. Fully-private blockchains' consist of blockchains where write permissions are kept centralized to one organization (even if spread among facilities), existing a closed group of known participants (e.g. a supply chain) [20]. Finally, consortium blockchains, are partly private in such way that the consensus process is controlled by a number of pre-selected set of nodes. In this type of blockchain, the right to query the blockchain may be public, or restricted to the participants (e.g. governmental institutions and partners).

2.2 Access Control

One of the more common approaches is the use of Access Control Lists (ACL), commonly used in modern operating systems. ACL consists of a list associated with an object that specifies all the subjects that can access it, along with the access level (or rights) [10]. Other systems use Access Control Matrix, in which, each row represents a subject, each column an object and each entry is the set of access rights for that subject to that object [10].

Specifically in healthcare, Role-Based Access Control (RBAC) and Attribute-Based Access Control (ABAC) have been applied [13]. RBAC defines the user's access right basing itself on his/her roles and the role-specific privileges associated with them. The ABAC system extends the RBAC role-based features to attributes, such as properties of the resource, entities, and the environment [11]. Policies in ABAC can be expressed resorting to the eXtensible Access Control Markup Language (XACML), defined by the OASIS consortium [9]. The XACML standard also includes a reference architecture for designing and implement access control systems, defining the system components and usage flow, that can be used in multiple application domains.

Another used approach to access control is the Entity-Based Access Control (EBAC) [1], which allows the definition of more expressive access control policies. This is accomplished by supporting both, the comparison of attribute values, as well as traversing relationships along arbitrary entities. Moreover, Bogaerts et al. presents *Auctoritas* as an authorization system that specifies a practical policy language and evaluation engine for the EBAC system [1].

2.3   Blockchain Applied to Access Control

Several approaches to resolve the access control issue based on blockchain appeared, including in e-Health scenarios. Di Francesco Maesa et al. [7] proposes a blockchain-based access control, implementing ABAC on top of the blockchain technology, following the XACML reference architecture. This approach validates itself through a reference implementation on top of Bitcoin. However this solution does not encompass the particularities of using such in e-Health context, namely, the possibility of having different authorities and/or entities as resource owners.

In the application of blockchain for e-Health, Yue et al. [21] proposes the use of a *Healthcare Data Gateway* (HGD) to enable the patient to own, control and share their data while maintaining data privacy. This solution also encompasses that all the patient's e-Health record is stored in a blockchain. Although the novelty of such approach, it implies a disruptive change on the already-existent systems of storing and retrieving e-Health data, what would require a considerable effort to implement which may call into question its current applicability. Also, there are cases where it is needed to access data without the explicit agreement from the patient itself (e.g. due to the patient inability to allow the access or by some governmental requirements) and this solution does not provide the ability to do such (e.g. some family member allow the data access). Also, considering the growth of e-Health data, storing this data on the blockchain itself will result in a rapid growth on its size, exceeding publicly available hard drive capacity, requiring special hardware to full nodes and, further, could lead to the centralization of the blockchain [4].

3   Proposed Scheme

The approach consists of using Blockchain technology as a way to accomplish a more reliable and user-empowered solution for access control management in an e-Health environment. Such approach allows us to define fine-grained access control while maintaining the consensus in a distributed system, authenticity, immutability and auditability.

This solution is somehow similar to the Access Control Matrix, which allows the establishment of a correspondence between a subject, an object and a set of rights. However, this information is not stored as is, due to the inherent proprieties of the use of blockchain. As a transaction-based state machine, we store transactions corresponding to a pre-defined set of the state machine on the Access Policies repository.

3.1   Access Control Model

Before defining the model, it is needed to define all the entities and relationships enrolled in it. Such model is given in Fig. 1, and five classes can be identified in this approach.

First off, we have *Entities* that can be 3rd Party's or Data Keepers. Further, we have Policies and Records. Each Policy refers to a relation between one and only one 3rd Party and an e-Health Record, with the respective level of access, PermissionLevel. In turn, each Record can have one or more Data Keepers that have partial or total ownership over it.

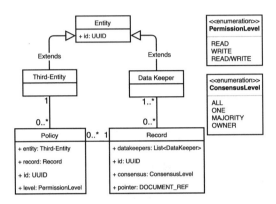

Fig. 1. UML diagram specifying the classes of the system and their relationships.

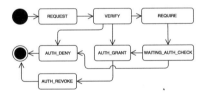

Fig. 2. State machine diagram representing the lifecycle of an access request by a 3rd Party. The composite state AUTH_CHECK represents the individual authorization requests needed by the Data Keepers of the record being queried. Each transition depends on external interaction by an actor of the system.

The information is preserved by the means of storing transactions in the blockchain. The transactions contain information about 3 different state machines, that have dependencies between them, being always related to the class model defined.

Access Policy State Machine, as represented in Fig. 2, is the state machine related to the main logic of creating Access Policies.

Record Life-cycle State Machine is responsible for the operations over an e-Health record. The record lifecycle begins with its creation, CREATE, then it can suffer diverse updates, UPDATE, until it is removed, REMOVE.

Individual Authorization State Machine describes the lifecycle for each user access grants over a given REQUIRE, which lead to a number of instantiations equal to the number of `Data Keepers` required. Each individual instantiation evokes a `REQUIRE_ACTION`, then the `Data Keeper` can allow (`AUTH_GRANT`) or deny (`AUTH_DENY`) the access. Eventually, before reaching the final state, the `Data Keeper` can revoke (`AUTH_REVOKE`) a previously granted authorization.

As stated, the main logic of attributing Access Policies for 3rd `Party` access an e-Health record is controlled by the *Access Policy State Machine*, detailed in Fig. 2. This state machine jumps the `init` state when an access request, `REQUEST` from a 3rd `Party`'s enters the PEP. Then, the `REQUEST` is verified, `VERIFY`, against the already existent information on the blockchain (by the means of a snapshot operation). If, and only if, the information about this particular access is present in the snapshot, the request can be granted, `AUTH_GRANT` or denied `AUTH_DENY`.

Additionally, if there is no information about an access request, the Access Policy must be required, `REQUIRE`, by the means of checking the number of permissions required form the `Data Keepers` in order to get a consensus (specific to the record), using for that the *Individual Authorization State Machine*. This means that there is no central authority authorizing requests from 3rd `Party`'s, and it is required that some set of `Data Keepers` allow the access. While this process is running, the state machine enters into a waiting state, `WAITING_AUTH_CHECK`.

At last, there can be a point in the future when it is needed to revoke a previously granted access, `AUTH_REVOKE`. The `final` state is, by this, reached by the existence of an `AUTH_REVOKE` or an `AUTH_DENY`.

3.2   Block Model

The basic structure of a block in the blockchain is the following:

- `index`: Corresponds to the index of the current block in the blockchain.
- `timestamp`: Timestamp corresponding to when the block was generated.
- `previousHash`: Hash of the previous block in the chain.
- `digitalSign`: Digital signature of the current block data.
- `data`: Content of the block. Corresponds to a set of transactions describing the access control policies, records information and individual authorizations.
- `nonce`: Value that is set so that the hash of the block will contain a run of leading zeros. This value is calculated iteratively until the required number of zero bits in the `hash` is found. This requires time and resources, making it so that the correct *nonce* value constitutes *proof-of-work*.
- `hash`: A SHA256 hash corresponding to the block data. This hash must have a leading *a priori* defined sequence being this leading sequence what defines the effort of the *proof-of-work*.

Additionally, the `data` field should be detailed, as it is the field that serves as transaction information storage. This `data` field includes in it three sub-fields, namely: (a) `records`: Transaction information relative to transactions of the state machine about creation, update or deletion of e-Health records of any kind. (b) `policies`: Transaction information relative to transactions of the state machine presented in Fig. 2, about creation and revocation of access policies. (c) `individualAuths`: Transactions about individual authorization by each one of a `Record Data Keeper` in relation to each `Policy`. The use of hashes allows us to maintain integrity along the immutable chain of transactions without a central authority, since any change in the data would result in a different hash, invalidating the next blocks in the blockchain. Additionally, as result, we can also achieve accountability and auditability. Authenticity is assured by the assign of a key-pair to each entity with access to the blockchain, identifying who write each block in the chain.

3.3   Architectural Design

At the system architectural level, the blockchain, being distributed by default, works as a peer-to-peer network connecting the different nodes, corresponding to the diverse facilities or organizations that can store, create or/and change e-Health data.

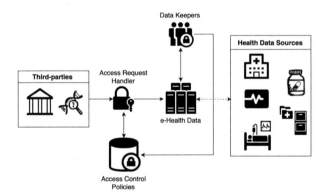

Fig. 3. Overview of the system architecture and interactions. The e-Health data access requests treated seamlessly by the handler. The e-Health data is associated with one facility or repository, but all the available resources are known by all nodes.

The system implements a consortium blockchain [3], which means that the blockchain is partly private, in the way that only several entities can carry the validation process. The advantage of using blockchain is that these entities can and will typically have distrustfulness between them (e.g. health clinics and insurance companies). The `Data Keepers` and 3rd `Party` can then interact with the system by the means of using public available Application Programming Interfaces (API's) or applications designed to do such.

Furthermore, despite the use of XACML standard for access control systems and blockchain as storage, there are some architectural choices due to the e-Health domain restrictions and more complex use cases. An overview can be observed in Fig. 3.

Focusing on the *e-Health Data*, this data is not typically aggregated in one storage, being spread by multiple institutions and organizations. As such, every time that any 3rd Party requests access to a specific record there is the need of locating this information, and then, proceed to check if the request is already approved or if there is the need to create a new access policy. So, as an improvement, information about the creation of new records must be kept and spread along all the organizations and facilities in such way that a request to a resource can be handled by any member of the private blockchain.

Additionally, aiming attention to the Data Keepers, it was noticed that there is a set of situations where the ownership of e-Health data records is not explicit to only one entity but shared among more than one entity or individual, as is the case of EMR *versus* PHR. Taking this into account, we set up a mechanism of consensus when creating new access policies. Each e-Health record has a level of agreement that must be achieved before allowing a 3rd Party the access to a Resource, being this level associated with the Resource itself (Fig. 1). Then, there is a number of executions of the individual authorization state machine corresponding to the number of that Record Data Keepers. Reaching the minimum number of individual authorization (that can be either AUTH_GRANT or AUTH_DENY), the access request state machine (Fig. 2) will create an access policy accordantly with the consensus reached (that can be, once more, either AUTH_GRANT or AUTH_DENY).

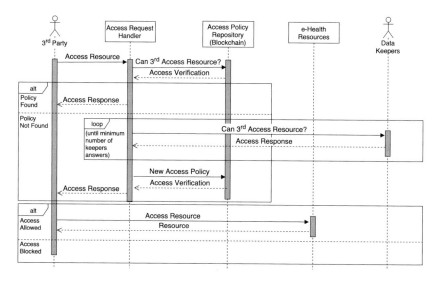

Fig. 4. Sequential view on how some 3rd Party can access or request access to an e-Health resource, detailing the communication between the inner modules of the architecture. It is also visible the process of creation of new access control policies.

From the functional architecture viewpoint, we can sum up the system interaction as stated in the sequence diagram of the Fig. 4. This diagram describes the process of a 3rd `Party` requesting access to an e-Health `Record`, owned partially or totally by one or more `Data Keepers`. Further, the diagram describes both the case of checking against an already existent access policy of the 3rd `Party` and the `Record` or the process of creating a new `Policy` by checking the necessary number of `Data Keepers`.

4   Conclusions

In this paper, we presented an approach to solving the problem of managing access control in the e-Health ecosystem. Access Control is a special complex task in e-Health since resources and data are distributed among different facilities and institutions. Further, this is even more problematic because in some cases, e-Health resources are not owned or managed by a single entity or individual. As a way of overcoming this complexity, we propose an approach that leverages the use of blockchain for store transactional information about e-Health records and access control policies.

For purposes of supporting the plausibility of the scheme proposal, a *proof-of-concept* was designed and implemented. This *proof-of-concept* allowed us to make some, even if preliminary, tests and validations over the sanity of the approach from a functional and applicational perspective.

Overall, we determine that the approach is viable, giving diverse advantages when comparing to the in-place systems. This advantages although, not limited to, includes the integrity, transparency, and authenticity of the access control policies in the system, being, this information distributed and synchronized by all the institutions and organizations that make part of the consortium.

Acknowledgment. This work was supported by Project "NanoSTIMA: Macro-to-Nano Human Sensing: Towards Integrated Multimodal Health Monitoring and Analytics/NORTE-01-0145-FEDER-000016" is financed by the North Portugal Regional Operational Programme (NORTE 2020), under the PORTUGAL 2020 Partnership Agreement, and through the European Regional Development Fund (ERDF).

References

1. Bogaerts, J., Decat, M., Lagaisse, B., Joosen, W.: Entity-based access control: supporting more expressive access control policies. In: Proceedings of the 31st Annual Computer Security Applications Conference, ACSAC 2015, pp. 291–300. ACM, New York (2015)
2. Boulos, M.N.K., Rocha, A., Martins, A., Vicente, M.E., Bolz, A., Feld, R., Tchoudovski, I., Braecklein, M., Nelson, J., Laighin, G.Ó., et al.: CAALYX: a new generation of location-based services in healthcare. Int. J. Health Geogr. **6**(1), 9 (2007)
3. Buterin, V.: On public and private blockchains, August 2015. https://blog.ethereum.org/2015/08/07/on-public-and-private-blockchains/. Accessed 06 June 2017

4. Chepurnoy, A., Meshkov, D.: On space-scarce economy in blockchain systems. IACR Cryptology ePrint Archive 2017, 644 (2017)
5. Intel Corporation: Sawtooth lake latest documentation (2015). https://intelledger.github.io/. Accessed 06 Feb 2017
6. Deloitte: Bitcoin, blockchain & distributed ledgers: caught between promise and reality. Technical report, Centre for the Edge, Australia (2015)
7. Di Francesco Maesa, D., Mori, P., Ricci, L.: Blockchain based access control. In: Distributed Applications and Interoperable Systems: 17th IFIP WG 6.1 International Conference, pp. 206–220 (2017)
8. Buterin, V.: Ethereum: a next-generation smart contract and decentralized application platform (2014)
9. Godik, S., Moses, T.: OASIS extensible access control markup language (XACML). OASIS Committee Specification cs-xacml-specification-1.0 (2002)
10. Hu, V.C., Ferraiolo, D., Kuhn, D.R.: Assessment of access control systems. US Department of Commerce, National Institute of Standards and Technology (2006)
11. Hu, V.C., Ferraiolo, D., Kuhn, R., Friedman, A.R., Lang, A.J., Cogdell, M.M., et al.: Guide to attribute based access control (ABAC) definition and considerations (draft). NIST Special Publication 800(162) (2013)
12. IDC: The digital universe: driving data growth in healthcare. Report, EMC Corporation and International Data Corporation (2014)
13. Li, M., Yu, S., Ren, K., Lou, W.: Securing personal health records in cloud computing: patient-centric and fine-grained data access control in multi-owner settings. In: Social-Informatics and Telecommunications Engineering. Lecture Notes of the Institute for Computer Sciences, pp. 89–106 (2010)
14. Lukowicz, P., Kirstein, T., Troster, G.: Wearable systems for health care applications. Methods Inf. Med.-Methodik der Inf. der Med. 43(3), 232–238 (2004)
15. Nakamoto, S.: Bitcoin: a peer-to-peer electronic cash system, p. 9 (2008)
16. Nijeweme-d'Hollosy, W.O., van Velsen, L., Huygens, M., Hermens, H.: Requirements for and barriers towards interoperable ehealth technology in primary care. IEEE Internet Comput. 19(4), 10–19 (2015)
17. Patrick, K., Griswold, W.G., Raab, F., Intille, S.S.: Health and the mobile phone. Am. J. Prev. Med. 35(2), 177 (2008)
18. Tan, L., Wang, N.: Future internet: the internet of things. In: 2010 3rd International Conference on Advanced Computer Theory and Engineering (ICACTE), vol. 5, pp. V5-376–V5-380, August 2010
19. Tang, P.C., Ash, J.S., Bates, D.W., Overhage, J.M., Sands, D.Z.: Personal health records: definitions, benefits, and strategies for overcoming barriers to adoption. J. Am. Med. Inform. Assoc. 13(2), 121–126 (2006)
20. Underwood, S.: Blockchain beyond bitcoin. Commun. ACM 59(11), 15–17 (2016)
21. Yue, X., Wang, H., Jin, D., Li, M., Jiang, W.: Healthcare data gateways: found healthcare intelligence on blockchain with novel privacy risk control. J. Med. Syst. 40(10), 218 (2016)

Blockchain-Based PKI for Crowdsourced IoT Sensor Information

Guilherme Vieira Pinto$^{(\boxtimes)}$, João Pedro Dias, and Hugo Sereno Ferreira

Faculdade de Engenharia da Universidade do Porto,
Porto, Portugal
{up201305803,jpmdias,hugosf}@fe.up.pt

Abstract. The Internet of Things is progressively getting broader, evolving its scope while creating new markets and adding more to the existing ones. However, both generation and analysis of large amounts of data, which are integral to this concept, may require the proper protection and privacy-awareness of some sensitive information. In order to control the access to this data, allowing devices to verify the reliability of their own interactions with other endpoints of the network is a crucial step to ensure this required safeness. Through the implementation of a blockchain-based Public Key Infrastructure connected to the Keybase platform, it is possible to achieve a simple protocol that binds devices' public keys to their owner accounts, which are respectively supported by identity proofs. The records of this blockchain represent digital signatures performed by this Keybase users on their respective devices' public keys, claiming their ownership. Resorting to this distributed and decentralized PKI, any device is able to autonomously verify the entity in control of a certain node of the network and prevent future interactions with unverified parties.

Keywords: Internet of Things · Blockchain ·
Public Key Infrastructure

1 Introduction

The Internet of Things concept gained popularity in the last couple of years. The convergence of the Internet with the RFID capabilities constituted, from the beginning, a powerful tool that provides great solutions for a wide variety of problems. With an interconnected network of smart devices and sensors, a large number of intelligent and autonomous services have been developed to improve personal, professional and organizational activities [7].

Besides confidentiality and privacy, trust became an important factor for any IoT system. The overall information shared through the network may, sometimes, require each smart device to properly identify the origins of the received packages as well as the recipients that it pretends to communicate with. Thus, defining a consistent identity management system, capable of satisfying the authenticity

© Springer Nature Switzerland AG 2020
A. M. Madureira et al. (Eds.): SoCPaR 2018, AISC 942, pp. 248–257, 2020.
https://doi.org/10.1007/978-3-030-17065-3_25

of each device towards the network, stands as a necessity to the evolution of trustworthy systems.

In this paper, it is presented an IoT solution that focuses on a blockchain implementation adapted to the Public Key Infrastructures roles on linking identities to public keys. In this case, it is pretended to link every device public key to a specific person and ensure that every action performed by a certain node of the network can be assigned to a proper entity.

The document begins by introducing the relevant concepts for this research, on Sect. 2, followed by the description of the problem to be addressed and which scenarios should be approached, in the third section. Further, in Sects. 4 and 5, the implementation of the blockchain-based PKI is presented and the results of the experiments recreated are discussed, respectively. Ending the paper, the Sect. 6, concludes with some general observations over the research presented and which improvements could be done in the future.

2   Background

2.1   Blockchain

Blockchain came up as an implementation of a distributed ledger. The term traces back to the Satoshi Nakamoto's original whitepaper from 2008, where he applied it as the core component of the digital currency named Bitcoin [8]. A blockchain is defined as a decentralized database, structured as a continuously growing list of ordered blocks, identified by a cryptographic signature. These blocks are linked in the chain by referring to the signature of the immediately previous record in the list. The blockchain contains an immutable collection of records of all the transactions processed, presenting a transparent, decentralized and secure solution for many applications [5].

A blockchain usually starts with a *genesis block*, an initial entry that marks the beginning of the collection of information. Following it, every block in the blockchain should contain the information respective to a specific transaction. Each of these transactions must be digitally signed by the entity that is emitting it, constituting a *block*. Upon a new block generation, the hash of the last block is retrieved and this new entry ends up referencing that previous record, becoming immediately linked to it [9].

2.2   Public Key Infrastructure

Public key cryptography requires users to hold a key pair composed by a public and a private keys and it is important to assure that a certain pair of keys is linked to a specific entity. The Public Key Infrastructures can be interpreted as systems that properly manage thh link between these keys and their respective owners. Usually, these records are based on digital certificates that verify the ownership of a public key (and its corresponding pair, the private key) by some entity. Furthermore, it is expected from a PKI that it supports a set of functionalities comprising registration and update of public keys, as well as revocation or backup of certificates [3].

2.3 Pretty Good Privacy

Pretty Good Privacy is an encryption standard that provides cryptographic privacy and authentication for communications. It can be used for signing, verification, encryption and decryption of digital data such as emails, files, text or entire directories [10].

PGP operations combine data compression with hashing, symmetrical and asymmetrical key cryptography to provide integrity and confidentiality of data. When information is exchanged between two actors, it is encrypted with a symmetric encryption algorithm. The symmetric key, known as *session key*, is used only once for encryption and decryption, and is refreshed as a new random number in each cycle of communication. The data is sent to the receiver together with session key, encrypted with the receiver's public key. Every public key is, therefore, linked to a single user and will be utilized by other entities anytime they pretend to send him confidential information. Upon obtaining the data, a receiver must first decrypt the session key with is own private key, which is kept personal, and decrypt the message with the provided session key.

Web of Trust is commonly introduced in systems that implement the Pretty Good Privacy (PGP) standard, aiming to establish the authenticity of the link between a specific public key and the respective owner, through a decentralized trust model.

In order to understand how web of trust works lets consider that a user named Alice witnessed that Charlie is in the possession of a pair of PGP keys and signed his public key with her own private key in order to vouch for him. If Charlie intend to email another user, called Bob, who doesn't know him, he might not trust Charlie right away. However, if Bob had previously verified Alice and signed her key, thus trusting her, then he can indirectly assume that Charlie is also trustworthy upon acknowledging that Alice signed his key. With this model, the more people sign each others keys, the shorter the trust paths between parties in a Web of Trust become [4].

3 Problem

With the number of identities in the IoT environment tending to grow, it becomes urgent for any platform to have the resources capable to manage them. According to [6], in order to implement a consistent Identity Management System for the Internet of Things, some properties like privacy, security, mobility and trustworthiness must be ensured.

The issues that this paper pretends to target is the centralized data collection of identities and management of devices, which is a method that doesn't scale in the context of IoT. In fact, managing billions of devices constantly exchanging messages between themselves in a smart and dynamic network, cannot be efficiently implemented in a centralized system [2].

The attention for this research focuses on specific kinds of networks, described as public, where multiple entities may participate while being globally distributed, if necessary. In this case, with a more open environment and free participation of unknown parties, it becomes necessary to determine the identity behind each node and assign the responsibility of its actions to a specific user.

Presenting a real example of this kind of networks, we can look at the **uRAD-Monitor** case, a network composed by IoT devices equipped with sensors for environmental monitoring cities, offices and homes, spread in more then 40 countries to generate uniform and comparable environmental data to be used on the analysis of current global pollution [1].

In scenarios like this, if participants decide to provide fake information, the services become untrustworthy and unreliable. However, if a system detects which device contributed with false information and the responsible entity gets identified, the system could apply proper punishments to this malicious actor.

4   Blockchain-Based PKI

The goal of the research is to develop a distributed infrastructure capable of registering the devices admitted in a network into a safe and verifiable data structure. However, this same infrastructure must be supported by an external system for identity management, where every user is supported by proofs that authenticate himself and assign an absolute level of trust that link that user account to the person that it belongs to.

4.1   High-Level Overview

Keybase can be summed up as a collection of tools that establish a Web of Trust, associating their users' accounts to the most common social networks, such as Facebook, Twitter or Reddit. In order to achieve this, Keybase implements Pretty Good Privacy (PGP) policies, assigning keys to each account which will be used to support a set of proofs that link the same user to other external accounts from distinct networks. This is done through signed statements posted in accounts that a user wants to prove ownership of. These constitute a publicly verifiable collection of identity proofs that can be individually verified to ensure trust on the interactions established with each account.

Assuming this support for PGP encryption and identity management of users, Keybase holds a set of capabilities useful for the objective in study. The signature of artifacts with the keys associated to each account create a direct link between the respective user and his own digital properties. Additionally, the publicly verifiable proofs that the platform administers are sufficient to ensure a truthful entity behind each user account on Keybase. However, considering the relevance of this platform for the system in mind, it becomes crucial to find a way to interact with the tools held by Keybase and integrate them into the solution to be developed.

For the infrastructure to be developed, Keybase supplies the enough resources to support it. Two of the actions required from the encryption operations that PGP offers are the signature of data and the respective verification of these originated statements. In order to implement them, the following must be ensured: the private key of a Keybase account must be securely exported from the platform, so that the respective owner can use it to sign external data; in the other hand, Keybase must be able to provide a user's public data, with its corresponding public key and account information.

Fortunately, the service provides a Keybase Command Line through the desktop application, with a specific set of commands to extract both keys assigned to an account on the platform. Furthermore, the public API that the network holds allows anyone to retrieve any information associated to a specific user account, with special urgency for the public key, for signature verifications. This way, all the requirements are satisfied to rely on this Keybase service.

A PKI Built on the Blockchain constitutes the second component of the system. It aims to hold a set of digital signatures that link each device to a single Keybase account. It must be guaranteed that any party on a network of devices must always be owned by some entity. The blockchain will act as the collection of records that will connect both ends in order to assign each device action to a specific person or organization.

In order to define how blockchain and PKI's can complement each other, it is convenient to expose their functions. While the blockchain supports the distribution of transactions and registration of the blocks in a secure and reliable way, the Public Key Infrastructure deals with the registration and revocation of digital certificates that are proof of the ownership of a public key by a specific identity. In this case, the blocks' structure must be adapted to the objective, displaying the necessary data to implement a protocol that easily verifies the data associated to both devices and Keybase users.

Each block should contain a meta data field, responsible to hold the order of the block in the chain, the hash of the immediately previous block and its timestamp. The second field is an object with the relevant information on the device to be introduced to the network. It contains the identification of the node, the Keybase username of the owner and the public key generated for the device. The username is essential to retrieve the identity information on Keybase, via the public endpoint of the API mentioned earlier. The signature property will provide a digital signature of the data object, produced with the private key of the owner, exported from Keybase. Finally, the hash field displays a fingerprint of the complete record, based on the three previously mentioned fields.

4.2   The Protocol

After the overview of the solution, it becomes important to clarify each step of the protocol to be developed. The following sections describe each of the steps that constitute the developed work and how they connect each of the components of the system to achieve the objective in mind (Fig. 1).

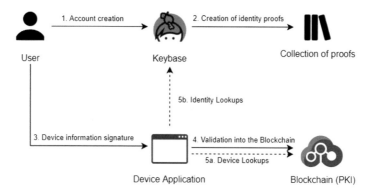

Fig. 1. Abstraction of the solution's relevant components and how they should interact to ensure the linkage between users and devices.

User Registration and Requirements: The initial stage of the procedure consists on creating a Keybase account. Every account must be complemented with a set of digital proofs that the platform supports. These proofs are public and can be validated by anyone in order to vouch for the user behind the account. The most relevant proof for a profile is the creation of a pair of PGP keys. This pair will provide means for the user to sign public statements that will prove his ownership of other social network accounts. These same statements also act as proofs on Keybase and are also a requirement in this protocol, as they are fundamental to guarantee the trust on a specific user account.

However, Keybase API doesn't provide any endpoint to submit a custom package of information to be digitally signed by the authenticated user. In order to gain control of these keys, Keybase enables any user to export any of his PGP keys through Command Line calls.

Device Signature: The second stage of the protocol consists on composing a structure of information relating a certain device to its public key and Keybase owner, who should digitally sign it and submit the output as a new transaction to the blockchain, which is acting as a Public Key Infrastructure.

It is considered a total of three fields to be required in order to create the pretended association: the device unique identification number, the Keybase username of the owner and the public key of the device.

The user should only be required to upload the *private key* extracted in the first phase together with the pass-phrase that unlocks it and the Keybase username. A new record is generated and the data is introduced to the blockchain as a new block that can be verified by any other device that pretends to interact with the signed equipment.

Ownership Verification: When an interaction between two unknown devices occurs, it needs to be established a set of verifications by both endpoints that provide information about the entities involved in the communication. For a certain device to validate another, it must first check on its owner's identity, verify the proofs that he presents and conclude whether they are trustworthy or not.

In the protocol, the verification can be done through look-ups on the blockchain and through queries to the Keybase API. Upon receiving data or a request from an unknown device B, the device A must verify for a registry in the blockchain that refers to B. This block will contain the signature of the information about this device and provide its respective public key and the username of its Keybase owner. Gathering the username, it is easy to query for the user on the public endpoint of the Keybase API. This call will provide a set of data concerning the Keybase account of the owner. Amongst this information is the public key of the user that can be applied to verify the block signature. If such is validated, the device A can then check on the number and type of proofs held by the account and decide whether or not to trust and continue to communicate with B.

5   Experiments and Observations

In order to take conclusions on the implemented solution, few experimental use cases are required to explore the most common situations that the system could face during execution. With it, it is intended to put the solution to the test and observe how it reacts towards each of them. Also, this practical method provides the opportunity to identify weaknesses on the implementation and target a few goals for future development.

Use Case #1 - Device Signature Verification. An interaction between two devices implies that a certain gadget is requesting to communicate with another device.

The implemented solution is simulating this request as a single *ping* to the network, where a device notifies every other connected nodes with a simple and unencrypted message containing its *nodeId*. Receiving this packet, a device can then lookup for the device that emitted the message and verify its ownership.

Upon retrieving the information of the sender's owner, the receiver can then proceed with the required verifications. From this instance, it is possible to face one out of three situations:

1. The public key retrieved from the user's Keybase account doesn't validate the block signature;
2. The public key retrieved from the user's Keybase account validates the block signature but the user doesn't provide enough identification proofs;
3. The public key retrieved from the user's Keybase account validates the block signature and the number of associated proofs complies with the minimum level imposed by the protocol.

The receiver is able to easily verify the signature of the block after collecting the public key and conclude whether or not the entity that signed the record really owns the appropriate secret key. If he doesn't, then it is assumed that he isn't the rightful owner of the sender device.

In the situations 2 and 3, the signature is successfully validated with the public key collected. However, depending on the configuration on the receiver device, it may require a higher or lower number of proofs in order to accept the communication with the requester. These proofs can be analyzed together with the information retrieved and the experiments proved that the implementation is able to respond correctly to these requirements.

Use Case #2 - Unreliable User Proofs. It is possible to create multiple fake accounts on Facebook, Reddit or Twitter and sign statements from a Keybase account to claim ownership over them. The most certain is that these fake accounts in Keybase will not be followed by any other users.

However, the implemented system is not prepared to consider how many followers a certain account has for a very simple reason: there is no API call nor resources provided by Keybase that return the number of followers that a given account has. The support of these complementary proofs would guarantee and extra parameter to consider and judge more accurately the identity behind the devices, being sure that their public statements were verified by other real entities. If a certain device is receiving interactions from two other nodes of the network, it may discard legitimate entities and approve malicious ones due to lack of complementary user data.

Use Case #3 - Revocation of the User's PGP Keys. The third use case contemplates the possibility of a Keybase user revoking his own PGP keys and replace them by another generated pair, associating it into the account. The Keybase platform is prepared to handle these actions and propagate the necessary changes on the user's chain of proofs.

However, these actions of revocation and update of the PGP keys of a certain device's owner brings a negative impact to the implemented solution. As result of the experiments made, the association of a new pair of PGP keys to a certain device owner leads to unsuccessful identity checks on the Keybase platform. This happens because the protocol will look up to the user profile, gathering his current public key and verify the device signature with it. If this signature was created with the previous pair of keys, the result will be to consider it as a bad device signature and it would be impossible to associate the devices to a rightful owner.

5.1   Results Evaluation

From the observed interactions with the conceived protocol, blockchain-based PKI's also show the potential to overcome the Web-of-Trust PKI's. While the WoT model requires a significant effort to produce a web capable of proving the trustworthiness of a node to a considerable portion of the network, blockchain-based PKI's don't require such an interconnected structure of authenticated entities.

The presented approach is built on top of a platform that already implements this Web of Trust definition through the PGP operations inherent to its functionalities. The creation of public and verifiable statements that support the identity of each user allow them to be validated by other real life entities and vouched through follower statements. Assuming the trust between a set of entities, the devices owned by them could also be able to interact with each other. Adopting these WoT characteristics from Keybase, the effort required to implement the Web-of-Trust PKI would be minimized and the result of this research could be different.

As denoted in the second use case, Keybase does not provide these follower statements neither a qualification of the trust between two given entities. Consequently, the intention to develop a WoT PKI becomes somehow more complex. However, Keybase allows anyone to access a public endpoint of the API to retrieve the ownership statements of external social networks accounts and digital assets, which became the source of trust in the blockchain-based PKI that is presented.

The implementation described during the previous chapters aims to transparently display a collection of signatures over the devices participating in a network. Resorting to this collection of signatures, securely appended to a blockchain that is distributed amongst the devices, they can check on the ownership of every machine they interact with and assign the actions of that device to a specific entity that claimed its ownership. The first use case proves that the protocol is successful and that the devices take the proper actions to prevent or allow the interaction with other unknown devices, based on the proofs provided by their owners on Keybase.

Considering the third use case, there are still a few operations that could be integrated and experimented, with special attention to the revocation of users' PGP keys in Keybase. The revocation and update of keys are common operations in Public Key Infrastructures and the protocol that was designed could be complemented with this functionality, providing more flexibility to the system and a more close approach to what PKI's should provide to their environments.

6   Conclusions and Further Work

The research presented in this paper provides a different approach for the Internet of Things segment, adopting the distributed ledger technologies as bridge between the interaction amongst smart devices and the importance of social networks in today's society.

With the described protocol, it is pretended to provide another feasible and secure implementation for the identity management of entities in the IoT systems, preventing malicious actors from anonymity and impersonation when introducing devices in a network that can't get their respective owner properly verified.

The most urgent improvement to this protocol would be the revocation and update of users' PGP Keys, from Keybase. This issue represents a fundamental functionality in every Public Key Infrastructure. Implementing it would prevent a device from being untraceable to the respective owner after the entity renewing his Keybase PGP keys.

Acknowledgment. This work was supported by Project "NanoSTIMA: Macro-to-Nano Human Sensing: Towards Integrated Multimodal Health Monitoring and Analytics/NORTE-01-0145-FEDER-000016" is financed by the North Portugal Regional Operational Programme (NORTE 2020), under the PORTUGAL 2020 Partnership Agreement, and through the European Regional Development Fund (ERDF).

References

1. Global radiation map. https://www.uradmonitor.com/. Accessed 27 June (2018)
2. Identity management for the Internet of Things. https://www.linkedin.com/pulse/identity-management-internet-things-george-moraetes/. Accessed 1 Feb 2018
3. An overview of public key infrastructures (PKI). https://www.techotopia.com/index.php/An_Overview_of_Public_Key_Infrastructu-res_(PKI). Accessed 15 May 2018
4. Fromknecht, C., Velicanu, D., Yakoubov, S.: CertCoin: A NameCoin Based Decentralized Authentication System 6.857 Class Project, pp. 1–19 (2014)
5. Jacobovitz, O.: Blockchain for Identity Management. Technical report, December 2016. https://www.cs.bgu.ac.il/~frankel/TechnicalReports/2016/16-02.pdf
6. Kaffel-Ben Ayed, H., Boujezza, H., Riabi, I.: An IDMS approach towards privacy and new requirements in IoT. In: 2017 13th International Wireless Communications and Mobile Computing Conference, IWCMC 2017, pp. 429–434 (2017)
7. Miorandi, D., Sicari, S., De Pellegrini, F., Chlamtac, I.: Internet of Things: vision, applications and research challenges. Ad Hoc Netw. **10**(7), 1497–1516 (2012)
8. Nakamoto, S.: Bitcoin: A Peer-to-Peer Electronic Cash System. www.bitcoin.org, p. 9 (2008). https://bitcoin.org/bitcoin.pdf
9. Samaniego, M., Deters, R.: Blockchain as a service for IoT. In: Proceedings - 2016 IEEE International Conference on Internet of Things, IEEE Green Computing and Communications, IEEE Cyber, Physical, and Social Computing; IEEE Smart Data, iThings-GreenCom-CPSCom-Smart Data 2016, pp. 433–436 (2017)
10. Shaw, D., Thayer, R.: RFC4880 - OpenPGP Message Format, pp. 1–6, November 2007. https://tools.ietf.org/html/rfc4880

The Design of a Cloud Forensics Middleware System Base on Memory Analysis

Shumian Yang$^{(\boxtimes)}$, Lianhai Wang, Dawei Zhao, Guangqi Liu, and Shuhui Zhang

Shandong Provincial Key Laboratory of Computer Networks,
Shandong Computer Science Center (National Supercomputer Center in Jinan),
Qilu University of Technology (Shandong Academy of Sciences),
Jinan 250014, China
yangshm@sdas.org

Abstract. The rapid development of cloud computing has not only brought huge economic benefits, but also brought the issue of computer related crimes. In this paper, a design method of cloud forensics middleware was proposed to obtain credible and complete digital evidence from the cloud in a comprehensive and convenient manner. The design method includes three parts: the remote control side evidence display, the server-side evidence analysis and monitoring management, and client-side memory acquisition and analysis. Compared with the traditional online forensics methods, this method was more in line with the requirements of traditional physical evidence technology, greatly improving the efficiency of the forensic staff and the credibility of the evidence. The method has been verified on Windows 10 (the client) and Centos 7.0 (the server) and was proved to be effective and reliable.

Keywords: Cloud forensics · Middleware · Physical memory · Remote control

1 Introduction

It has brought huge economic benefits and more convenience, but it has also brought computer crime problems with the rapid development of cloud computing, communication technology and computer technology. More and more malicious attacks are being implemented through cloud computing and computer networks. Cloud forensics is the most important means of combating cybercrime. It increased the difficulty of event emergency response and digital forensics because of different deployment and service modes. So it is a difficult problem in current cloud forensic research to obtain complete and reliable evidence data from the cloud environment [1].

In addition, the number of forensic cases was increasing year by year with the increase of cloud computing users, but the number of forensic personnel was very limited, and it was difficult to go to the scene for evidence collection in time, and it had different methods of forensics for each case which required a variety of analytical methods. To obtain full-scale evidence by own personal experience forensics officers was very difficult [2].

© Springer Nature Switzerland AG 2020
A. M. Madureira et al. (Eds.): SoCPaR 2018, AISC 942, pp. 258–267, 2020.
https://doi.org/10.1007/978-3-030-17065-3_26

Based on the above questions, domestic and foreign experts have given a lot of solutions to the data collected and whether they were credible and complete. Xie et al. proposed a forensic framework [3] under the infrastructure-as-a-service (IaaS) cloud model. ICFF, Wang proposed an online forensic model and method [4] based on physical memory analysis, which solved the problems of traditional online forensic methods and made the credibility of online evidence be possible. Wang [5] proposed a physical memory analysis method for windows System based on Kernel Processor Control Region (KPCR) structure; Yang proposed a remote forensics system [6] based on physical memory analysis. All of these methods had put forward solutions to the credibility of evidence.

In view of the current lack of forensic tools in cloud computing, Deng [7] proposed the middleware from video surveillance to cloud computing, introduced the background of middleware and the role of middleware; Pei [8] proposed the design and implementation of VPN system based on cloud middleware; Luo proposed the design and implementation of health management cloud platform [9]; Li et al. proposed remote monitoring OPC client [10] based on C/S model(Client/Server); Liu et al. proposed an intelligent control system [11] based on public cloud platform for information interaction; Cui et al. proposed a transparent message channel design [12] based on KVM virtual desktop. All of these have proposed certain solutions in the design of cloud middleware, but they could only transmit data of traditional types, and could not guarantee that the evidence of transmission was credible and complete.

Based on the credible and complete problems of the evidence, this paper designed a remote cloud forensics middleware system based on physical memory analysis, which could real-time analyze the physical memory of a large number of computing devices such as servers, personal computers and intelligent terminals in the cloud computing environment. On the one hand, in Client it mirrored real-time the physical memory, calculated the hash of the memory image files to ensure the authenticity of the evidence and ensure the evidence was not tampered with in the transmission process. On the other hand, it analyzed real-time the mirrored memory files to ensure the credibility and integrity. Finally, it transmitted the memory analysis results and the mirrored physical memory files which were encoded by base64 to the server which could analyzed further. Its core components included client physical memory acquisition and analysis software, server evidence analysis and monitoring management, remote control service software. This article could find evidence according to the needs of customers, and further in-depth analysis.

2 Design Method

The system collected the client information through the mode of the remote management service. The server could remotely obtain the client's physical memory image file, the physical memory analysis result and the log information of the client, and provide the function of retrieving the physical memory analysis result and analyzing the memory in parallel to acquire the valuable information. The system mainly consisted of three parts including the remote control terminal, the server and the client. Compared with the existing technology, the advantages of the design method proposed in this paper are: it could real-time obtain the physical memory mirror files, system status information, registry information, open file information, network connection information, hook information and other valuable information from the physical memory mirror files, because of the physical memory information was not been tampered which make the result of physical memory analysis be great credibility and effectiveness. The design method could help judicial officers to find the necessary evidence in a large amount of memory analysis results, and greatly improve the efficiency of work.

2.1   The Remote Control Management

The remote control terminal obtained the information of the client according to the requirements of the server, and displayed the log information of the client. The log information mainly included the IP address of the client, the port, the name of the transferred image file, the result of the memory analysis file, and the corresponding md5 value. There was also a retrieval function. If the user wanted to retrieve a keyword from a table, the memory image analysis result of the client could be obtained by retrieving. And provide export function, it could export any or all of the search results and memory analysis results for the forensic personnel to further analysis. The remote console function was shown in Fig. 1:

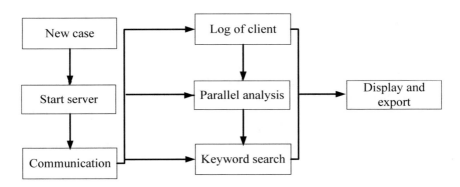

Fig. 1. The flow chart of remote control side

The remote control terminal contained five modules: new case, communication module, log management module, key file retrieval module, and parallel analysis module. The main modules were designed as follows:

The communication of the remote client and the server were used TCP protocol to establish a connection and communicate after the authentication was passed. The log management module mainly analyzed the log for windows OS including application logs, security logs, and system logs, and extracted and analyzed keywords in a log. The main key fields were log name, source, recording time, time ID, and user, keyword, detailed information, etc.; The retrieval part was completed by retrieving the memory analysis result in Hadoop hive database, and display the qualified information in the list box. The retrieval part included eight tables: basicinfo, process, driver, net, registry, syslog, hookinfo, winlogon. Each table corresponded to its own primary key, the key fields in the basicinfo table ClientIP, MemoryTime, registerowner, productname, systemname; the key fields in the process table: ClientIP, MemoryTime, name, pid; the key fields in the driver table: ClientIP, MemoryTime, name; and the net table: ClientIP, MemoryTime, remoteaddress; Registry table: ClientIP, MemoryTime, name, type; syslog table: ClientIP, MemoryTime, systemloginfo; hookinfo table: ClientIP, MemoryTime, proname, procPID; winlogon table: ClientIP, MemoryTime, Username, rid. For example Table 1: "process" table, Table 2: "net" table and Table 3 "hook" table:

Table 1. Process

ClientIP	MemoryTime	name	pid	parentcid	createtime	openfile

Table 2. Net

ClientIP	MemoryTime	createtime	remoteaddress	localaddr	tablepa	tableva	protocol

Table 3. Hook

ClientIP	MemoryTime	Proname	proPID	chain	proarch	threadsearch	csrsshandle

2.2   The Server

The server started multi-threading and listened to multiple clients at the same time. If there was a client connection request, it send a fixed string "INF#Physi-calMemoryString#" to the client. The client transmitted different information to the server according to the received string, mainly including collecting the client's physical memory image files and the memory analysis results of different image file. Because multi-threading had been opened, it could collect physical memory image files and memory analysis results of several clients at the same time, and store the results of memory analysis into Hadoop hive database; at the same time, it established a connection to send the client's log information to the remote client according to the needs of the remote client. According to the retrieval conditions of the remote control end, it could find the qualified retrieval information from the Hadoop hive database which was sent to the remote control end to display in the list control. The server function was shown in Fig. 2.

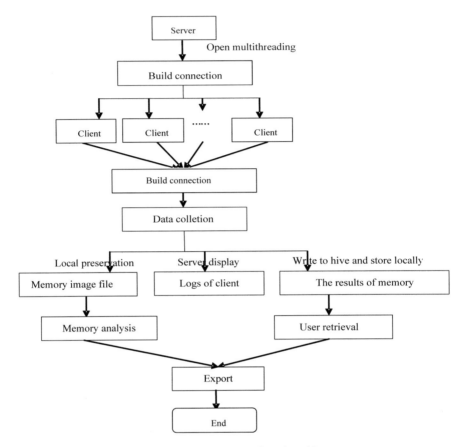

Fig. 2. The flow chart of service side

The server stored data into the hive database. According to the different fields received, the data was inserted into the database by using the hive −e to load data, for example "local inpath 'table name' into table". Parallel analysis: According to the needs of the remote console, the server started multi-threading and simultaneously processed and analyzed the multiple physical memory image files.

2.3   The Client (Intelligent Terminal and Client Terminal)

The client established a connection with the server, and selected different method of mirror physical memory according to the system version, stored the image file locally for analysis and translation, and calculated the hash value of the image file, verified the sent physical memory mirror file whether had been tampered by the hash value. The physical memory analysis line program was invoked to analyze the memory image file based on KPCR, and the memory analysis result was transmitted to the server along with the physical memory image file, and the memory analysis result was written to the Hadoop hive database. The data collection analysis processing flow of the client part was shown in Fig. 3.

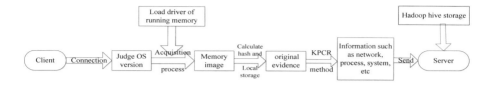

Fig. 3. The flow chart of client side

2.3.1 The Method for Loading Driver Program

Firstly, the system obtained the complete physical memory information on the target computer, obtained the memory image file, and finally obtained the required online evidence by analyzing the physical memory image file. The system must load the driver before imaging a physical memory. The method of loading the driver was as follows: SCMLoadDeviceDriver(str.GetBuffer(str.GetLength()), "64BitMemDump") and SCM LoadDeviceDriver(str.GetBuffer(str.GetLength()), "32BitMemDump"), of which 64Bit MemDump and 32BitMemDump were the drivers for obtaining physical memory image files under different versions of the system respectively.

2.3.2 Memory Analysis Program

Memory analysis technology could extract lots of online evidence from physical memory image files. If the client host was a Linux system, the Virtual Machine Control Structure (VMCS) analysis method was used; if the client host was Windows operating system, the KPCR-based physical memory analysis method was used. The physical memory based on KPCR mainly solved the problem of physical address and virtual address translation. Only when the physical memory address was obtained, the physical memory could be analyzed.

This online forensic method can clearly divide the stages of online forensics acquisition, fixation, analysis, and fusion presentation, so that each stage can be repeated and tested, and the evaluation of forensic techniques at each stage is also clear and feasible. Moreover, the analysis process relies on the obtained physical memory image file, and also facilitates the correctness of the verification process, which is more in line with the requirements of the physical evidence technology than the traditional online forensic method.

This online forensics method could clearly divide the stages include of acquisition, fixation, analysis and fusion presentation [4], so that each stage could be repeated and tested, and the evaluation of forensics technology in each stage is also clear and feasible. Moreover, the analysis process depended on the physical memory image file which was obtained from the client, which also verified the correctness of the process. The method was more in line with the requirements of the physical evidence technology than the traditional online forensic method.

3   The Design and Implementation of Middleware System

The remote control terminal was developed by using VS2010, which required running in the Windows operating system, and supported Windows XP, Windows 7, and Windows 10 systems. The remote control end was an interface program. It didn't need to be installed and open the .exe file directly, including call the acquisition and parsing of the client log information.

The server was a command line program. It was developed in C language under Centos 7.0, which required the installation of Hadoop hive database. The server executed the command line program: gcc -o tcp -lpthread -D_FILE_OFFSET_BITS=64 tcp_rev_server-new-remote.c (./tcp port), and the port could be set arbitrarily, such as 5000. After the service was started, the server listened to the client and the remote console. On the one hand, it established a connection with the client, received the physical memory image file and the physical memory analysis result from the client, and stored the memory analysis result to the hive database which was supported by hadoop. The transmission process supported encoding and decoding calculation of the file. On the other hand, a connection was established with the remote console, and the server program provided the log information and the physical memory analysis results of the client according to the retrieval requirement. The transfer of more than 4G of memory image file required the addition of a parameter -D_FILE_OFFSET_BITS=64 in the Linux kernel system to ensure normal transmission between the client and the server.

The client program was a process-based command line program. The directory of the client process contains physical memory mirror and physical memory analyzer programs, which supported Windows XP, Windows 7 and Windows 10 systems respectively. The client was a command line program under Windows. Opening tcp_client_send.exe file, the client obtained the physical image file and saved it locally and then called the memory analyzer to analyze the physical memory image file whose analysis results were transmitted to the server. The mirroring interface was shown: the memory page size of each physical memory, the number physical memory page (decimal), minimum physical memory page, maximum physical memory page (decimal) and the ratio of physical memory which has been mirrored.

Network was needed between the remote control and the server, the client and the server before they could connect and transmit. To ensure the integrity and credibility of the evidence, the system used the TCP protocol as the communication protocol. Take the Windows 7 system as an example, it run the remote_tcp_client.exe program and started the interface is as shown in Fig. 4.

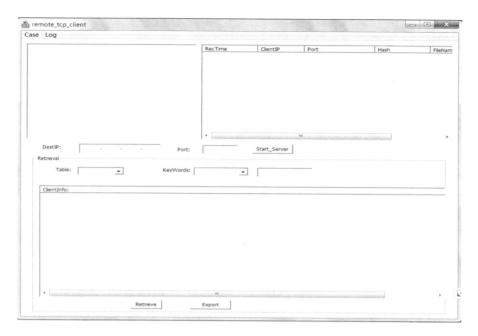

Fig. 4. The main interface of remote control

The remote control terminal mainly consisted of three parts: starting services, logging information, and retrieving information. The tree control on the left side of the upper part showed the IP address of the server, the IP address of the client and the memory image files which had been transmitted by the client to the server. The right list box showed the log information of the client, including the receive time, MD5 value, file name, IP address, port, and the memory image file when the client memory image file was generated. The middle part of the main interface was the IP address and the port of the server. The server was connected to open the service. The lower part of the main interface list box mainly displayed the memory analysis results and the qualified memory search results.

Select any line or several lines of the client log information in the upper part of the list box, right click "send", then extract the memory image file to the server, according to the image file name, the server will analyze the memory image file. The memory analysis process needed to take about 2 min. After the analysis was completed, the memory analysis results would be displayed in the list box in the lower part of the main interface is as shown in Fig. 5.

The design method proposed and implemented the cloud forensics middleware, which satisfied the needs of IT industry users in forensics and security auditing, expanded the application field, and extended the industrial chain of the computer forensics.

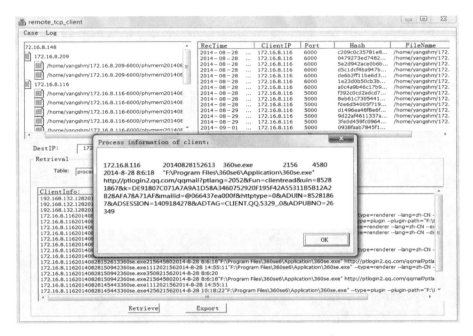

Fig. 5. The memory analysis result of client

4 Conclusion

Aiming at the shortcomings of the traditional forensics tools, such as the lack of online forensics tools and the difficulty of dealing with a large number of cases, this paper designed a cloud forensics middleware, which solved the problems of the credibility of evidence and the difficulty of dealing with a large number of cases in time. Firstly, this paper introduced the current forensics methods and cloud middleware products, and then introduced the design method of cloud forensics middleware. Using this method, a cloud forensics middleware system was developed. The client software of the system could realize the real-time processing of local evidence according to the requirements of remote server including evidence preservation and evidence analysis and could remotely control forensics tools and other third-party applications to achieve real-time collection and processing of evidence, and send remote evidence and analyzed results to the server for evidence preservation, analysis and processing. The remote control end of the system could realize the client's security audit, including collecting the application log, security log, system log and other audit information of the client's operating system in real time. At the same time, it could monitor and record the whole process of the witness, and send it to the remote server for analysis and processing, so as to achieve timely detection of network security risks or illegal criminal activities. The greatest advantage of this design method was that it relied on the physical memory image file, so that the whole forensics process could be divided into different stages, ensuring the credibility and integrity of evidence. In the future, we will further expand the development of versions, capture and analyze the physical memory of different systems, and serve more and more cloud users.

Acknowledgments. This work is supported by the National Natural Science Foundation of China (Grant Nos. 61572297, and 61602281), the Shandong Provincial Natural Science Foundation of China (Grant Nos. ZR2016YL014, ZR2016YL011, and ZY2015YL018), the Shandong Provincial Outstanding Research Award Fund for Young Scientists of China (Grant Nos. BS2015DX006), the Shandong Academy of Sciences Youth Fund Project, China (Grant Nos. 2015QN003), the Shandong provincial Key Research and Development Program of China (2018CXGC0701, 2018GGX106005, 2017CXGC0701, and 2017CXGC0706).

References

1. Garfinkel, S.L.: Digital forensics research: the next 10 years. Digit. Invest. **7**, 64–73 (2010)
2. Wang, X., Xiong, X., Zhang, X., et al.: Methods and systems for collaborative forensic analysis of remote forensics target terminals, China, CN1044629A, 03 December 2014
3. Xie, Y., Ding, L., Lin, Y., et al.: ICFF: a cloud forensics framework under IaaS model. J. Commun. **34**(05), 200–206 (2013)
4. Wang, L.: Research on online forensics models and methods based on physical memory analysis. Shandong University, Jinan (2014)
5. Guo, M., Wang, L.: Windows physical memory analysis method based on KPCR structure. Comput. Eng. Appl. **45**(18), 74–77 (2009)
6. Yang, S., Wang, L., Han, X., et al.: A remote Forensics System Based on physical memory analysis, CN105138709A, 09 December 2015
7. Deng, Y.: From video surveillance middleware to cloud computing middleware. China Secur. (Z1), 60–63 2014
8. Pei, Z.: Design and implementation of VPN system based on cloud inter parts. Dalian University of Technology (2012)
9. Luo, G.: Design and implementation of cloud platform for health management. Beijing Jiaotong University (2017)
10. Li, G., Li, Y., Yuan, A.: The OPC client based on C/S model realizes remote monitoring. Microcomput. Inf. (12), 25–26, 189 (2007)
11. Liu, Y., Qin, C.: Intelligent control system for information interaction based on public cloud platform. Exp. Technol. Manag. **33**(08), 149–151, 155 (2016)
12. Cui, J., He, S., Guo, C., et al.: Design of transparent message channel based on KVM virtual desktop. Comput. Eng. **40**(09), 77–81 (2014)

Privacy Enhancement of Telecom Processes Interacting with Charging Data Records

Siham Arfaoui$^{(\boxtimes)}$, Abdelhamid Belmekki, and Abdellatif Mezrioui

INPT, Rabat, Morocco
{arfaoui,belmekki,mezrioui}@inpt.ac.ma

Abstract. Telecommunication is the foundation stone of the current economic context which allows an opening in all areas and facilitates the interaction using different platforms and technologies (from 2G to 5G). In such context, personal data is massively collected, stored and processed for different aims which is considered as big privacy concerns. Personal data should not be known by no matter whom. It is thus important to raise the question of what can be communicated and to which. The protection of personal data needs a deep attention from companies that process with this data such telecommunication operator. The dilemma of privacy in telecommunication operators is that telecommunication employees, suppliers and subcontractors need personal data access but the privacy requirements consist on knowing less personal data and protects such data. The goal of this article is to deal with this dilemma by providing a privacy approach for protecting personal data in telecommunication operators processes interacting with Charging Data Records CDR in such way to ensure that the illegible employees whom access to personal data cannot use them for purpose other than authorized one.

Keywords: Privacy · Anonymization · Charging Data Records · Telecommunication

1 Introduction

Personal data designates all that make possible to identify a person directly (such as last name, sex, and photos) or indirectly (such as number of social security, place and date of birth, the national identifier, the phone number…).

We should notice that end user's personal data are used in telecommunication processes to deliver data and voice services. This personal data is needed in telecommunication operators also for avoiding churn.

Privacy is considered as human right; laws and regulations oblige all companies using personal data to make all necessary mechanism to ensure protecting of personal data. We focus on this article on providing a privacy approach for protecting personal data in telecommunication operators processes interacting with Charging Data Records by highlighting in Sect. 3, according to 3GPP Third Generation Partnership Project, the meaning of Charging Data Records CDR. In Sect. 4, we summarize all anonymization techniques that allow to protect personal data, afterwards we propose the approach of

© Springer Nature Switzerland AG 2020
A. M. Madureira et al. (Eds.): SoCPaR 2018, AISC 942, pp. 268–277, 2020.
https://doi.org/10.1007/978-3-030-17065-3_27

protecting personal data. In Sect. 5, we implement our approach and we study its performances which help us to conclude that this approach is applicable in real production environment of telecommunication operators.

2 Motivation

We have chosen to focus in this article on protecting personal data of telecommunication operators processes interacting with Charging Data Records because the CDRs are one of the big links between operator telecom network and its information system in term of conveying personal data. CDRs allow knowing our localization, our phone and Internet activities like called number, visited web sites and key words typed in the search engines which can be recorded and monitored for different aims. The figure below Fig. 1 shows the link that CDRs establish between telecom network and its information system:

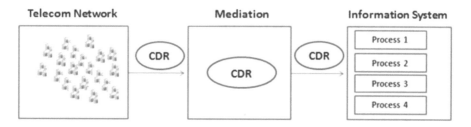

Fig. 1. Charging data records location

In this article we will focus on improving data privacy in Charging Data Records.

3 Charging Data Records

Charging Data Records are a formatted collection of information about a chargeable event for use in billing and accounting [1]. The billing domain is a part of the operator network which is outside the core network and receives CDR file from the core network charging function by the intermediate of the mediation [2]. The mediation is the first step after generating CDR; it is a process that converts call data to predefined layout that can be imported by specific billing system or other application. Not all the data transferred via billing mediation platforms is actually used for billing purposes. For instance, the CDR can provide information for other processes like statistics, revenue assurance; Quality of Service (QoS) and Business Intelligence. Figure 2 summarize the interaction of CDR with the different processes cited above:

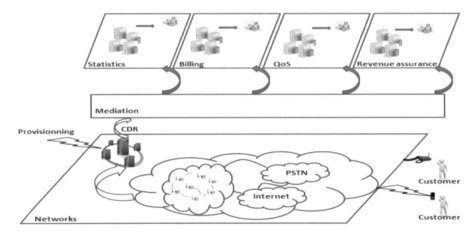

Fig. 2. Processes interacting with CDR

Each CDR includes all charging data relevant to the call [2], some of this data are considered as personal data such as [1, 3, 4]:

- Calling party number/Called party number
- Calling party IMSI: International Mobile Subscriber Identity
- Called party IMSI: International Mobile Subscriber Identity
- Calling location area code party/Called location area party
- Calling Cellule ID/Called Cellule ID
- IMEI: International Mobile Equipment Identity/URL: Uniform Resource Locator

4 Privacy Approach

We should notice that interface principals such as security and performance of the processes seen in the figure above (Fig. 2) are dependent on vendor implementation and operators network design. They are not covered by the 3GPP Third Generation Partnership Project specification [5]. That is why we will work in the following on increasing the security of personal data use helping by this way the network operator to secure personal data for their clients in processes interacting with CDR.

In the following section we focus on the different techniques that can be used in our approach to protect personal data.

4.1 Anonymization

The anonymization is the process of ensuring the non-identification of individuals from its personal data using several techniques. We established a categorization of the anonymization techniques which we divided into reversible anonymization and irreversible one. The irreversible anonymization is characterized by the fact that the changes made on the data are final and is not reversible.

This category is divided into 3 sub categories: the suppression, the generalization and the modification. The irreversible anonymization consists on changing the original dataset by another data. Concretely, that means that all data directly and indirectly identifying individuals are removed and it makes impossible any deduction of people identity. The reversible anonymization consists on replacing an identifier by a pseudonym. This technique allows the lifting of anonymity or the study of correlations where necessary. All the techniques of anonymization that will be cited are defined in the Fig. 3 where we represent the categorization of anonymization techniques according to the reversibility criterion:

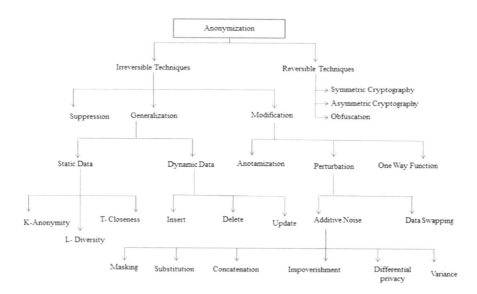

Fig. 3. Global diagram of anonymization techniques

Irreversible Anonymization Techniques

Suppression
The suppression consists on removing the data concerned by anonymization and replaces them by the value "Null" or downright to delete it [6]. It is an effective way to ensure that these data are no longer visible. However, it is one of the options the least desirable because it has a large impact on the usefulness [7, 8], it is used generally on the direct identifiers [9] such as the last name and the first name.

Generalization
The generalization can be described as the grouping of the data in greater category [10–12]. Its purpose is to reduce the number of record with a unique combination of variables allowing potentially the re-identification [8] usually the dates, sex, localizers and the phone numbers so that we can achieve an acceptable level of risk of re-identification. It is therefore possible to reduce the dates of birth as cited in [8] by the year of birth. If the year of birth does not provide enough protection against

re-identification, the age at a given time may be considered [12]. The generalization can be carried out on the static databases and also on dynamic databases [9].

Modification

The techniques of modification manipulate the data in more radical way and reduce greater the degree of data information comparing to the generalization [8].

- Anatomization Fig. 3: the technique of the anatomization proceeds by the separation of the quasi-identifiers and sensitive data by modifying the column of sensitive data and adding columns ensuring the usefulness of sensitive data such as shown in [13].

Perturbation [11] can be divided in two categories such as shown in Fig. 3: additive noise and data swapping, under additive noise we find 6 methods:

- Masking: the masking data [13] replaces some character by a mask character (such as an X), which hides the content of data while maintaining the same formatting of the data
- Substitution: technique that consists on replacing the content of the data of a column in a random manner or to select data that resembles of the contents of the origin database but which is totally without any report with the first ones. For example, the names of the family in a customer database could be anonymized by the replacement of family names with names drawn from a random list [13].
- Concatenation: The concatenation [6] consists on replacing the data by a value from the combination of the records contained in the database in question. It is one of the anonymization techniques used to keep the format and the validity within a database generally used to test a computer application.
- Impoverishment: Impoverishment [14] consists on making the data less precise, conducting selective changes. For example; we can remove the detail of the location of the call made during a certain period, subsequently retaining the age range of the calling party [14]. This method is applied to images, and sound to make impossible the identification of persons included in a film or on a photo.
- Differential privacy: It is one of the methods that secures against the attacks oriented database and not the record. It is based on a calculation of probability of Laplace [15]. The attacker has an arbitrary knowledge of the presence of the person sought in the database in question.
- Variance: This technique [13] is useful when it comes to numerical data or dates. It is to change the numerical values or the dates with a percentage increase or decrease of actual values. This technique has the advantage of providing a reasonable disguise for the data while maintaining the distribution of values in the column [13]. For example, a column of details of salary could have a statistical variance of $\pm 10\%$ applied on it.

Under perturbation category as shown in Fig. 3, we have a technique which called data swapping. This technique is similar to a substitution with the exception that the swapped data are derived from the column itself. Essentially, the data in a column are displaced in a random manner between the lines of database table, in this case there is more correlation with the remaining information in the column. There is a certain danger in this technique. The original data are always present and sometimes relevant questions can always find answers [13].

Reversible Anonymization Technique

The reversible anonymization, allows a return to the data not anonymized from anonymized data if necessary. Under this category we have symmetric anonymization, asymmetric anonymization and obfuscation:

- The symmetric cryptography uses the same key in the encryption and decryption process. The algorithms of this category are based on substitutions, the transposition and the product. Two sub categories of the symmetric cryptography exist: by block where the encryption operation is performed on blocks of text. The algorithms the most known are 2DES, 3DES and AES. The second sub-category is the encryption by flow where the encryption operation is performed for each element of the text in the clear (a bit or character at a time). The algorithms the most well-known are the RC4, LFSR, and ARC4 (which is a lighter version of the RC4 used by the public contrary to RC4).
- Asymmetric cryptography uses a key for encryption and another for decryption. The Algorithms the most known are the RSA and El Gamal, ECC. Note that El Gamal and RSA are multiplicative homomorphic algorithms. The asymmetric and symmetric cryptography have for purpose to preserve the confidentiality.
- Obfuscation: is a technique used by some developers of computer programs to make the understanding of their code very difficult for a human, while ensuring perfect compilation by a computer [7]. This technique is to add to the entry another data completely fictitious.

4.2 CDR Protection Approach

Because of mediation is the first step after getting the CDR and we want to protect personal data in CDR, we have three options: add the layer of personal data protection just after generation the CDRs, so before the mediation or in mediation or after mediation. If we opt for the addition of the personal data protection layer before or after the mediation, we will have to use a personal data protection tool, but if we opt to add the protection layer in the mediation, we have the opportunity not only to use a personal data protection tool but also the possibility of adding a layer of anonymization in the mediation. We will study in the following the effectiveness choice of the two options: either use a personal data protection tool or the addition of a layer of anonymization in the mediation. Starting by analyzing the addition of the layer of protection before/after the mediation, so we will need to add a personal data protection tool. No organization or institution systematically evaluates the tools of anonymization [16]. Isolated comparisons are found in the academic literature, where authors compare their algorithm or method to other published techniques. However, they cannot be considered as an objective evaluation of software tools, for three reasons:

- Although peer-reviewed, in most cases these evaluations compare old methods with the methods proposed by authors and are therefore not really impartial;
- Evaluations generally focus on algorithms or conceptual models rather than toolboxes.
- Software tools may not be commercialized or in a rudimentary form that does not fully meet the objective criteria of software maturity (robust development method, documentation, user interfaces, etc.). We deduce that:

1. These tools do not contain a panoply of anonymization methods choices (especially as regards the presence of reversible methods) [16], which represents a real hindrance of these tools, since for all the data to be anonymized in the CDR, we will need some reversible methods except for the case of the URL which must be anonymized by an irreversible method.
2. The expensive price of these tools [17] which are all paying and those which are open source are not mature enough [16] for use in a production environment.
3. The majority of these tools are inefficient for a production environment since they are designed for the test environment.

Let Analyze now the choice to add a layer of anonymization in the mediation. This solution has the advantage:

- Performing for a production environment as we will demonstrate in Sect. 5.
- No need to add another tool.
- The existence of panoply of choice of anonymization method.

Thus, the suitable choice is the addition of the protection layer in the mediation. Figure 2 of Sect. 2 becomes as follows (Fig. 4):

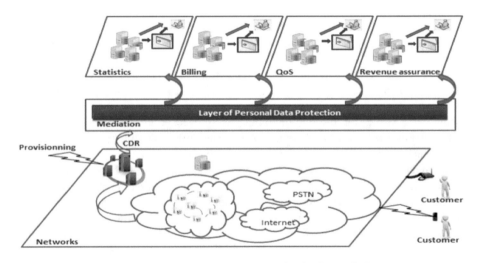

Fig. 4. Layer of personal data protection in the mediation

As mentioned above, all field containing personal data in the CDR must be anonymized with a reversible method, the choice of such method is the result of the fact that all processes will need to access to data in clear mode. Except for the case of the URL which is a sensitive data, it must be anonymized by an irreversible method because access to this data determinate the political, religious or customs of the client.

In the table below, we summarize the anonymization techniques to be adopted for the CDR personal data (Tabel 1):

Table 1. Anonymization techniques for personal data in CDR

Personal data	Anonymization techniques
Calling party number	Reversible techniques
Called party number	
Calling party IMSI	
Called party IMSI	
Calling location area code	
Called location area code	
Calling cell ID	
Called cell ID	
IMEI	
URL	Irreversible techniques

5 Performance

In this section, we focus in the performance aspect of adding a layer of anonymization and their impact on the production environment if a telecommunication operator decides to include it. The performance parameter that we will study is run time.

The CDR sample we worked on contains 110 records separated by "|" [18]. The tool used for the tests is Microsoft BI, on a 2 GB memory and 4 CPU 1.8 Ghz. The algorithms we applied is a symmetric one: AES-256 whose private key is protected by a certificate. The results of the tests are shown in Fig. 5.

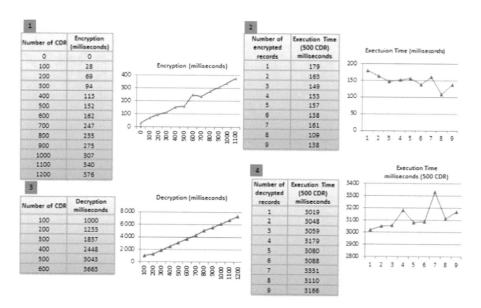

Fig. 5. Performance of the implementation

We study 4 varieties of tests: varying the number of CDR encrypted, varying the number of encrypted records simultaneously, varying the number of CDR decrypted and varying the number of decrypted records simultaneously.

6 Discussion

The results of the tests are shown in Fig. 5 as follow:

1. We studied encryption by varying the number of encrypted CDRs and setting the number of records encrypted to 1 value. The execution time is in the order of milliseconds, so it can be used in production environment.
2. We studied encryption by varying the number of encrypted records simultaneously and fixing the number of CDRs at 500. The execution time is in the order of milliseconds, so it can be used in production environment.
3. We studied decryption by varying the number of decrypted CDRs and setting the number of records decrypted to 1 value. The execution time is in the order of milliseconds, so it can be used in production environment.
4. We studied decryption by varying the number of decrypted records simultaneously and fixing the number of CDRs at 500. The execution time is of the order of milliseconds, so it can be used in production environment.

7 Conclusion

We deduce from all what we have seen about protecting personal data in CDR that securing them in the mediation is the efficient approach. Also we conclude that the number of encrypted/decrypted CDRs and number of encrypted and decrypted records simultaneously has a minimal influence on the execution time. So this approach of protecting personal data in mediation is applicable in a production environment of a telecommunication operator.

References

1. Technical Specification 32.298: Telecommunication management, charging management, Charging Data Record (CDR) parameter description, 3GPP, version 15.3.0 (2018)
2. Technical Specification 32.250: Telecommunication management, charging management, Circuit Switched (CS) domain charging, 3GPP, version 15.0.0 (2018)
3. Technical Specification 32.276: Telecommunication management, charging management, Voice Call Service (VCS) charging, 3GPP, version 15.0.0 (2018)
4. Technical Specification 32.274: Telecommunication management, charging management, Short Message Service (SMS) charging, 3GPP, version 15.0.0 (2018)
5. Technical Specification 32.297: Telecommunication management, charging management, Charging Data Record (CDR) file format and transfer, 3GPP, version 15.0.0 (2018)
6. AFCDP, Glossary anonymization of data (2008)

7. Mivule, K., Turner, C.: A review of privacy essentials for confidential mobile data transactions. In: International Conference on Mobility in Computing-ICMiC 2013, pp. 36–43 (2013)
8. Benjamin, C.M., Fung, M., Wang, K., Chen, R., Yu, P.S.: Privacy-preserving data publishing: a survey of recent developments. ACM Comput. Surv. **42**(4), 1–53 (2010)
9. Sweeney, L.: Achieving k-anonymity privacy protection using generalization and suppression. Int. J. Uncertain. Puzziness Knowl.-Based Syst. **10**(5), 571–588 (2002)
10. Islam, M.R., Islam, M.E.: An approach to provide security to unstructured big data. In: IEEE 8th International Conference on Software, Knowledge, Information Management and Applications (SKIMA), pp. 1–5 (2014)
11. Byun, J.-W., Sohn, Y., Bertino, E., Li, N.: Secure anonymization for incremental datasets, pp. 48–63. Springer, Heidelberg (2006)
12. Fraser, R., Willison, D.: Tools for anonymization of personal health information prepared for the pan-Canadian group of protection of personal health information (PRPS), September 2009
13. Xiao, X., Tao, Y.: Personalized privacy preservation. In: Proceedings of the 2006 ACM SIGMOD International Conference on Management of Data, pp. 229–240 (2006)
14. Križan, T., Brakus, M., Vukelić, D.: In-situ anonymization of big data. In: Information and Communication Technology, Electronics and Microelectronics (MIPRO), pp. 292–298 (2015)
15. Wong, W.K., Mamoulis, N., Cheung, D.W.: Nonhomogeneous generalization in privacy preserving data publishing, pp. 747–758 (2010)
16. Data Masking: What you need to know before you begin, A Net 2000 Ltd. White paper
17. https://www.quora.com/how-much-does-it-cost-tu-use-informatica
18. Call Detail Record Reference, V300R003C01LMA101, pp. 22–229. HUAWEI (2013)

Warning of Affected Users About an Identity Leak

Timo Malderle[1(✉)], Matthias Wübbeling[1,2], Sven Knauer[1,2],
and Michael Meier[1,2]

[1] University of Bonn, Endenicher Allee 19a, 53115 Bonn, Germany
{malderle,wueb,knauer,mm}@cs.uni-bonn.de
[2] Fraunhofer FKIE, Endenicher Allee 19a, 53115 Bonn, Germany

Abstract. Identity theft is a typical consequence of successful cyber-attacks, which usually comprise the stealing of employee and customer data. Criminals heist identity data in order to either (mis)use the data themselves or sell collections of such data to fraudsters. The warning of the victims of identity theft is crucial to avoid or limit damage caused by identity misuse. A number of services that allow identity owners to check the status of used identities already exist. However in order to provide proactive timely warnings to victims the leaked identity data has to be on hand. In this paper we present a system for a proactive warning of victims of identity leaks.

Keywords: Identity theft · Identity Data Leaks · Privacy ·
Password Security · Early Warning System

1 Introduction

Normally, Internet users use multiple web services like forums, social media platforms, or online shops [1,15]. Each user owns one or more accounts for each of these web services, which at least includes his email address and a chosen password. The user account describes a digital identity that is representing the user himself within the specific service domain. Every user can own different identities for different services. Digital identities are of big interest for criminals not only because they are traded with a high value in cyber space. There is a growing volume of illegally copied identities, which might be used for several kinds of fraud. In most cases the affected person notices too late that he fell victim to a criminal act. Usually, stolen identities are sold or published when the criminal intends no further use. Publication increases the risk for affected identity owners because the number of potential attackers is raised.

A person affected by an identity theft can be harmed by four different negative effects: financial, emotional, physical or social damage [14]. Therefore, it is desirable to inform affected identity owners about the theft of their digital identity so that preventive or reactive steps can be taken. To ensure that the

A. M. Madureira et al. (Eds.): SoCPaR 2018, AISC 942, pp. 278–287, 2020.
https://doi.org/10.1007/978-3-030-17065-3_28

majority of affected people will be notified, it is necessary to develop a service that automates the notification of affected people regarding recent identity leaks.

This work contributes an innovative design of an identity theft early warning service to reach a wider range of affected users than existing services.

The paper is organized as follows: Sect. 2 introduces the background and discusses related work. An analysis of the gathered identity leaks is provided in Sect. 3. In Sect. 4 our identity theft early warning service is presented and Sect. 5 concludes.

2   Background and Related Work

In the range of identity leakage and the warning of affected users there are different projects aiming similar goals. Additionally, there are services that a user can use to inform oneself about appearance in existing identity leaks. Most of these services require the user to enter his email address into a text field on the service's website. Afterwards, the service checks if the email address is contained in correlated leak databases. Such services are namely *have i been pwned* [10], the *HPI-Leak-Checker* [8], *vigilante.pw* [22] or *hacked emails* [3]. In addition there are companies that provide such a service commercially [4, 12, 16].

Some of these services display the result of the email review directly on the screen without further verification: *have i been pwned, vigilante.pw* or *hacked emails*. Such services have one big issue: everyone can check out an email addresses of another person. If a persons identity data is included in such a leak and the leak becomes public then it is impossible for him to delete or hide the leaked information. Moreover, it is published that a person has used a specific service. This leads to further problems for a person included in such leaks. Suicides are linked to data breaches, e.g. related to a leak of an online dating service [18]. Thereto, it can be seen that this is a high sensitive topic. Therefore, we think that the services mentioned above are ethically questionable regarding their process of such leaks and the decision to make them searchable for everyone.

It is better to report the result of the email review after an identity verification. The *HPI-Leak-Checker* sends the result via email to the user [6]. This approach ensures that the analysis result is handed out to the identity owner.

Nevertheless, all these services have a downside: a user has to know at least one of these services and needs to use it regularly to stay informed. Furthermore, not every service displays a list of identity leaks that are contained in the service's database. Thus, the user does not know whether he can use a service to find information about the existence of his personal data in the latest identity leaks. It is not visible to the user whether the identity leak database of this service is up-to-date. A further problem is that the results of the analysis will be sent to the user by email. This service is useless if the login credentials of their email accounts are leaked and the emails account is already hijacked by an attacker. Unfortunately, none of the services named here mentions how they keep their database up-to-date.

Our goal is to inform every affected user proactively, i.e. without the user's necessity of knowing about such a service. Therefore, we propose an early warning service aimed to contact a user proactively when his identity is leaked.

In our terms such a service is called **Identity Theft (Early) Warning System**. In contrast to the existing services, it operates on publicly available leaked identity data but proactively informs identity owners as long as they can be uniquely identified utilizing the sufficient identity features available in the leaks.

Onaolapo et al. provide a different perspective on this research area by exploring possible consequences for a user after his credentials are leaked [19]. They analyzed the identity leak ecosystem to get a better view on the real risks that occur for affected individuals. Therefore, they leaked login credentials of 100 registered *gmail* accounts. They demonstrate that accounts are really used by attackers for impersonation attacks.

A similar problem is the reuse of personal passwords. Han et al. and Subrayan et al. found out that around 33% of users, who have accounts for two different tested web services, reuse their passwords [7, 20]. In addition, if the password is not reused it is popular by users to use a prefix or suffix for passwords that is similar across different web services [7]. Therefore, it is necessary to inform users about leaked credentials and advise them on how to react accordingly.

The result of identity theft can be worse for a victim if the same password is used for several services as not only the leaked service can be attacked. For example, the attacker can try leaked credentials on multiple websites. Thomas et al. identified potential victims that are affected by identity leaks, phishing or malware [21]. They analyzed relevant data and which amount of data can be aggregated by passwords. Also Heen and Neumann studied the aggregation of identity data [9]. They tried to reveal the true identity of users by linking attributes of different identity leaks, e.g. hashes of passwords. Multiple privacy attacks are shown and an advice for users is given. It is shown that aggregating different identities from different leaks lead to high quality databases. Another study by Jaeger et al. found out that 20% of users affected by identity leaks reuse the same password for two or more different websites/services [13]. It can be seen that password reuse is relevant topic that intensifies the possible damage caused by identity theft.

Only few of the identity leaks contain information from which services the credentials are taken. Sometimes it would be useful to verify the authenticity of identity leaks. To verify found login credentials Casal "tested a subset of these passwords" at real service sites [2]. Most of the tested passwords are found valid and can be used for interactions with the services. We consider that there are ethical and legal concerns regarding this approach because it is a penetration in the personal privacy of the affected persons. In addition, it is legally questionable to do this.

It is a comprehensive task to notice new data leaks. DeBlasio et al. developed an identity leak detection service for external web services [5]. This service should notify about identity data breached from one of the supported services. This is done by registering honey pot accounts at the corresponding web services. Additionally, an email account is registered at a fake email provider having the same login credentials as used for the service. When the account data is robbed from a monitored service the attackers try to reuse the login credentials to log in to other services. Needless to say that the credentials are tried to log in to the fake email provider as well. If that happens successfully, a breach notification is triggered right away. However, this service does not provide user notification.

3 Leak Gathering and Analysis

In this section the gathering process and results of the identity leaks we have found so far will be presented.

Identity leaks are distributed through special forums, blogs or websites. It depends on the particular leak which attributes are included. Attributes can be of different types: email address, password, password hash, username, name, birth date, postal address, credit card number, etc. In most cases identity leaks are text files that consist of multiple lines. Each line contains a single digital identity separated into different attributes. The number of lines represents the size of a leak. The number of attributes per line may vary. Because pure spam lists are expelled from this analysis, the number of attributes varies from 2, typically an email address with another attribute (see below), up to ten, seldom more. A separator separates the different attributes per line. The most common formats are: [email:password], [email:hash], [email:hash:cracked-password], [email:username:hash:salt], [userid:username:email:ip:hash:salt], [entire SQL-Table as CSV], [Whole SQL-Dump]. Instead of the separator *colon* in the example every other character or string of characters can be used: *semicolon, comma, tab, \r, space* and many more. Most of the leaks have different attributes and no standardized format. In the period of one year we used parallel two different approaches to gather identity leaks. The first approach is a manual method. We searched through hacker forums and blogs. The second approach is an automated procedure. We implemented a fetcher that retrieves identity leaks with a defined structure from different sources automatically. Altogether, 521 identity leaks have been downloaded through manual gathering. These files contain 5,127,243,813 email addresses. Some of these email addresses can be found multiple times. This is due to multiple occurrence in the same or different leaks. When every unique email address is counted once, then there are still 1,563,002,958 email addresses. Related services store a comparable number of email addresses. The service *have i been pwned* owns 4.86 billion email addresses but email spam lists without any further information are included [10,11]. The service *vigilante.pw* owns 3.56 billion email addresses [22]. This comparison shows that this project gathered identity leaks in the same order of magnitude as other well established services. The multiple occurrence of email addresses might be caused by two reasons: First, some of the gathered leaks are combination lists that are an accumulation or enrichment of multiple leaks. Second, some users use the same email address for multiple services, which have been affected by data leaks.

With the second approach we gathered a lot more leaks with significant content. This approach supplies in the arithmetic mean 69 new files per day. In the period from April 2017 to July 2018 overall 18003 files were stored having 9.268.809 email addresses contained. The biggest part of the gathered files is scraped from the *pastebin* website, having 69.5% of the data. *Slexy* has 26.0%, *siph0n* has 3.9% and the rest 0.6%.

The data formats used in automatically collected leaks are similar to those in the manually identified leaks. However, there is a specialty about the first or last few lines of some files. Sometimes, there is a text giving some additional information about the leak. In some cases the text says that present data is only a part of the whole leak and the whole leak can be bought at a referred URL. Sometimes the leaks, manually collected ones as well as automatically gathered ones, include more information, which is given in a separate section with continuous text or so called ASCII-Art. This information, if present at all, varies a lot and differs from additional information about the data source, the targeted victims, the hacker, or still open security vulnerabilities, by which the identity data has been stolen.

When both approaches are compared it is shown that manually collected data lead to more data than the automatic approach. In the same time period the manual approach gathered 5.1 billion email addresses, the automatic approach only 9.3 million.

A development of an automatic parsing system is necessary. Therefore, a previous paper of the authors present an approach of how to automatically parse these files [17]. The software detects different blocks within a leak where the identity attributes have a distinct order or a varying separator. Following, in each block the separator is detected and in each line the attributes are separated. After that, it is tried to assign meaning to the column of attributes. Then, values of recognized attribute types are normalized and stored in the leak database. This database is the groundwork for the warning system which is presented next.

4   Early Warning Service

We collected a tremendous amount of digital identities that are processed and normalized. This is the basis for the intended notification of affected individuals. As shown before, services where users can enter their email addresses to check if they are affected by an identity leakage are not a satisfactory solution due to their reactive approach. It is likely that only a small percentage of individuals find out their consternation with the help of such a service. Therefore, a service is required that proactively notifies the victims with an alert including further instructions on how to react. We designed such a *warning service* as described in the following paragraphs.

The core question is how to contact the affected person behind the leaked identity. The most trivial solution is to send emails to the users, as in most cases an email address is included in identity leaks. However, this kind of emails will most probably be classified as spam mails by the receiver before being noticed.

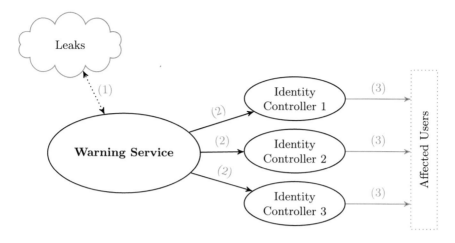

Fig. 1. Structure of the warning service

The problem is that a user would not trust the sender of such an email because he is unknown and has a low credibility for the user. Whom would the user trust when getting an identity leak notification? A better solution would be that a more credible company or institute can notify the user. A more suitable approach would be that a company or institution, which has a relation to the user, would inform him on behalf of the *early warning service*. Such a company could be a social network, email provider or online bank. That would be advantageous because these companies access additional communication channels for example a phone call, a letter, a login message or a push notification to the smartphone.

Our vision is that an *early warning service* continuously gathers and analyzes identity leaks. This service cooperates with different companies (*identity controllers*) and shares necessary data of an identity leak with those companies. At best, such companies have a high credibility for an affected user. So, when an *identity controller* warns a customer about an identity leak, he should trust this information. Thereby, privacy concerns are very important. As a result, transmitted attributes of an identity leak should be as restricted as possible but they should describe the person belonging to the leaked identity sufficient. These identifiers have to be shared with the *identity controller*. Therefore, different approaches are thinkable. One approach is that the *warning service* provides a Rest API that can be queried about the consternation of users by the *identity controller*. Hence, the *identity controller* must send user data to a third instance without any consent of the user. That would not work because of privacy concerns. The *identity controller* would release who is a user of this service.

Another approach is that the *identity controller* provides a REST API where the *warning service* can send prepared data out of a new identity leak. The collaboration between the *warning service* and the *identity controller* can be seen in Fig. 1. (1) The *warning service* gathers identity leaks from the mentioned sources and processes them. (2) Then, the *warning service* sends the processed

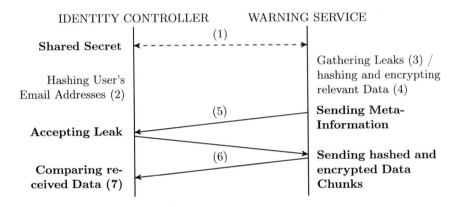

Fig. 2. Communication model of identity leak exchange

leaks to the API of the *identity controller*. (3) The *identity controller* checks whether a user is affected and in such a case a warning can be submitted to the matching user.

A problem is that a valid email address alone in an identity leak would lead into a user warning. Too many and irrelevant warnings should be avoided because it would decrease the awareness of security warnings and would end in a situation where the user ignores all warnings. In order to mind that it is necessary to proof if there is still a threat in form of impending identity theft. This is only given when a combination of login name and password included in a leak are valid login credentials for a service at the specific point in time. As a result, the *warning service* should share email addresses and passwords, so that an *identity controller* can review the received data. Though, the *identity controller* should not get any information about unknown individuals. Thus, email addresses and passwords have to be delivered in a protected form. Hashing is the first idea and it works well for email addresses. However, an *identity controller* should not store passwords as plain text but in a hashed format. So that the *identity controller* can compare the received password with the stored user passwords there are two options: First, the *identity controller* publishes his hashing process with the included salts and the number of iterations. Then this process can be used by the *warning service*. The publication of the hashing process would not work for the most companies because there will be huge security concerns. Second, the *identity controller* receives the passwords in an encrypted form that can be decrypt into plain text. So, the *identity controller* can hash the passwords himself to compare them with the stored user hashes.

The communication between the *early warning service* and the *identity controller* usually happens in multiple steps which are explained below and plotted in Fig. 2 (detailed view of (2) in Fig. 1):

1. A shared secret S is agreed between the *warning service* and the *identity controller*. It is generated during the first connection of these communication partners using the diffie hellman algorithm.

2. The *identity controller* hashes every users' email address with the hash algorithm h() and the shared secret S as a salt and stores the result in its database.
3. When the *warning service* gets a new identity leak then it will be analyzed onto relevant attributes like email address, password and additional meta information.
4. The *warning service* hashes the found email addresses with the hash algorithm h() and the shared secret S. Additionally the password is encrypted with a symmetrical encryption method e(). The email address in plain text combined with the shared secret is used as the key.
5. After that, the *warning service* sends some meta information to the *identity controller* like the number of contained data records and the approximate age of the leak.
6. If the *identity controller* accepts the leak then the *warning service* sends the hashed email addresses and the encrypted passwords to the *identity controller*. The whole data is separated in chunks for a better handling.
7. The *identity controller* compares the received hashes with its own hashed user email addresses. If there is a match the *identity controller* can lookup the plain text of the email address. This email address and the shared secret can then be used to decrypt the encrypted password with the decryption method d(). Now the *identity controller* can check if the user provides the same password for their services.
8. If there is a match then the *identity controller* warns the affected user.

For each cooperating *identity controller* this communication protocol is used with a different shared secret. The advantage of this approach is that the *warning service* does not come aware of the registered user of the *identity controller*.

The legal assessment is not part of this paper but it is considered in parallel work by lawyers and data protectionists. Our whole approach and the implemented system complies with European law. We involved different organizations in the design of our *early warning service* that are able to contact many users, such as social media platforms and telecommunication providers. With growing maturity, email providers and online shops should be incorporated as well. The participation in such a *warning service* has different benefits for an *identity controller*. In case of leaked user data this system would help to realize the obligation to inform every affected user that is regulated by article 34 of the EU GDPR. It is also a tool which improves the reactive IT security of a company. Such a participation also helps to reduce spam or fraud which arises through misused user accounts. Possibly, it is a successfully marketing feature that a company protects their users against identity theft.

5   Conclusion

Accounts, associated information and login credentials represent service specific digital identities of users. Several ways exist, e.g. by successful attacks on computers of service providers, how this identity data falls in unauthorized hands and is misused for criminal activities or sold by data dealers. In the course of these activities extensive collections of identity data become publicly available on the Internet. Publicly accessible internet-based data storage providers or exchange services are (mis)used as data sinks for huge collections of identity data. Typically all this happens unnoticed by the affected identity owners who are then exposed to the resulting threats. Following the vision to resolve this deficiency we systematize the processes of identity leakage, involved services and components around such activities. We propose a process for systematic collection and analysis of identity data for the purpose of proactive warning of affected identity owners. Different types of data sinks as well as possible strategies for data discovery and gathering have been investigated and challenges regarding varying formats of identity data collections have been discussed. It can be supposed that only few of these victims notice the consternation by the use of existing identity leak information services like *have i been pwned*. Therefore, we present our novel and innovative approach to notify a wider range of affected users in contrast to the existing services. For this purpose, a *warning service* is designed.

Future work will focus on identifying the preferable communication channel for each affected user through a deeper inspection of available data. Additionally, the design and contents of warning messages will be regarded. Further, a process has to be developed to measure the general authenticity and threat level of leaks. This is crucial to choose optimal countermeasures and provide users necessary and valuable information on how to react.

The work presented here was funded by the German Federal Ministry of Education and Research under contract no 16KIS0696K.

References

1. Bras, T.L.: [infographic] online overload - it's worse than you thought, July 2015. https://blog.dashlane.com/infographic-online-overload-its-worse-than-you-thought/. Accessed 27 Sept 2018
2. Casal, J.: 1.4 billion clear text credentials discovered in a single database. A Medium Corporation - 4iQ, December 2017. https://medium.com/4iqdelvedeep/1-4-billion-clear-text-credentials-discovered-in-a-single-database-3131d0a1ae14. Accessed 27 Sept 2018
3. Chia, J.M.: Hacked-Emails (2017). https://hacked-emails.com/. Accessed 27 Sept 2018
4. Corp, S.: LifeLock (2018). https://www.lifelock.com/education/4-lasting-effects-of-identity-theft/. Accessed 27 Sept 2018
5. DeBlasio, J., Savage, S., Voelker, G.M., Snoeren, A.C.: Tripwire: inferring internet site compromise. In: Proceedings of the 2017 Internet Measurement Conference, IMC 2017, pp. 341–354. ACM, New York (2017)
6. Graupner, H., Jaeger, D., Cheng, F., Meinel, C.: Automated parsing and interpretation of identity leaks. In: Proceedings of the ACM International Conference on Computing Frontiers, CF 2016, pp. 127–134. ACM, New York (2016)
7. Han, W., Li, Z., Ni, M., Gu, G., Xu, W.: Shadow attacks based on password reuses: a quantitative empirical view. IEEE Trans. Dependable Secur. Comput. **15**(2), 309–320 (2016)

8. Hasso-Plattner-Institut für Digital Engineering gGmbH: HPI Leak Checker (2017). https://sec.hpi.de/leak-checker. Accessed 27 Sept 2018

9. Heen, O., Neumann, C.: On the privacy impacts of publicly leaked password databases. In: Polychronakis, M., Meier, M. (eds.) Detection of Intrusions and Malware, and Vulnerability Assessment, pp. 347–365. Springer International Publishing, Cham (2017)

10. Hunt, T.: have i been pwned? (2017). https://haveibeenpwned.com. Accessed 27 Sept 2018

11. Hunt, T.: Inside the massive 711 million record onliner spambot dump (2017). https://www.troyhunt.com/inside-the-massive-711-million-record-onliner-spambot-dump/. Accessed 27 Sept 2018

12. IdentityForce, I.: IdentityForce (2018). https://www.lifelock.com/education/4-lasting-effects-of-identity-theft/. Accessed 27 Sept 2018

13. Jaeger, D., Pelchen, C., Graupner, H., Cheng, F., Meinel, C.: Analysis of publicly leaked credentials and the long story of password (re-)use. In: Proceedings of the 11th International Conference on Passwords (PASSWORDS2016). Springer, Bochum (2016)

14. Johansen, A.G.: 4 lasting effects of identity theft (2018). https://lifelock.com/education/4-lasting-effects-of-identity-theft/. Accessed 27 Sept 2018

15. Lord, N.: Uncovering password habits: are users' password security habits improving? (infographic), September 2017. https://digitalguardian.com/blog/uncovering-password-habits-are-users-password-security-habits-improving-infographic. Accessed 27 Sept 2018

16. Experian Ltd.: Experian (2018). https://www.experian.co.uk. Accessed 27 Sept 2018

17. Malderle, T., Wübbeling, M., Knauer, S., Sykosch, A., Meier, M.: Gathering and analyzing identity leaks for a proactive warning of affected users. In: Proceedings of the 15th ACM International Conference on Computing Frontiers, CF 2018, pp. 208–211. ACM, New York (2018)

18. Malm, S.: Two suicides are linked to Ashley Madison leak: Texas police chief takes his own life just days after his email is leaked in cheating website hack (2015). https://www.lifelock.com/education/4-lasting-effects-of-identity-theft/. Accessed 27 Sept 2018

19. Onaolapo, J., Mariconti, E., Stringhini, G.: What happens after you are pwnd: understanding the use of leaked webmail credentials in the wild. In: Proceedings of the 2016 Internet Measurement Conference, IMC 2016, pp. 65–79 (2016)

20. Subrayan, S., Mugilan, S., Sivanesan, B., Kalaivani, S.: Multi-factor authentication scheme for shadow attacks in social network. In: 2017 International Conference on Technical Advancements in Computers and Communications (ICTACC), pp. 36–40 (2017)

21. Thomas, K., Li, F., Zand, A., Barrett, J., Ranieri, J., Invernizzi, L., Markov, Y., Comanescu, O., Eranti, V., Moscicki, A., Margolis, D., Paxson, V., Bursztein, E.: Data breaches, phishing, or malware?: Understanding the risks of stolen credentials. In: Proceedings of the 2017 ACM SIGSAC Conference on Computer and Communications Security, CCS 2017, pp. 1421–1434. ACM, New York (2017)

22. vigilante: vigilante.pw (2017). https://vigilante.pw/. Accessed 27 Sept 2018

Network Security Evaluation and Training Based on Real World Scenarios of Vulnerabilities Detected in Portuguese Municipalities' Network Devices

Daniel José Franco[1](✉), Rui Miguel Silva[2](✉),
Abdullah Muhammed[1], Omar Khasro Akram[3], and Andreia Graça[2]

[1] Faculty of Computer Science and Information Technology,
University Putra Malaysia (UPM), 43400 Serdang, Selangor, Malaysia
luopdf@gmail.com
[2] Lab UbiNET/LISP, Polytechnic Institute of Beja, 7800-295 Beja, Portugal
ruisilva@acm.org
[3] Faculty of Design and Architecture, University Putra Malaysia (UPM),
43400 Serdang, Selangor, Malaysia

Abstract. Nowadays, public and private organizations have demonstrated some sensibility in maintenance and security updates of their equipment. However, their main focus are servers and workstations, leaving network devices, such as routers and switches often forgotten in this process. This research addresses the vulnerabilities on network equipment, intending to evaluate their dimension, in Portugal's City Halls and, after that, analyze and rate their impact according to taxonomies, such as CAPEC. This study also aims to set of vulnerabilities to reply, elucidate and sensitize not just City Halls ITs, but also other type of public and private organizations about the risks related to outdate network devices. The vulnerability demonstrations were done through the design of different scenarios, with real devices, installed in a mobile rack, called "Hack Móvel" and using network simulators. Each scenario was documented with multimedia contents, allowing teaching hacking techniques in network devices. As methodology, the study adopts the quantitative method, through the application of questionnaires applied to each City Hall, in order to collect relevant information about the device models and brands, as well as the firmware version they are really using. It is also adopted the quantitative method in order to perform tests with real users and evaluate the scenarios that were designed. Results show a really good acceptance of the "Hack Móvel" by users and their motivation to increase their knowledge on the computer security and hacking techniques field.

Keywords: Pentest · Audits to information systems · Network devices ·
Network security · Network vulnerabilities · Security training

1 Introduction

Public and private organizations are nowadays sensible to the need of maintaining their devices updated to the last security software versions, however, when focusing on devices, they are usually related to just servers and users' workstations, where network

© Springer Nature Switzerland AG 2020
A. M. Madureira et al. (Eds.): SoCPaR 2018, AISC 942, pp. 288–297, 2020.
https://doi.org/10.1007/978-3-030-17065-3_29

devices, such as routers and switches, are most of the time left behind, during the update process.

It is a common practice among operating system programmers and developers to promote a set of periodic updates that not just increases some features to the systems but also increases their security level. In contrast to servers and workstations, network devices, such as routers and switches, require a bigger attention and careful by IT professionals, due to the lack of alert features related to the release of new patches and versions. It is periodically necessary to consult the device and system providers in order to obtain detailed information about new updates, as well as to manually download them in case of existence. Also, the installation of new updates is not as user friendly as the installation done to servers and workstations. Network devices require an advanced knowledge in order to perform a correct system update. Taking Cisco devices as an example, users must have advanced knowledge on network concepts as well as on IOS programming and configuration, for instance the different memories existing on the devices and the specific procedures for updating the system, where it is needed IOS commands, copying files from and to the device, backup procedures, between others. Not being an easy task that requires careful and time, it is most of the time left for later, with the possibility of never being done, increasing as a consequence, the risk of successful hacking attacks to organizations.

Network devices, namely routers, are mainly used to provide Internet access, connecting the Local Area Network (LAN) to the Internet Service Provider (ISP). Having at least one of its interfaces connected to the outside, these devices are accessible through the Internet, allowing its analysis and exploitation by malicious attackers. Using an outdate system makes possible for an attacker to exploit IOS and firmware vulnerabilities, taking advantage over the device and resulting, for instance, in the network isolation by a simple DoS attack, or even gaining access to the device and consequently to the organization local area network, obtaining private and sensitive information. Being devices that are accessible from the outside (Internet), their management and update should be constantly performed, in order to prevent future attacks and private or sensitive information loss.

Among the years a large number of vulnerabilities has been found, not just on software level, but also on hardware side. Due to this constantly growing number, efforts are being done in order to better understand such vulnerabilities and present possible corrections to be applied, as well as their classification in order to better evaluate the associated risks. Robert Seacord and Allen Householder [1] present on their study, about security vulnerabilities classification, that it should be done based on a solid engineering analysis, in order to allow the possibility of determining the threats for each vulnerability and, consequently, prevent future ones. The vulnerabilities classification and analysis allow studies on the frequency, trend analysis and correlation with incidents and exploits, providing a better way of sharing information among organizations from different countries.

The correct way on classifying vulnerabilities is still not fixed nowadays, however there are a large number of research groups and organizations that keep different classification between them. Among these classifications it is important to highlight the following:

- CAPEC – Created to answer the community needs in relation to the attackers' perspective and their adopted ways on vulnerabilities exploitation. It consists in a list of common standards associated to a wide scheme of taxonomical classification [2];
- CWE – Worldwide available this classification provides, for free, a measurable and unified set of weak points related to the software, turning the discussion, description, selection and usage of security tools more effective [3];
- CVE – This classification is a list of standardized software vulnerability and weaknesses names and was developed to standardize the name of all known vulnerabilities and weaknesses. The CVE assigns a unique identifier to each vulnerability, composed by the year of its discovery, followed by a four-digit number which increments on every new discovered vulnerability [4].
- CVSS – The CVSS classification consists on a system for assigning a severity score to each one of the existing vulnerabilities and it was developed to promote an open and standardized method for computer vulnerabilities classification. The score is calculated using different metrics and may present a value between 0 and 10, being 10 the most serious one [5];

There is a lack of attention to the security of network devices in Portuguese Municipalities (and probably in other countries, as well), besides that, it is good and usefull case study for network security attacks due to its importance in society life. Therefore, there is a dire need to evaluate the current risk of attack and use that information and results, to develop scenarios with real equipment for awareness and training on network security hacking techniques and that way contribute to a better conscientiousness on network security domain.

So, as objectives this research aims to first summarize and identify the main existing vulnerabilities' taxonomies and classifications, second to analyze its current risk of impact in the Portuguese City Halls network devices and third to replicate some vulnerability scenarios integrated in a mobile lab also to be develop on this study, to better demonstrate and make available a hands-on environment for awareness and training on network security hacking techniques.

2 Literature Review

As it was told before, security is still a growing area and IT professionals tend to pay more attention to servers and workstation, neglecting most of the times the network devices that are installed on their local networks. The study handled by Liebmann [6], about the real SNMP vulnerability, presents the real protocol security issue as being its architecture. The SNMP is mostly based on an architecture named by the author as "in-hand" that is focused on the network internal level, where the protocol is working on the same network segment as the monitored devices. This condition brings, to the author, two severe problems, namely the case when a flaw happens on the network, making it impossible to establish the communication with the devices, compromising the monitoring process and if the flaw happens in one of the remote network devices, making almost impossible for a technician to quickly access it in order to correct the issue, being, most of the times, the physical access the only possible way.

The study addressed by Agarwal and Wang [7] about the Mesh networks performance on DoS attacks showed that the existence of medium network errors, presents a significant impact on the network performance, being also compounded by the retransmissions that happen on the MAC level and showing the weakest links. The study shows also that a "path-based DoS" attack is most of the times more efficient if done through weak links, when compared to the ones done through strong links and, by the physical diversity, an attacker with a far physical location may be more dangerous than a closer one. As conclusions, the authors agreed that collisions between attacks do not necessarily represent a risk to the network performance, being the physical local of the attackers the most important factor.

The study conducted by Shivamalini and Manjunath [8] about the "Man-in-the-Middle" attack shows a possible security method where each node has one key. When communications are made between the nodes, it first starts with an authentication attempt, where the keys must be validated. In other words, when a node sends data to the other, a server compares its key to the one registered on the database. The data transmission just occurs if the authentication was successfully completed.

The study made by Stasinopoulos, Ntantogian and Xenakis [9] came to show that routers usually used on ADSL connections could once more be considered the weakest link on the network. Typically, these devices are provided by ISPs and, on their majority, are managed by users that have almost no technical knowledge. Incorrect configurations leave them vulnerable and an easy target for network attacks. This study shows that the security level of some of the routers used in residential services have vulnerabilities that may compromise their management Web pages.

Research Hypothesis

The analysis of the current risk of attack to network devices in the Portuguese Municipalities and its correspondent level of impact, makes it possible to develop real world scenarios to be used for awareness and training on network security hacking techniques.

3   Methodology

The adopted methodology makes use of a quantitative method, taking the Portuguese City Halls a survey objects, identifying their network devices and respective IOS and firmware versions in order to reproduce scenarios and achieve the initial objectives of the study. As part of the survey methodology, questionnaires are also applied in order to collect all information, while keeping the answers anonymous, not allowing the correlation between them and the participant Municipalities. This method allows the study to evaluate the developed scenarios, through test with users, in order to calculate their performance. A questionnaire needs also to be applied in order to study the population who will be testing the scenarios and procedures [13].

4   Municipalities and the Collected Data

The contacts made with the 308 Municipalities of Portugal were done through their email addresses available on their official Webpages, mainly to the Head of the IT Department of each City Hall (Table 1).

Table 1. Portuguese municipalities' contact statistics

	Total	Percentage
Total number of Municipalities	308	100%
Number of emails sent	290	94%
Emails that reached to the destination	275	89%
Emails that were read by the receiver	143	46%
Number of answers	89	29%

Confronting the obtained results with the information provided by the National Institute of Statistics (INE), in relation to the Portuguese population [10], Portugal had by the date of this study, 10.636.979 inhabitants, where the total number of people covered by the Municipalities that have answered the questionnaires is 55.337.603 inhabitants. By this, the results cover around 50.2% of the whole Portuguese population, which gives real world significance to this study.

With a total of 65%, which correspond to around 58 of the obtained answers, Cisco is the most used brand of routers on the Municipality networks, followed by Draytek, with 10%, which corresponds to just 9 of the obtained answers. In accordance with the study handled by the Infonetics Research, Cisco, Alcatel-Lucent and Huawei are the most recognized brands among consumers, being Cisco the leader one [11].

Focusing on the Cisco models, 40% of the Municipalities are using the 800 Serie Cisco routers. It is also important to highlight that 26% of the them are using a more robust device, the Cisco 2811, followed by the Cisco 1700 present in 11% of the Municipalities.

With the constantly technology growth, appearing new software and hardware each day, it is hard for the Municipalities and other organizations do follow all the new trends, being the market study in accordance to their need or most advantage way. Having, most of the times, reduced budgets City Halls find themselves buying new devices just when the old ones are no longer working or completely obsoletes.

Questionnaires show that 48% of the Municipalities are using devices with more than 4 years of usage, which if not regularly updated, monitored and managed may provide a large number of vulnerabilities easily exploitable by attackers.

Questionnaires also show that almost half of the responders are sensible to the need of performing regular updates to their network devices, identifying security, performance improvement and new features as main reasons. However, on the other hand, and this time a bit more perturbing, the great majority (52%) of the City Halls do not perform any type of update on network devices and their firmware, using the lack of time as the main reason for such a dangerous choice. Another disturbing fact is that even existing Municipalities that perform regular updates to their devices, the great majority of identified IOS versions present old release dates.

When asked about suffered attacks and vulnerabilities, 88% of the Municipalities do not perform any type of analysis, not knowing the high risks they are going through. A more positive fact is that just 12% of them have already suffered a security attack, where in 50% of them, there were no negative consequences neither any systems nor device damages. Among the identified security issues caused by attacks, it is possible to identify DoS attacks, which added to present communication flaws a total of 38% of the attacks. It is also important to highlight the attacks where the device was compromised (12%), being necessary a password recovery procedure.

By the identified router brands, models and IOS versions, it was possible to develop a survey using the CVE Details website [12], detecting around 152 vulnerabilities between DoS, Remote Code Execution, Information Gathering, Bypass and XSS (Fig. 1).

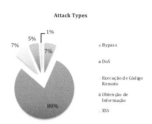

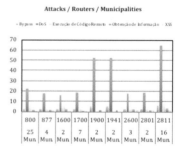

Fig. 1. Percentage of the identified attack types

Fig. 2. Comparison between attacks, routers and Municipalities

In a more specific way, Fig. 2 presents a comparison between attack types for each of the identified routers and the number of Municipalities that are using that device.

5   Analysis of Vulnerabilities

Confronting the obtained vulnerability survey with the attack pattern CAPEC classification, it is possible to identify the following main categories:

- 118 – Collect and Analyse Information;
- 152 – Inject Unexpected Items;
- 156 – Engage in Deceptive Interactions;
- 172 – Manipulate Timing and State;
- 223 – Employ Probabilistic Techniques;
- 225 – Subvert Access Control;
- 255 – Manipulate Data Structures;
- 262 – Manipulate Systems Resources.

It is possible to find a specific CVE in more than one CAPEC category, due to the nature of this classification that focus on the objectives and service types as attack patterns. However, each CVE belongs just to one specific type of exploited weakness CWE. An important aspect to observe is the potential impact of the founded vulnerabilities specified in the CVSS classification. In analysis (Fig. 3) it is possible to observe that the majority of the identified vulnerabilities (80 out of 151) have a CVSS score of 7.8 out of 10, representing a really high risk of impact. A score of 7.8 indicates severe system flaws that may result in confidentiality, integrity and availability loss.

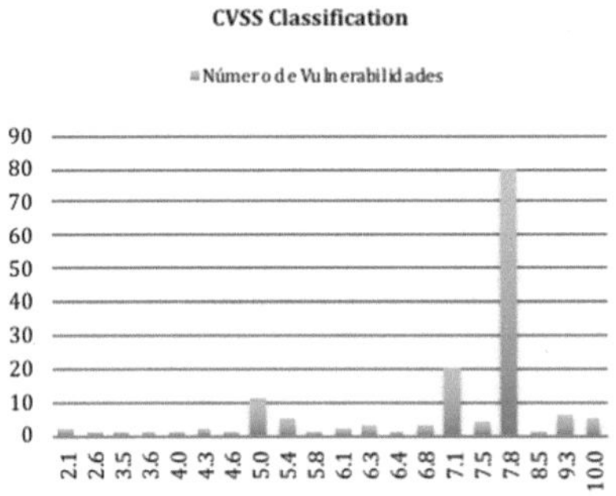

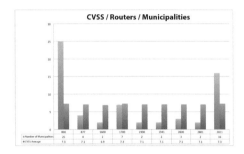

Fig. 3. Vulnerabilities' CVSS classification

Fig. 4. Average CVSS classification by devices and number of Municipalities

At this point it was also important to understand the risk of impact that each router represents and how many Municipalities are using those devices on their networks (Fig. 4).

Focusing on the 800 and the 2811 models, which are the most used by the Municipalities, they also present the highest CVSS average score. It is true that these are really high values, however, it is not possible to claim that all the Municipalities that are using those devices have exactly 7.3 of risk of impact, being this score just a reference for the average risk. In order to correctly obtain the CVSS score it is necessary to take some factors into consideration, like the types of active services on the devices and what kind of vulnerabilities may in fact be exploited. Though, it is possible to assume 7.3 on CVSS score as the highest average score after the vulnerability's analysis and, due to its high value, consider that probably a large number of Municipalities are at a high risk of suffering network attacks. 6.9 is the lowest obtained CVSS score, corresponding to the Cisco 1600, which is also the less used device by the Municipalities. Despite of being the lowest observed value, it is still a high score, where it is also probable that the user may be at a high risk of losing availability, confidentiality and integrity on their network routers.

6   Implementation of Scenarios and "Hack Móvel"

In order to implement tests as real as possible, to allow a better awareness and training with hands-on possibility and using real world equipment and with a wide range of techniques, some factors was taken into consideration:

- Attack environment as internal or external to the routers' LAN;
- Exploited vulnerability in CVE classification;
- Exploited vulnerability impact in CVSS classification;
- Exploited protocol;
- Attack consequence.

Table 2, bellow, presents the selected four implemented scenarios.

Table 2. Selected implemented scenarios

Environ.	CVE	CVSS	Protocol	Consequence
Internal	1999-0517	7.5	SNMP	Information gather
External	2001-0537	9.3	HTTP	Bypass authentication
Internal	2003-0567	7.8	IPv4	Denial of service
External	2008-3821	4.3	HTTP	Cross site scripting

One of these study objectives was to develop a mobile lab to allow a better demonstration of the founded vulnerabilities, using real world equipment in a hands-on perspective for a better awareness impact. Figure 5 illustrates the developed "Hack Móvel", as a Mobile Rack containing the following equipment's:

- LCD Screen, KVM console (Keyboard/VGA/Mouse) Switch;
- Keyboard and Mouse;
- 2 Cisco Router 2811, 1 Cisco Router 2600 and 1 Cisco Router 800;
- Cisco Switch Catalyst 2960;
- Fujitsu Siemens PC (3x);
- Cisco Rollover cable (4x);

Fig. 5. The "Hack Móvel"

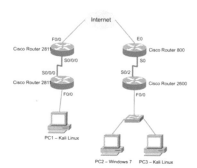

Fig. 6. Scenarios – General network topology

Devices were chosen in accordance to the Municipalities most used ones. In the specific case of the Cisco Switch Catalyst 2960, it was installed in order to provide network connections to the PCs, not having any influence on the scenarios, neither on the obtained results. Using different types of operating systems was also a factor taken into consideration and, by this, on the three installed computers there is a Kali Linux, provided with a large number of attack tools, and two Windows 7. It is also important to highlight that these systems may be easily changed in accordance to the implemented scenario.

The mobile lab allows a large number of possible configurations and scenarios, however, in this specific study, it was adopted the following base network topology (Fig. 6).

7   User Tests

Tests with real users were performed in order to better evaluate the quality of the developed scenarios and "Hack Móvel" platform. Tests were majority performed in an individual way. After the implementation and configuration of each one of the four scenarios, users have also answered a small questionnaire [13], in order to demographically study the responders, classify the implemented security network hacking techniques and the provided multimedia materials.

The users' questionnaire analysis shows that 50% of the users are between 26 and 35 years old, being half men and half women. Focusing on the education level, 50% of the responders have completed their Bachelor Degree and the other were performing it. With 62%, responders have some knowledge about network devices and their management and configuration, however, when focusing on hacking skills, 75% assume to not have any type of knowledge about this field.

The obtain results on the implemented scenarios showed that 100% of the users felt really satisfied and motivated performing the configurations and the attacks on the "Hack Móvel" platform, as well as the supporting material that was provided in the beginning of the tests.

In general, 50% of the responders classified the scenarios, support materials and multimedia materials with 5 points on a scale from 0 to 5, where 0 is the lowest classification and 5 the highest. 88% of the responders classified the scenarios as very good.

8   Conclusions

From the resulting analysis of this research, there were found different taxonomies and vulnerability classifications, each one with its own specifications and characteristics, being important to highlight the CAPEC, CVE and CVSS classifications that allow us to identify the vulnerability nature and its potential of impact to organizations. When mixing some of the studied classifications, it is possible to obtain a deep approach from each one of the existing vulnerabilities, proving a better understanding on the issue, in order to administrators, developers and IT professionals to take the finish actions and correct or prevent future attacks.

Highlighting the Municipalities average risk of impact, it is once more a factor that must be taken into consideration. A score of 7.8 on the CVSS classification, where the maximum value is 10.0 is considered a high and dangerous classification. The "Hack Móvel" development was considered a great help to awareness about network security hacking techniques and methodologies. Having, on this first stage, just Cisco devices, the "Hack Móvel" allows, in future evolutions, to include new devices from different brands and models, providing a dynamic configuration and adaptation to the needed scenarios. Its mobility feature makes it a smart choice to educational and training classes, where it is needed to move devices from classroom to classroom, as well as to demonstrate attacks in real time, with real world equipment, promoting a more dynamic and efficient learning.

Acknowledgments. Thanks to "Fundação para a Ciência e a Tecnologia" for grant through "UID/CEC/04668/2016-LISP.

References

1. Seacord, R., Householder, A.: A Structures Approach to Classifying Security Vulnerabilities (2005). http://oai.dtic.mil/oai/oai?verb=getRecord&metadataPrefix=html&identifier=ADA4 30968. Accessed June 2018
2. CAPEC: Common Attack Pattern Enumeration and Classification. https://capec.mitre.org. Accessed Aug 2018
3. CWE: Common Weakness Enumeration. http://cwe.mitre.org. Accessed Aug 2018
4. CVE: Common Vulnerabilities and Exposures. https://cve.mitre.org. Accessed Aug 2018
5. CVSS: Common Vulnerability Scoring System. http://www.first.org/cvss. Accessed Aug 2018
6. Liebmann, L.: SNMP's Real Vulnerability. Communication News, p. 50, April 2002
7. Agarwal, A.K., Wang, W.: An experimental study of the performance impact of path-based DoS attacks in wireless mesh networks. Mob. Netw. Appl. **15**(5), 693–709 (2010)
8. Shivamalini, L., Manjunath, S.: An approach to secure hierarchical network using joint security and routing analysis. Int. J. Comput. Appl. **28**(8) (2011). http://www.ijcaonline.org/volume28/number8/pxc3874752.pdf. Accessed July 2018
9. Stasinopoulos, A., Ntantogian, C., Xenakis, C.: The weakest link on the network: exploiting ADSL routers to perform cyber-attacks. IEEE (2013). http://ieeexplore.ieee.org/xpl/article Details.jsp?reload=true&arnumber=6781868&sortType%3Dasc_p_Sequence%26filter%3D AND(p_IS_Number%3A6781844). Accessed July 2018
10. National Statistical Institute. https://www.ine.pt/xportal/xmain?xpid=INE&xpgid=ine_ destaques&DESTAQUESdest_boui=83328022&DESTAQUESmodo=2&xlang=en. Accessed July 2018
11. WSilicon Week. http://www.siliconweek.es/noticias/y-los-fabricantes-mas-valorados-de-routers-y-switches-son-52916. Accessed July 2018
12. CVE Details: The Ultimate Security Vulnerability Datasource. http://www.cvedetails.com. Accessed June 2018
13. Akram, O.K., Mohammed Jamil, N.F., Franco, D.J., Graça, A., Ismail, S.: How to Guide Your Research Using ONDAS Framework (2018)

A Novel Concept of Firewall-Filtering Service Based on Rules Trust-Risk Assessment

Faouzi Jaïdi[✉]

ESPRIT School of Engineering, Tunis, Tunisia
faouzi.jaidi@gmail.com, faouzi.jaidi@esprit.tn
http://www.esprit.tn

Abstract. The importance given to firewalls as a security mechanism for protecting sensitive and private infrastructures has been well justified in literature. Nowadays, we consider firewalls as one of the most important security mechanisms that is widely deployed and highly approved. The main goal of this fundamental security component is to provide a filtering service by blocking or providing access to specific areas and segments of a network based on a set of filtering rules defined with regards to the global security policy. Hence, the effectiveness of the protection provided by a firewall is governed by the efficiency of the filtering policy deployed in that firewall. To enhance the quality of the filtering service provided by firewalls, we propose a novel filtering technique that integrates a risk assessment approach to evaluate the risk associated to firewalls rules. Our goal is to strengthen the filtering service with pertinent information relative to rules risk values that allows (i) changing the actions associated to critical rules in specific/critical contexts or (ii) dynamically injecting new rules in the firewall that refine other rules (by giving precision or reducing domains) to reduce the risk or (iii) changing the behavior of the firewall by changing its configuration (the set of rules) to avoid malicious scenarios.

Keywords: Data and networks security · Security policies ·
Firewalls · Risk assessment

1 Introduction

Nowadays, security approaches and solutions rely on firewalls as a fundamental access control mechanism for ensuring the protection of resources (computers, networks, data, services, devices, etc.) in critical infrastructures. Security Architects consider firewalls as a key node in Information Systems (IS) security that constitutes the backbone of every IS security solution. This main component provides a filtering service by blocking or providing access to specific areas and segments of a network based on a set of filtering rules. Rules configured and implemented in the firewall are defined with regards to the global security policy.

© Springer Nature Switzerland AG 2020
A. M. Madureira et al. (Eds.): SoCPaR 2018, AISC 942, pp. 298–307, 2020.
https://doi.org/10.1007/978-3-030-17065-3_30

Nevertheless, it is clear that the effectiveness and reliability of a firewall protection depends on the correctness and coherence of its configuration. Indeed, an oversight in the configured policy may have dramatic consequences on the security and/or the functionality of the infrastructure. Moreover, several common problems may arise from firewall misconfiguration that may have dramatic effects on IS operations.

On another hand, it is commonly agreed that, over time, firewall rule bases may rise in size and complexity. This situation, inevitably, can lead to misconfiguration problems, non-compliant rules, excessive permissions and other issues that expose the IS to a serious risk. Therefore, enhancing the quality of the filtering service is an unavoidable task. To address this issue, several research works proposed a variety of approaches for: (i) the discovery and removal of firewall misconfigurations; (ii) the conduct of firewall reviews for compliance reasons; and (iii) firewalls rules analysis. Nonetheless, existing approaches do not provide an effective solution to clear up and dismiss this risk and have some drawbacks: (a) no complete analysis of relations and correlations between all rules that may cause uncharted classes of configuration errors; (b) no distinction between syntactic unintended anomalies and influential misconfigurations; and (c) the main issue concerns firewall autonomy and automatic application of the solutions since, even, anomalies resolution is a tedious and error prone task, it is generally done manually by the administrator.

In order to improve the effectiveness of firewall policies, we explore in the current paper the thematic of enhancing the quality of firewalls filtering services. We propose a novel filtering technique that integrates a risk assessment approach for evaluating the risk associated to firewall rules. By examining pertinent information relative to rules risk values, the solution helps in identifying firewall misconfigurations and allows strengthening the filtering service via incorporating new actions:

- modifying all or particular actions associated to critical rules (rules evaluated as risky rules) in specific/crucial contexts;
- activating or injecting in a dynamic manner new rules in the firewall rule base that refine other rules (via giving precision or reducing domains of some rules) to minimize the risk of attacks or malicious operations;
- changing the firewall behavior via changing its configuration (resetting the firewall rule base or applying a set of stored predefined rules in order to authorize only vital/necessary services) to avoid malicious scenarios;
- keeping a real time view on the situation (criticality and risk) of the firewall policy since firewall misconfigurations or policy alterations will engender different (mainly highly) risk values;
- improving the ability (based on quantified risk values) to spot weaknesses in the network security posture and identify where the policy need to be changed.

As a main property of our proposal, we should satisfy firewalls autonomy via working on automatic application of our solution and avoiding (otherwise reducing to the minimum) manual interventions of architects and administrators.

The rest of this paper is structured as follows. In Sect. 2, we introduce the work backgrounds and review related works. In Sect. 3, we introduce the main concepts of our solution. We present future directives to enhance the effectiveness of firewall policies. Finally, Sect. 4 concludes the paper.

2 Background and Related Works

2.1 Firewall Techniques

A firewall is a device (soft or hard) used to enforce security policies within or across networks. This component behaves as a security guard of network gateways which is responsible for checking incoming and outgoing packets and treating them in accordance with a set of rules like described in Fig. 1. Hence, a firewall has the ability to block, permit, or drop incoming or outgoing traffic. More, firewalls can also provide details about network traffic and may assist in identifying security threats [1]. A firewall may perform a set of features such as managing and controlling network traffic, authenticating access, protecting resources, filtering packet content and may support the analysis of VPN (Virtual Private Networks), NAT (Network Address Translation), cryptographic protocols, etc.

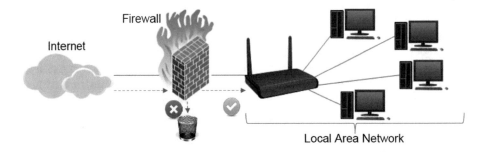

Fig. 1. Integration of firewalls in networks.

2.2 Firewalls Evolution

The importance of a firewall in IS security attracted researchers to enhance its quality and integrate more features for a better protection [17]. This led to several firewall generations:

First Generation - Packet Filter Firewall: the firewall is considered as a simple packet filter (operating at layer three of the OSI model) that forward, block or drop a packet based on its IP addresses (source and destination), port numbers (source and destination) and protocol type. This type of firewalls is not efficient and has some drawbacks such as: no users and applications authentication, no sessions tracking, no content filtering, performance problems in scalable networks that has huge number of users and devices (to much rules), can't resist to IP Spoofing, IP Flooding and DoS attacks.

Second Generation - Application Proxy Filter Firewall: this kind of firewalls (operating at layer seven of the OSI model) allows filtering communications based on applications type. It allows verifying the authenticity of individual packets. Actually, proxy server are considered as software based firewall (application level firewall) that examine incoming and outgoing packets within the network. Even a proxy firewall is more efficient than other firewalls, but in obverse, a fine analysis of application data often results in a slowing.

Third Generation - Stateful Inspection Firewall: called also dynamic filtering, this type of firewalls keeps track of packets, the state of every network connections and sessions. It maintains an internal state table to follows communication sessions until they close. Whens a packet is permitted by the firewall, it stores in a dynamic table the log (incoming/outgoing packet details) concerning this communication.

Forth Generation - Adaptive Response Firewall: this generation represents firewalls that can communicate with active security components such as Intrusion Detection Systems (IDS) and Intrusion Prevention Systems (IPS) to obtain an adaptive response to network attacks. The firewall may perform a self-reconfigure to block ports or reset connections. Nevertheless, such a firewall may accidentally disable sensitive services to engender a deny of service.

Fifth Generation - Kernel Proxy Firewall: this type of firewalls creates a dynamic virtual network layer stack (that contained only required protocols) whence a packet needs to be evaluated. Hence, the packet evaluation and transformation tasks take place in the kernel without moving them to a higher layer that ensures a better speed of the filtering service.

2.3   Related Works

Several research works addressed the thematic of enhancing security solutions provided by firewalls. Authors in [1] worked on the implementation of a solution that integrates firewalls with intrusion detection systems in order to enhance network security. Others, worked on verification of firewalls policies with the aim to identify misconfigurations and rules anomalies. Authors in [2] proposed an approach for formally verified static analysis of firewall rule sets and focused particularly on a formal semantics of the behavior of iptables firewalls. In [3], the authors propose a framework for firewall policy anomaly management that adopts a rule-based segmentation technique for detecting anomalies and deriving possible and effective resolutions. Authors in [4] proposed an approach for the optimization and clean-up of firewall rule-set by removing superfluous rules, detecting and fixing misconfigurations. In [5], authors introduced a quick method for managing firewalls configuration files based on a so called firewall anomaly tree (FAT). Several other works treated the problematic of identifying anomalies and conflicts in firewall rules such as [6–11]. A main drawback of those efforts is related to firewall autonomy and automatic application of the solutions since anomalies resolution requires manual intervention of the network and security administrators.

The importance of risk assessment in IS security attracted researchers to focus on risk-assessment based security solutions. A variety of solutions have been proposed, as examples: authors in [13, 16] proposed a risk assessment approach for deploying and monitoring access control policies in critical systems; authors in [14] proposed a fuzzy risk assessment approach for distributed intrusion prediction and prevention systems; and in [15], authors addressed risk assessment solutions for SCADA and DCS networks. As for the risk assessment of firewall rules, the only work that we found in literature is defined in [12] which started parallel to our work. The author evaluates the risk associated with firewall rules based on the risk of the remote address from which the traffic can originate (the remote address risk value) and the kind of the traffic (the traffic risk value). Our approach proposes a different and comprehensive solution for the assessment of the risk values associated to firewall rules and provides various responsive methods to enhance the quality of the filtering service based on obtained risk values.

3   Firewall-Filtering Service Based on Rules Risk Assessment

3.1   Structure of the Approach

We introduce in Fig. 2 the general structure of our quantified rule (trust) risk-assessment approach that allows:

1. considering the criticality (sensitivity) levels associated with firewall rules;
2. considering a set of rules risk thresholds and a global risk threshold;
3. measuring the risk values of firewall rules taking into account a set of risk factors;
4. storing the risk values in a repository for logs and real time analysis;
5. classifying the set of rules based on the risk values;
6. defining the appropriate responsive actions face critical situations;
7. generating the log messages and notifying administrators with reports concerning the situation and the actions that was taken.

We consider that the risk associated to a firewall rule is evaluated as a metric where the higher value is considered more risky than the lower one. To do so, the structure of our approach defines the following components:

(01) A Risk Assessment Engine (RAE) - responsible for evaluating rule risk values taking into account a set of risk factors and managing the set of the risk factors.

(02) A Risk Repository (RR) - that stores the risk values associated to different rules as well as risk threshold values and rules sensitivity levels.

(03) A Responsive Engine (RE) - that ensures the analysis, the classification of firewall risk values with regards to the corresponding risk thresholds, and mainly automatically reacting regarding critical situations.

(04) A Rules Anomalies Manager (RAM) - that manages the set of rules to identify anomalies associated to high level risk values.

(05) A Risk Factors Depository (RFD) - that stores pertinent predefined risk factors like history events, context factors, etc.

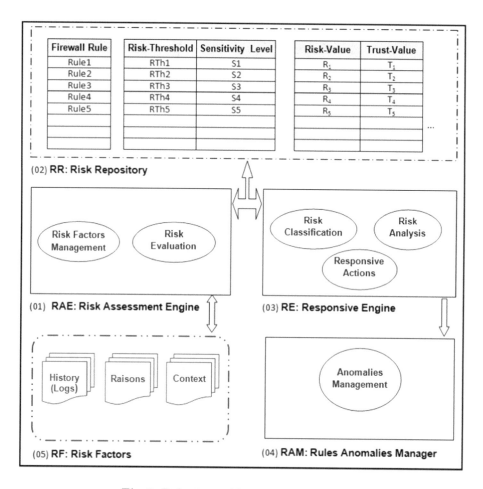

Fig. 2. Rules trust-risk assessment approach.

3.2   Overview on Rules Risk-Trust Assessment

For the assessment of a firewall rules risk, we opted for a quantified approach that allows computing rules risk values.

To explain this process, we need first to cast a glance on a sample of firewall rules. Like described in Fig. 3, the action of accepting, denying or dropping a network traffic is based on the specification of a set of parameters such as

IP addresses (source and destination), the type of protocol, the port numbers (source and destination), etc. The order of the rules (rules number) is important since the firewall considers the first rule that applies.

Rule	Action	IP Source	IP Destination	Protocol	Port Source	Port Destination
1	Accept	192.168.10.20	194.154.192.3	tcp	any	25
2	Accept	any	192.168.10.3	tcp	any	80
3	...	...	...	...	...	...

Fig. 3. A sample of firewall rules.

To evaluate the risk value associated to a rule, we consider the following parameters: the risk value (vs the trust level) of both IP addresses source and destination, the criticality (sensitivity) level of the service defined by the corresponding protocol, the usefulness of both ports source and destination (appropriate standard port numbers or defined by the user), and the traffic direction (ongoing or outgoing).

We consider that the risk value associated with each parameter varies in a scale between 0 and 1 (respectively 0% and 100%). This choice is not compulsory and the scale may be defined according to the *security architect* point of view. However, it is an appropriate choice when we treat with complex and confusing tasks to simplify the process. Dealing with risk values or trust values is not confusing because trust is coupled with risk since a parameter with a high risk has a low level of trust and vice versa. As an example, if we consider that the risk value of a parameter is 0.4, so the trust level of the same parameter is 0.6.

- *Risk values associated with IP addresses*: we evaluate in (1) the risk of an IP address $R(IP)$ as the sum of the probabilities $Pr(k)$ of occurrence of malicious usages $k; k = \{1, \ldots, m\}$; multiplied by the cost associated to each malicious usage $C(k)$.

$$R(IP) = \sum_{(k=1)}^{m} Pr(k) * C(k). \tag{1}$$

Hence, the risk value associated with **Rs** (source risk) is evaluated in (2) as the highest IP risk value associated with the range of IP source addresses.

$$Rs = \max_{1 \leq j \leq n} R(IP_j) | IP_j \in source - addresses. \tag{2}$$

By analogy, the trust level associated with **Ts** (source trust) is evaluated in (3) as the lowest IP trust level associated with the range of IP source addresses.

$$Ts = \min_{1 \leq j \leq n} T(IP_j) | IP_j \in source - addresses. \tag{3}$$

Idem, we compute in (4) (respectively (5)) the risk value associated with **Rd** (destination risk) (the trust level associated with **Td** (destination trust)).

$$Rd = \max_{1 \leq j \leq n} R(IP_j) | IP_j \in destination - addresses. \qquad (4)$$

$$Td = \min_{1 \leq j \leq n} T(IP_j) | IP_j \in destination - addresses. \qquad (5)$$

- *Service sensitivity level*: the *security architect* has to define the criticality (sensitivity) level of the service defined by the corresponding protocol with regards to the business goals and the system security policy. For example we consider that telnet protocol is more sensitive than FTP and FTP is more sensitive than HTTP, etc.
- *Appropriateness of ports numbers*: in general, using appropriate standard port numbers associated with the different services is less risky than others. For example, using the standard port number 23 with telnet is less risky than using another port that makes it more difficult to be identified. This is not a basic rule, since sometimes for security reasons, we may choose non standard port numbers with critical services. Hence, the *security architect* has to verify the predefined appropriateness levels of the different port numbers.

Hence we obtain the risk values associated with the different rules like shown in Fig. 4. The risk value R_i of the rule number i is computed according to the formula (6) with α, β, γ, ϵ and θ are coefficients that quantify the risk parameters and identified with regards to the set of risk factors and σ is the number of the considered parameters.

$$R_i = \frac{\alpha * Rs + \beta * Rd + \gamma * S + \epsilon * A1 + \theta * A2}{\sigma}. \qquad (6)$$

More, the security architect and administrator should define a risk threshold level and a sensitivity level relative to each rule. This task can be automated based on predefined parameters to avoid complexity in case of huge number of rules in scalable infrastructure.

		Rs: Source Risk	Rd: dest Risk	S: Sensitivity	A1, A2: appropriateness		
Rule	Action	IP Source	IP Destination	Protocol	Port Source	Port Destination	Risk
1	Accept	192.168.10.20	194.154.192.3	tcp	any	25	R_1
2	Accept	any	192.168.10.3	tcp	any	80	R_2
3	...	...	...	...	...	...	...

Fig. 4. Rules risk assessment.

Based on the risk values attributed to the firewall rules, we can easily compute a global risk value that describe the risk and the state associated to the firewall policy.

3.3   Overview on Responsive Actions

Whence the RAE evaluates the risk values of the firewall rules, it stores the obtained values in the RR and sends notification to the RE. The RE proceeds with classifying the set of rules based on obtained values. It compares, for each rule, the risk value with the risk threshold. In case the risk value exceeds the risk threshold and based on the sensitivity level of the rule, the RE chooses an appropriate action such as: (i) changing the action of the rule; (ii) deactivating temporarily the rule; (iii) reducing the source/destination address domain by allowing only particular addresses; (iv) moving source/destination address to white/black list; (v) applying a predefined set of rules; (vi) resetting the connection between the peers; (vii) sending notification alerts; etc.

3.4   Implementation and Preliminary Evaluation

To establish such a risk-trust assessment solution, our approach defines a set of components that can be integrated in the kernel of a firewall or implemented as a separate framework that defines some processes that communicate and interact with the firewall.

Preliminary obtained results with a simple prototype that defines a few number of rules in the context of a local network are satisfactory and motivating. Nevertheless, they does not guarantee the real behavior of the solution in large scale and complex networks. Hence, we address in future works the refinement of the proposal and the proof of its consistency, persistence and applicability in large scale and complex infrastructures.

4   Conclusion

We discussed in this paper the effectiveness of firewall filtering solutions. We introduced our approach to enhance the quality of the filtering service provided by firewalls. Our proposal defines a novel filtering technique that integrates a risk assessment approach to evaluate the risk associated to firewalls rules. Our goal is to strengthen the filtering service with pertinent information relative to rules risk values.

The paper defines a framework for organizing thinking about taking into account rules risk values in firewall filtering solutions. Our ongoing works address mainly the refinement of the different components of the approach as well as the proof of its consistency, correctness, persistence and applicability in scalable networks.

References

1. Kumar, D., Gupta, M.: Implementation of firewall & intrusion detection system using pfSense to enhance network security. Int. J. Electr. Electron. Comput. Sci. Eng. **5**(ICSCAAIT–2018), 131–137 (2018)
2. Diekmann, C., Hupel, L., Michaelis, J., et al.: Verified iptables firewall analysis and verification. J. Autom. Reason. (2018). https://doi.org/10.1007/s10817-017-9445-1
3. Hu, H., Ahn, G.J., Kulkarni, K.: Detecting and resolving firewall policy anomalies. IEEE Trans. Dependable Secur. Comput. **9**(3), 318–331 (2012)
4. Saâdaoui, A., Souayeh, N.B.Y.B., Bouhoula, A.: FARE: FDD-based firewall anomalies resolution tool. J. Comput. Sci. **23**, 181–191 (2017)
5. Abbes, T., Bouhoula, A., Rusinowitch, M.: Detection of firewall configuration errors with updatable tree. Int. J. Inf. Secur. **15**(3), 301–317 (2016)
6. Halle, S., Ngoupe, E.L., Villemaire, R., Cherkaoui, O.: Distributed firewall anomaly detection through LTL model checking. In: 2013 IFIP/IEEE International Symposium on Integrated Network Management (IM 2013), pp. 194-201 (2013)
7. Cuppens, F., Cuppens-Boulahia, N., Garcia-Alfaro, J.: Detection and removal of firewall misconfiguration. In: Proceedings of the 2005 IASTED International Conference on Communication, Network and Information Security, vol. 1, pp. 154-162 (2005)
8. Hu, H., Ahn, G.J., Kulkarni, K.: Fame: a firewall anomaly management environment. In: Proceedings of the 3rd ACM Workshop on Assurable and Usable Security Configuration, pp. 17-26 (2010)
9. Mayer, A., Wool, A., Ziskind, E.: Offline firewall analysis. Int. J. Inf. Secur. **5**(3), 125–144 (2006). https://doi.org/10.1007/s10207-005-0074-z
10. Yuan, L., Chen, H., Mai, J., Chuah, C.N., Su, Z., Mohapatra, P.: FIREMAN: a toolkit for firewall modeling and analysis. In: IEEE Symposium on Security and Privacy, pp. 199–213 (2006)
11. Bodei, C., Degano, P., Focardi, R., Galletta, L., Tempesta, M., Veronese, L.: Firewall management with firewall synthesizer. In: CEUR Workshop Proceedings, vol. 2058, no. 16 (2018)
12. Phillips, I.: Assessing risk associated with firewall rules. U.S. Patent Application No 15/215,792 (2018)
13. Jaidi, F., Ayachi, F.L.: A risk awareness approach for monitoring the compliance of RBAC-based policies. In: Proceedings of the 12th International Conference on Security and Cryptography, (SECRYPT 2015), pp. 454–459 (2015)
14. Haslum, K., Abraham, A., Knapskog, S.: Fuzzy online risk assessment for distributed intrusion prediction and prevention systems. In: Proceedings of the Tenth International Conference on Computer Modeling and Simulation, pp. 216–223 (2008)
15. Ralstona, P.A.S., Grahamb, J.H., Hiebb, J.L.: Cyber security risk assessment for SCADA and DCS networks. ISA Trans. **46**, 583–594 (2007)
16. Jaidi, F., Ayachi, F.L., Bouhoula, A.: A methodology and toolkit for deploying reliable security policies in critical infrastructures. Secur. Commun. Netw. **2018**, 22 (2018)
17. Scarfone, K.J., Hoffman, P.: Guidelines on firewalls and firewall policy. In: National Institute of Standards and Technology, NIST Special Publication 800-41, Revision 1 (2009)

A Survey of Blockchain Frameworks and Applications

Bruno Tavares[1(✉)] ⓘ, Filipe Figueiredo Correia[1,2] ⓘ,
André Restivo[1] ⓘ, João Pascoal Faria[1,2] ⓘ, and Ademar Aguiar[1,2] ⓘ

[1] Department of Informatics Engineering, Faculty of Engineering,
University of Porto, Porto, Portugal
{up201700372, filipe.correia, arestivo, jpf,
ademar.aguiar}@fe.up.pt
[2] INESC TEC, Porto, Portugal

Abstract. The applications of the blockchain technology are still being discovered. When a new potential disruptive technology emerges, there is a tendency to try to solve every problem with that technology. However, it is still necessary to determine what approach is the best for each type of application. To find how distributed ledgers solve existing problems, this study looks for blockchain frameworks in the academic world. Identifying the existing frameworks can demonstrate where the interest in the technology exists and where it can be missing. This study encountered several blockchain frameworks in development. However, there are few references to operational needs, testing, and deploy of the technology. With the widespread use of the technology, either integrating with pre-existing solutions, replacing legacy systems, or new implementations, the need for testing, deploying, exploration, and maintenance is expected to intensify.

Keywords: Blockchain · Applications · Distributed ledger · Framework

1 Introduction

Research groups are looking for new applications for the blockchain technology. Although this technology has achieved popularity in the implementation of decentralized crypto currencies, the technology offers many capabilities that can be advantageous in different applications. This study aims to find what research works exist for blockchain frameworks that assist in the development of blockchains or blockchain-related applications. As for new technologies, sometimes the academic world and the industrial world work at different paces, so it is also important to find out where the innovation is coming from, to find out what approaches are still missing [27].

This survey intends to find out what is the types of scientific contribution each paper contains. The analysis of research works about blockchain frameworks over the past few years, and the analysis of the metadata of academic literature can occasionally reveal important information about the technology.

© Springer Nature Switzerland AG 2020
A. M. Madureira et al. (Eds.): SoCPaR 2018, AISC 942, pp. 308–317, 2020.
https://doi.org/10.1007/978-3-030-17065-3_31

To understand the impact blockchain can create, it is necessary to understand what does blockchain provide. Understanding from a high-level perspective how the technology works and what capabilities it possesses can help the analyses of the literature.

The second section of this paper presents the blockchain technology from a high-level perspective and some blockchain applications. The third section presents meta-data information about blockchain-related academic literature. The forth section contains the scientific contribution for each paper and contains the analysis of the research works. In the fifth and last section the author presents his conclusions.

2 Blockchain

2.1 Blockchain Concept

A blockchain is a series of blocks connected to each other, where the last block connects to the previous one and so on. This distributed ledger leverages cryptographic techniques, that can time-stamp information in a system that is immune to fraud or central control if the processing power of the blockchain is distributed by different nodes. Each block holds information; when a node creates a block it will try to add the block to the blockchain. The other nodes need to verify each block. When the validation of the block is complete, the nodes will accept the new block, extending the blockchain. The algorithms, used to handle the verification and consensus between multiples entities, can be different for each blockchain implementation.

Figure 1 shows how the blocks connect to each other in a blockchain [29].

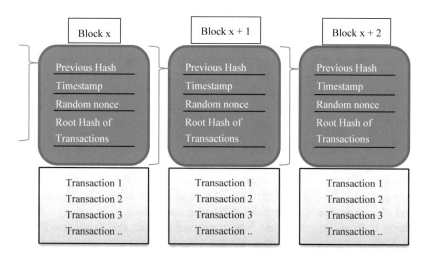

Fig. 1. Blockchain scheme

The top part of the block is the header and the lower part is the body of the block. The body contains the transactions of that block. The block header contains the previous hash; the previous hash is a hash of the header of the previous block. Adding the hash of the previous block to a new block creates a chain. The header also contains the timestamp, a random nonce, and the root hash of the transactions in the body. The random nonce is an arbitrary number used once. The root hash of transactions is the top hash of a Merkle tree demonstrated in Fig. 2 [31].

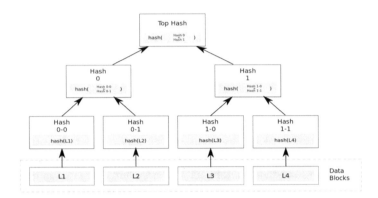

Fig. 2. Merkle tree for root hash of transactions

2.2   Blockchain Capabilities

The blockchain technology offers different capabilities, but the main benefits that the technology offers are the ability to mitigate the need for a trusted third party. When dealing with other companies or individuals, some level of trust is necessary. In today's world people trust the good name of the company or third-parties' certifications. Blockchain offers an alternative, moving the trust needed from an entity to the technology itself. In summary, blockchain is a shared ledger that provides a consensus mechanism with a certain degree of anonymity and privacy in an untrusted network. Blockchain technology can also be the ground for smart contracts that allow transactions with different types of validations where the verification occurs in a transparent system. In the end, the transactions are immutable and it is not possible to tamper the record [27]. The main benefits the technology has to offer are the following [13]:

- **Trustless** – The entities are in control of the information and two entities can make an exchange without the need of a trusted third party.
- **Integrity** – The execution of transactions is exactly as the protocol commands.
- **Transparency** – Modifications to public blockchains are public and can be analyzed and accessible to all entities creating transparency.
- **Quality data** – The data in the blockchain is complete, consistent and immutable.
- **Reliability** – Blockchain uses a decentralized network; this eliminates the central point of failure problem and improves survival from attacks.

2.3   Blockchain Challenges

Blockchain is a technology that contains several challenges. Transaction speed, limit of information each block can contain, and the verification processes need validation in different scenarios to make the technology widely accepted. Depending on the consensus algorithm used a distributed cryptographic ledger can also consume a large amount of energy. Regarding control, security, and privacy, it is necessary further research, because, although all data can be encrypted, the data is shared, and if the encryption protocol is broken, then the information will be exposed [27].

For the technology to be able to replace existent systems, it is necessary to overcome, at least, the following problems: integration issues, cultural adoption, and uncertain regulation. The integration concerns exist because the blockchain application solutions requires significant changes and, most of the times, complete replacement of current systems. It is necessary for the entities to organize and define a strategy for a technology evolution. The cultural adoption is dependent on the robustness and strength of current applications of the blockchain. Most people see the technology as something new but not as a disruptive technology, and there is some skepticism around the true capabilities of blockchain [27]. For a technology shift, and a widespread adoption of the blockchain technology, regulation is necessary. Although blockchain promises to solve the trusted third party issue, it is still required that the regulation and authorities enforce by legal means any dispute that may arise [12].

2.4   Blockchain Applications

The blockchain technology has been used to put proof of existence in legal documents, health records, supply chain, IoT E-business, energy market, E-Governance, decentralized registry, stock exchanges and smart cities [27]. The economic, legal and political systems deal with contracts, transactions, and records. They protect assets and set organizational restrictions. They create and authenticate information. This information allows interactions among countries, societies, business, and individuals. Any system or organization that wants to share information with transparency and safe immutability between different parties, can leverage the capabilities offered by the blockchain technology [12].

There are several applications of the technology in the finance world. Fintech companies are starting to explore cryptocurrencies and exchange markets. In addition, to explore the blockchain technology for business-to-business transactions, prospective benefits include cyber risk reduction, counterparty risk, and increase of transparency [8, 13]. In healthcare the information from patients and medical research data need to be shared, in a way where sensitive personal information is not revealed but also cannot be tampered [10]. The technology can also be applied in organizations responsible for: civil registries, deaths, marriages, land registries, etc. [28].

The energy market is exploring the blockchain technology to reduce the cost of existent centralized solutions, where maintenance and management is too high. It is demanding to support the collection, transaction, storage and analysis of substantial data in the energy market and a peer-to-peer solution can help to mitigate this problem [7].

Internet of Things (IoT) is a new phenomenon predisposed to deliver several services and applications. Developing technologies at the same time can promote work synergies, methods, and frameworks, capable of enabling the best features from each technology [12, 23].

3 Research Works About Blockchain

The meta information that exists about a technology can indicate the progress and interest of a subject in the academic community. Figure 3 illustrates the number of papers related with the blockchain keyword found in Scopus when this survey was made.

Fig. 3. Number of Blockchain related papers in Scopus

Another interesting metric is the one presented in Fig. 4. There are still unknown applications of blockchain. However, a search of the related papers by subject area can reveal that many different areas are studying a way to apply the technology.

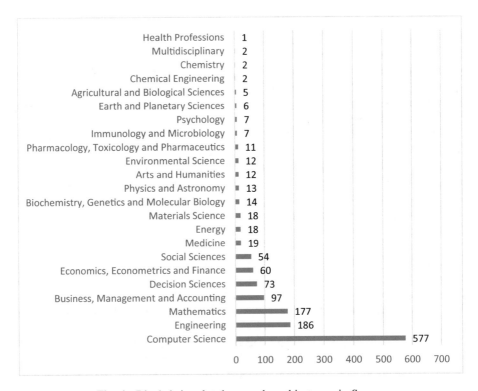

Fig. 4. Blockchain related papers by subject area in Scopus

The interest in blockchain is increasing exponentially in the last couple of years. However, the academic interest is related to a commercial interest. Many businesses are trying to leverage the blockchain capabilities to solve real-world problems [19, 21].

In Fig. 4 chart, the same paper may relate to more than one subject area. Nevertheless, this information indicates that there are many potential applications of the technology in very distinct fields.

Figure 5 classifies blockchain related papers by document type. This information does not vary from subject to subject. Usually there are more conference papers and articles indexed by Scopus.

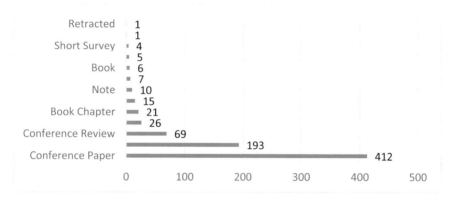

Fig. 5. Blockchain related papers by document type in Scopus

Figure 6 shows the number of citations that the blockchain papers have received. This information shows that the more relevant papers are a couple of years old.

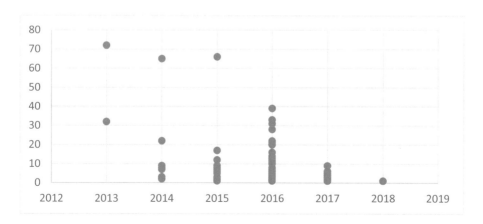

Fig. 6. Blockchain related papers by citations in Scopus

4 Research Works About Blockchain Frameworks

This research is based in a Scopus search with the keywords: blockchain and frameworks; the search query was also limited to conference papers and articles. This analysis focuses on frameworks created for blockchain technology (Table 1).

Table 1. Research works about blockchain frameworks

Author	Framework	Area	Scientific contribution
Cha et al. [5]	Smart Contract-based Investigation Report Management framework for smartphone applications security (SCIRM)	Certification	Architecture Interfaces
Allen et al. [1]	Institutional Possibility Frontier framework	E-Govern	Algorithm
Ateniese et al. [2]	New framework that makes it possible to re-write or compress the content	No specific area	Algorithm
Baracaldo et al. [3]	Framework for cryptographic provenance data protection and access control	IoT	Architecture
Chen et al. [6]	Theoretical framework for evaluating a PoET based blockchain system	Not specific to any area	Analysis
Cheng et al. [7]	Transaction framework is based on the blockchain technology in the distributed electricity market	Energy Market	Analysis Algorithm
Dinh et al. [9]	BLOCKBENCH, the first evaluation framework for analyzing private blockchains	Fintech	Analysis
Dubovitskaya et al. [10]	Framework for managing and sharing EMR data for cancer patient care	Healthcare	Architecture
Fabiano [12]	Global privacy standard framework	IoT	Analysis
Geranio [13]	Common framework and a proper management of risks	Fintech	Survey
Goyal et al. [14]	Non-interactive zero-knowledge (NIZK) system	Fintech	Algorithm
Hou [17]	Chancheng's project called "The Comprehensive Experimental Area of Big Data in Guangdong Province"	E-Govern	Analysis
Kinai et al. [20]	Blockchain backend built on hyperledger fabric and mobile phone applications for stakeholders to engage with system	Fintech	Architecture

(continued)

Table 1. (*continued*)

Author	Framework	Area	Scientific contribution
Li [22]	Conceptual framework of blockchain	Academic	Analysis
Liu et al. [23]	Blockchain-based framework for Data Integrity Service	IoT	Architecture
Ouyang et al. [25]	Framework of large consumers direct power trading based on blockchain technology	Energy Market	Architecture
Wang et al. [30]	Blockchain router, which empowers blockchains to connect and communicate cross chains	Communication	Architecture
Ye et al. [34]	Customized reputation system framework (DC-RSF) to evaluate the credibility of cloud service vendors	Cloud	Architecture
Yuan et al. [35]	Parallel blockchain aims to offer the key capabilities including computational experiments and parallel decision-making to blockchains via parallel interactions	Academic	Analysis
Hari et al. [16]	Blockchain based mechanism to secure the Internet BGP and DNS infrastructure	Communication	Algorithm
Lemieux [28]	General evaluative framework for a risk-based assessment of a specific proposed implementation of Blockchain technology	E-Govern	Architecture
Yasin et al. [33]	Systematic framework for aggregating online identity and reputation information	E-Business	Survey
Zhang et al. [36]	Universal Composability (UC) framework, authenticated data feed system called Town Crier (TC)	E-Business	Architecture
Zhang et al. [37]	Blockchain in Energy Internet were analyzed from three dimensions- function, entity and property	Energy Market	Analysis

5 Conclusions

The blockchain technology is a novel technology that gained considerable momentum in the last couple of years. It is a distributed ledger and a possible disruptive technology; the way blockchain manages information, offers a new way of doing things. Creating new foundations and giving meaning to smart contracts, the application of the technology is still in the beginning. There are several frameworks related with the technology; some frameworks solve problems related with specific areas such as healthcare, IoT, energy market and so on. Other frameworks are developed with the type of blockchain in mind [9].

There are still open challenges and replacing legacy systems with new implementations is a process that takes a lot of time and preparation. Although many problems are known, whenever the technology is applied in new environments, new difficulties will arise. Being able to define a framework concentrated in the lifecycle of a blockchain-based project, can probably identify problems before hand and deliver a tested and reliable solution [17].

References

1. Allen, D., et al.: The Economics of Crypto-Democracy (2017)
2. Ateniese, G., et al.: Redactable blockchain-or-rewriting history in bitcoin and friends. In: Proceedings of IEEE European Symposium on Security and Privacy, pp. 111–126 (2017)
3. Baracaldo, N.B., et al.: Securing data provenance in internet of things (IoT) systems. In: Service-Oriented Computing – ICSOC 2016 Workshops, ICSOC 2016. Lecture Notes in Computer Science, pp. 92–98, vol. 10380. Springer, Cham (2017)
4. Buterin, V.: A next-generation smart contract and decentralized application platform. Ethereum, pp. 1–36, January 2014
5. Cha, S.-C., et al.: On design and implementation a smart contract-based investigation report management framework for smartphone applications. Presented at the Proceedings of the Thirteenth International Conference Advances in Intelligent Information Hiding and Multimedia Signal Processing (2018)
6. Chen, L., et al.: On security analysis of proof-of-elapsed-time (PoET). Lecture Notes in Computer Science (including subseries Lecture Notes in Artificial Intelligence and Lecture Notes in Bioinformatics), vol. 10616 (2017)
7. Cheng, S., et al.: Research on application model of blockchain technology in distributed electricity market. In: IOP Conference Series: Earth and Environmental Science, vol. 93, no. 1 (2017)
8. Dimbean-Creta, O.: Fintech - already new fashion in finance, but what about the future? Qual. Access Success **18**, 25–29 (2017)
9. Dinh, T.T.A., et al.: BLOCKBENCH: a framework for analyzing private blockchains. In: Proceedings of the 2017 ACM International Conference on Management of Data, SIGMOD 2017 (2017)
10. Dubovitskaya, A., et al.: How blockchain could empower ehealth: an application for radiation oncology: (Extended abstract). Lecture Notes in Computer Science (including Subseries Lecture Notes in Artificial Intelligence and Lecture Notes in Bioinformatics), vol. 10494, pp. 3–6 (2017)
11. Exonum: Exonum - A framework for blockchain solutions (2017)
12. Fabiano, N.: The Internet of Things ecosystem: the blockchain and privacy issues. The challenge for a global privacy standard. In: Proceedings of Internet of Things for the Global Community, IoTGC 2017, vol. 2060 (2017)
13. Geranio, M.: Fintech in the exchange industry: potential for disruption? Masaryk Univ. J. Law Technol. **11**(2), 245–266 (2017)
14. Goyal, R., Goyal, V.: Overcoming cryptographic impossibility results using blockchains. In: Lecture Notes in Computer Science (including Subseries Lecture Notes in Artificial Intelligence and Lecture Notes in Bioinformatics), vol. 10677, pp. 530–561 (2017)
15. Greenspan, G.: MultiChain Private Blockchain - White Paper, pp. 1–17 (2013)

16. Hari, A., Lakshman, T.V: The internet blockchain: a distributed, tamper-resistant transaction framework for the internet. In: HotNets Proceedings of 15th ACM Work. Hot Top. Networks, pp. 204–210 (2016)
17. Hou, H.: The application of blockchain technology in E-government in China. In: 2017 26th International Conference on Computer Communication and Networks, ICCCN 2017 (2017)
18. Hyperledger, Linux Fundation: Hyperledger Whitepaper v2.0.0, pp. 1–19 (2016)
19. Khaqqi, K.N., et al.: Incorporating seller/buyer reputation-based system in blockchain-enabled emission trading application. Appl. Energy **209**, 8–19 (2018)
20. Kinai, A., et al.: Asset-based lending via a secure distributed platform. In: Proceedings of Ninth International Conference on Information and Communication Technologies and Development, ICTD 2017. Part F1320, pp. 1–4 (2017)
21. Kraft, D.: Difficulty control for blockchain-based consensus systems. Peer-to-Peer Netw. Appl. **9**(2), 397–413 (2016)
22. Li, M.-N.: Analyzing intellectual structure of related topics to blockchain and bitcoin: from co-citation clustering and bibliographic coupling perspectives. Zidonghua Xuebao/Acta Autom. Sin. **43**(9), 1509–1519 (2017)
23. Liu, B., et al.: Blockchain based data integrity service framework for IoT data. In: Proceedings of 2017 IEEE 24th International Conference on Web Services, ICWS 2017, pp. 468–475 (2017)
24. Microsoft: The Coco Framework - Technical Overview (2017)
25. Ouyang, X., et al.: Preliminary applications of blockchain technique in large consumers direct power trading (2017)
26. Schuh, F., Larimer, D.: BitShares 2.0: General Overview. 8 (2017)
27. Tavares, B.: The contribution of blockchain technology for business innovation. In: Proceedings of the 13th Doctoral Symposium in Informatics Engineering, DSIE18, pp. 1–10 (2018)
28. Lemieux, V.L.: Trusting records: is Blockchain technology the answer? Account. Audit. Account. J. **2**(2), 72–92 (2016)
29. Wander, M.: Bitcoin Block Data, September 2018. https://commons.wikimedia.org/wiki/File:Bitcoin_Block_Data.png
30. Wang, H., et al.: Blockchain router: a cross-chain communication protocol. In: Proceedings of the 6th International Conference on Informatics, Environment, Energy and Applications, IEEA 2017, pp. 94–97 (2017)
31. Wikipedia: Merkle tree, September 2018. https://en.wikipedia.org/wiki/Merkle_tree
32. Wind River: Navigating the open source legal maze, pp. 74–77, April 2005
33. Yasin, A., Liu, L.: An online identity and smart contract management system. In: Proceedings of International Computer Software and Applications Conference, vol. 2, pp. 192–198 (2016)
34. Ye, F., et al.: DC-RSF: a dynamic and customized reputation system framework for joint cloud computing. In: Proceedings of IEEE 37th International Conference on Distributed Computing Systems Workshops, ICDCSW 2017, pp. 275–279 (2017)
35. Yuan, Y., Wang, F.-Y.: Parallel blockchain: concept, methods and issues (2017)
36. Zhang, F., et al.: Town crier: an authenticated data feed for smart contracts. In: 23rd ACM Conference on Computer and Communications Security, CCS 2016, pp. 270–282 (2016)
37. Zhang, N., et al.: Blockchain technique in the Energy Internet: preliminary research framework and typical applications (2016)

Filtering Email Addresses, Credit Card Numbers and Searching for Bitcoin Artifacts with the Autopsy Digital Forensics Software

Patricio Domingues[1,2,3(✉)], Miguel Frade[1,3], and João Mota Parreira[4]

[1] ESTG, Polytechnic Institute of Leiria,
Leiria, Portugal
patricio.domingues@ipleiria.pt
[2] Instituto de Telecomunicações, Coimbra, Portugal
[3] CIIC – Computer Science and Communication Research Centre,
Leiria, Portugal
[4] Void Software, SA, Leiria, Portugal

Abstract. Email addresses and credit card numbers found on digital forensic images are frequently an important asset in a forensic casework. However, the automatic harvesting of these data often yields many false positives. This paper presents the *Forensic Enhanced Analysis* (FEA) module for the Autopsy digital forensic software. FEA aims to eliminate false positives of email addresses and credit card numbers harvested by Autopsy, thus reducing the workload of the forensic examiner. FEA also harvests potential Bitcoin public addresses and private keys and validates them by looking into Bitcoin's blockchain for the transactions linked to public addresses. FEA explores the report functionality of Autopsy and allows exports in CSV, HTML and XLS formats. Experimental results over four digital forensic images show that FEA eliminates as many as 40% of email addresses and 55% of credit card numbers.

Keywords: Digital forensics · Email addresses ·
Credit card numbers · Bitcoin

1 Introduction

The ubiquity and omnipresence of digital devices in our lives, either as computers, smartphones or IoT devices, has made digital forensics almost unavoidable when a abnormal situation, such as physical, financial or cyber crime, occurs. An email address is often an high value asset in a digital forensics casework. Indeed, finding email addresses in a digital device may allow to establish connections between individuals or to detect email accounts that were unknown to the examiners [15]. Also, many online services, such as social networks, rely on email addresses for authentication. All of this highlights the importance of email

© Springer Nature Switzerland AG 2020
A. M. Madureira et al. (Eds.): SoCPaR 2018, AISC 942, pp. 318–328, 2020.
https://doi.org/10.1007/978-3-030-17065-3_32

addresses in a digital forensic casework. When harvesting data in a digital forensic image, the digital forensic Autopsy software[1] resorts to regular expressions (*regex*) to collect strings and create a list of potential email addresses. However, the regex-based list often has a high percentage of false positives, which adds noise and burden to the forensic examiner.

Credit cards have, since their inception, been the target of frauds and theft. This trend continues in the digital age, with credit cards being an important method of payment [2]. Credit cards are involved in several types of crimes. For instance, fraudulently obtained credit card numbers (spied, illegally bought, *etc.*) can be cloned to bogus cards, or simply abused for online transactions. Credit cards can also have a relevant role in an investigation by placing a given individual in a place at a given time, or having performed an action such as buying an item of interest (*e.g.* ammunition). Therefore, in some cases, it might be important in the context of a digital forensic analysis to detect credit card numbers that exist in a digital forensic image. Similarly to email addresses, Autopsy relies on *regex* to harvest credit card numbers, which yields numerous false positives that might obfuscate the forensic analysis.

Since the inception of Bitcoin in 2009, cryptocurrencies have gained significant traction and valuation. Its anonymity has attracted malicious actors to use cryptocurrencies to hide financial transactions, collect ransom from ransomware schemes [10], just to name a few. Additionally, valuation of cryptocurrencies has attracted malicious users with the intent of robbing funds, or to steal CPU and GPU cycles to mine cryptocurrencies [5]. Bitcoin was the first cryptocurrency and is still the most used and valued one. Currently, no support exists within Autopsy to collect Bitcoin-related artifacts.

We present the *Forensic Enhanced Analysis* (FEA) report module for the digital forensic software Autopsy. The main goal of FEA is to filter out false positives from the lists of email addresses and credit card numbers, as well as, presenting a list with Bitcoin public addresses and private keys, if any are found. To detect invalid addresses, FEA resorts to various techniques such as validating the employed set of characters, verifying the existence of the domain name, and checks its existence in the Internet Archive archive.org. The credit card numbers are verified through the validation of their check digit. Additionally, the FEA is able to harvest potential public addresses, and private keys, of Bitcoins and then to search them in the transactions ledger. Since FEA[2] is a report generator for Autopsy, all of its output is available through reports.

The main contributions of FEA are the ability to filter out (*i*) invalid email addresses, (*ii*) invalid credit card numbers and (*iii*) to report on Bitcoin artifacts, namely public addresses and private keys found in digital forensic images. Features (*i*) and (*ii*) aim to reduce noise, contributing to reducing the volume of data to be analyzed by the forensic team, while (*iii*) aim to harvest and filter Bitcoin related data, if any, that can be relevant for some investigations.

---

[1] www.sleuthkit.org/autopsy/.

[2] FEA is available at https://doi.org/10.5281/zenodo.1006703 (GPLv3 license).

The remainder of this paper is organized as follows. Section 2 reviews related work. Section 3 briefly presents the digital forensic software Autopsy, while Sect. 4 describes the main characteristics of the FEA module. Section 5 describes the experimental scenario and the main experimental results of FEA/Autopsy. Finally, Sect. 6 concludes the paper and points out possible future work.

2 Related Work

Validating an email address is, at first glance, simple: attempt to contact the destination email server. If the destination mail server cannot be contacted, for instance, because the domain does not exist, or no MX records pointing to email servers exist on DNS records, then the domain, and consequently the email address is invalid [4]. On the contrary, if an email server for the destination can indeed be contacted, then it suffices to ask the email server whether the email address under scrutiny exists or not. RFC 5321 [9] defines that a request for a non existent email address should trigger from the server the answer *550 "no such user"*. Nonetheless, the 550-response is not used by mail servers, since in the past it had been abused by spammers to validate email addresses and to test randomly generated addresses or common names. Instead, when receiving a mail request for a non existent email address, the mail server just acts as the address is valid, pretending to accept the email. Even if a remote email server was honoring RFC 5321 by properly answering for email addresses validation, this most certainly could not be used in the context of a digital forensic examination. Requesting the validation of a given email address could alert the targets of the investigation and/or modify their behavior. Additionally, the elapsed time between an event under investigation and the digital forensic examination can be quite long, ranging from days to months, if not years. Within this time range, a given DNS domain, which was valid when the facts under investigation occurred, might have expired and no longer be valid when the forensic analysis is performed.

Rowe et al. [15] resort to Bayesian networks to detect false positives in the lists of potential email addresses when forensically analyzing digital storage. In their tests, Rowe et al. point out that 73% of the email addresses were classified as false positives and were thus removed from the analysis. Note however, that their main goal was to detect connection on social media accounts based on the email addresses.

As reported by Garfinkel [7], several digital forensic tools resort to *regex* to search for telephone numbers, email addresses, credit card numbers and some other similar artifacts. The Autopsy software follows this approach. It also allows users to define their own *regex* to harvest the content of digital forensic images. However, results produced by *regex* include a high percentage of false positives. FEA aims precisely to remove false positives from the list of email addresses and credit card numbers returned by Autopsy. Moreover, FEA also resorts to *regex* to build a list of potential Bitcoin public addresses and private keys. FEA further filters out the list, applying validation rules proper to Bitcoin public addresses and private keys, to eliminate false positives.

3   The Autopsy Software

The Autopsy software aggregates within a graphical user interface a wide set of functionality and tools for the analysis of digital forensic images. To access the file systems, files and directories of the forensic images under analysis, Autopsy resorts to the Sleuth Kit (STK)[3]. STK can process several types of file systems and file formats used to archive disk forensic images, such as Encase file format (E01), AFF [6], raw/dd, vmdk, vhd, just to name a few. Besides STK, Autopsy relies on other software to perform some forensic actions. For instance, to recover and carve deleted data, Autopsy resorts to the external tool *PhotoRec*, while data from Windows OS registry is parsed and interpreted by the *RegRipper* tool [12]. An ingest operation consists in parsing and processing the data from the forensic disk images. Autopsy can be extended through modules, allowing for the addition of new features. These modules can be switched on/off as per user request at the start of any ingest operation.

Two of the services of Autopsy are the creation of lists holding (*i*) email addresses and (*ii*) number of credit cards. However, both lists present a high percentage of false positives. From the forensic point of view, it is important to filter out the lists, to reduce the workload of the forensic examiner. FEA does not intervene on the methodology used by Autopsy to harvest and create these lists, instead, FEA filters them out since both of them are made available as artifact lists. This approach can better withstand future internal changes of Autopsy that can affect the creation of the lists.

4   The FEA Module

FEA is a module for Autopsy, developed in Jython and with a Java Swing configuration interface. The module follows the report paradigm, one of the three module paradigms available within Autopsy. A report module is used to organize data in a report format defined by the code of the module. FEA can produce reports for the three main types of data that it processes: email addresses, credit card numbers and Bitcoin addresses. Non valid data are labelled as false positives. The filtering techniques for each category of data are described next.

4.1   Validating Email Addresses

Checking the Syntax. The syntax of an email address obeys the complex rules defined in RFC 5322 [14]. Many of the extreme cases defined in RFC 5322 are not supported by traditional email systems. For instance, RFC 5322 allows for the inclusion of a comment in the first part of the email – before the @ symbol – with the comment being delimited by round brackets. For example, `name(comment)@example.com` and `(comment)name@example.com` are, from the point of view of RFC 5322, accepted as valid, with both being interpreted as the

---

[3] www.sleuthkit.org.

address `name@example.com`. However, in practice, the email services only allow, for an email address, alphanumeric symbols, dot (.), underscore (_) and dash (−). To validate the syntax of email addresses, FEA follows the approach of the main email systems. For instance, there is a maximum of 64 characters for the local part of the address and a total of 254 characters for the whole address.

Validating Top Level Domain. The top level domain (TLD) of an address corresponds to the last section of the domain. For instance, `COM` in the case of `example.com`. The top level domains are defined by IANA and include, besides the traditional `EDU`, `COM`, `NET`, `ORG`, `GOV`, `MIL` and `INT`, the domains of countries. These domains use the two-character code that corresponds to the ISO-3166 [13] standard. Lately, the number of top level domains has increased significantly [1]. FEA uses the TLD list maintained by IANA[4]. Since the list is updated every day, FEA attempts to download the updated list before processing the list of email addresses harvested by Autopsy. For every address in the list, FEA checks whether the top level domain exists on the TLD list, removing the non-existent ones. This way, email addresses whose TLD is invalid are discarded without the need to query the Domain Name System (DNS) service.

Validating Domain with DNS. The domain of an email address is the section that is on the right of the @ symbol. The validation of the whole domain is done by querying the DNS service for its existence. Thus, for every potential domain, a DNS query is performed. To hide the latency of the DNS service, FEA supports multithreading, launching several queries in parallel, one per thread. The number of simultaneous DNS queries can be defined by the forensic examiner when launching FEA. Therefore, it is up to the user to decide whether he/she wants a faster execution at the cost of an higher load on DNS, or restrict the parallelism to control the load on DNS, thereby lengthening execution times. Note that some applications such as browsers are known to be DNS-intensive [3], without impacting much the DNS service, which is a highly scalable service [8].

History Analysis at *Internet Archive*. A currently nonexistent domain might have existed in the past. Since the registration of a domain is valid for a given amount of time, if the registration is not extended, nor is the domain taken by another entity, the domain ceases to exist. For the purpose of the validations performed by FEA, it would be convenient to be able to consult the history of DNS domains. However, this approach is not viable for FEA due to two main reasons: (*i*) existing services do not maintain history records for all existing top level domains and (*ii*) they are not free. To circumvent these limitations, FEA resorts to the *Internet Archive*(See Footnote 1) to assess whether a given domain existed in the past. Thus, for syntactically correct domain names whose top level domain exists, but which are currently not registered, a lookup is performed through the Internet Archive. The project has been active for more

---

[4] Available at data.iana.org/TLD/tlds-alpha-by-domain.txt.

than 20 years and thus has a significant amount of data. FEA queries the Internet Archive through the online API, and if the queried domain exists within the archive, FEA returns the date of the last time the domain was indexed. To avoid overloading the service, FEA only queries the Internet Archive in sequential mode, having no more than one query pending. Note that a positive answer from the Internet Archive – the domain has existed in the web – does not mean that the domain had an associated email service. Similarly, a domain might have existed in the past, but may have never been captured by the Internet Archive.

4.2   Credit Card Numbers

The credit card module of FEA aims to validate the credit card numbers harvested by Autopsy. If instructed to do so at ingest time, Autopsy builds a list of potential credit card numbers. These potential credit card numbers are available under the result set of artifacts. To filter the list of harvested credit card numbers, FEA simply verifies, for each candidate number, whether the number obeys the LUHN's checksum[5], as it is required for any valid credit card number. The LUHN checksum is a simple and public domain validation mechanism that appends a check digit at the end of the number that it aims to protect from input errors. The check digit is computed such that the validation yields a modulo 10 number, when the whole number is correct, hence its another names, "modulo 10" or simply "mod 10".

LUHN's checksum is used with credit card numbers and with International Mobile Equipment Identity (IMEI) of mobile devices, to cite just the most important domains of usage. LUHN's algorithm only protects against certain accidental errors, such as single mistyped digit, where, for example, a 2 is entered instead of a 9 [16]. Other detected errors include some, but not all, transposition of digits, namely where two digits are swapped (e.g, 12 instead of 21). Nonetheless, and despite certain errors that are not detected, the LUHN verification allows to swiftly filter out some invalid candidate credit card numbers, thus removing false positives.

4.3   Bitcoin

Cryptocurrencies have emerged in the past decade with the main goal of allowing financial transactions to be performed without the control of a central authority. While traditional transactions rely on intermediary agents such as banks or other types of financial institutions, cryptocurrencies allow for transactions to flow directly between two entities, relying on a network of peer-to-peer nodes and on public/private cryptographic keys to authenticate transactions. From the point of view of cryptocurrencies supporters, the main advantages are the (i) anonymity that can be achieved by having non-centralized transactions and the (ii) lower costs of performing transactions. Since transactions can be done

---

[5] ISO/IEC 7812-1:2006. Identification cards – Identification of issuers – Part 1: Numbering system.

anonymously, cryptocurrencies are increasingly linked to crime, like for instance ransomware attacks [10], cryptojacking abuses where CPU cycles are stolen to engage in cryptocurrency mining [5], to mask illicit incomes, and, more broadly, to engage in malicious activities.

Bitcoin was the first cryptocurrency, being introduced through the seminal white paper "Bitcoin: A peer-to-peer electronic cash system" authored by a mysterious Satoshi Nakamoto [11]. The paper proposed a novel distributed infrastructure – blockchain – that comprises a sequential set of numerically identified blocks, where each block is chained to its predecessor by a cryptographically secure link, and holds the transactions accepted by the Bitcoin network. A transaction in Bitcoin requires (*i*) funds, (*ii*) a public address of the receiving side of the transaction and (*iii*) the private key needed to authorize the transfer of funds.

Public Addresses. Public addresses can either be regular addresses or compressed ones. The former comprise 65 bytes, while compressed public address are smaller, with 33 bytes, hence the name. A public address is the public part of the private/public key pair, and it is derived from the private key. In Bitcoin, public addresses are represented for human purposes as hash string using the base58 character set. Base58 corresponds to the well known base64 sets, stripped of the visually ambiguous characters 0 (zero), O (capital o), I (capital i) and l (lower case L). Additionally, the slash (/) and the plus (+) characters are not part of base58. Format-wise, the hash string of a public address has a length that ranges between 26 to 35 base58 characters. A public key hash must obey the following format: the last four bytes of the public address must match the first four bytes of a double SHA256 hash computed over the first 21 bytes of the address. This way, the last four bytes act as a checksum. FEA uses this property to validate potential public Bitcoin addresses.

Private Keys. In textual representation, a private key has 50 or 51 base58 characters and starts either with a K, a L, or a 5. Validation of a potential private key is done indirectly through the associated public key/address. For this, the Elliptic Curve Digital Signature Algorithm is used to obtain the corresponding public key. Then, this public key is validated as above, with the last 4 bytes compared to the double SHA256 of the first 21 bytes. The rationale is that if a valid public address is generated, then the Blockchain API will return a result, even if the public key has never participated in a transaction. Even though private keys obtained with this approach may lead to unused public keys, they are still relevant information, as they may provide proof of intent to engage in illicit activities within the broader scope of an investigation.

Harvesting Potential Bitcoin Keys. To harvest potential Bitcoin public addresses and private keys, FEA follows Autopsy's approach with potential email addresses and credit card numbers: it resorts to the *keyword search* module,

Table 1. Regular expressions to harvest Bitcoin artifacts

Public addresses:	`^[13][a-km-zA-HJ-NP-Z1-9]{25,34}$`
Private keys:	`^[5KL][1-9A-HJ-NP-Za-km-z]{50,51}$`

creating a custom list with Java-based *regex*. Two *regex* for harvesting Bitcoin keys are used: one for public addresses, another one for private keys. Both are shown in Table 1. Similarly to the email addresses and credit card numbers, prior to running FEA for a report on potential Bitcoin artifacts, the user needs to run the keyword search module during the ingest stage with the Bitcoin option activated. This triggers text extraction and indexing by the Apache Solr engine, which is used by Autopsy to implement keyword search. As these *regex* are not part of Autopsy, they need to be installed by the user through the manage lists option. This can be achieved via the "Options" menu in Autopsy, under the "Keyword Search" tab.

Online Search for Transactions. Besides the local validation of both public addresses and private keys, if instructed to do so, FEA can also research within Bitcoin's blockchain the existence of transaction data involving the harvested and validated Bitcoin public addresses. As all valid Bitcoin transactions are recorded in the blockchain and often Bitcoin is used to hide identities, the search might reveal valuable data.

5   Experimental Results

FEA was assessed on four digital forensic images identified as *Win7*, *Win8.1*, *Win10* and *macOS*. Each name matches the installed OS. Besides having different OS, the forensic images have different levels of activity. Specifically, Win8.1 and macOS are from systems with low activity and few installed applications, while Win7 and Win10 images have a larger number of installed applications and have seen many more hours of usage. The tests were conducted with Autopsy software 4.8, which was the latest available version at the time of the experiments.

The adopted modus operandi for processing each forensic image was as follows. First, the PhotoRec Carver ingest module of Autopsy was run to recover content, namely deleted files, from unallocated space in disk. Then, the keyword ingest module was run, with the *Email Addresses*, *Credit Card Numbers* and *Bitcoin* options activated. Finally, the FEA report module was run, first to produce the report with the filtered out email addresses and a second time for credit card numbers. The Bitcoin report was also run, but no Bitcoin artifacts were found on the images. Thus, only results for email addresses and credit card numbers are considered.

Table 2. Filtered out emails addresses by FEA

	Email addresses	Non valid syntax	Non valid TLD	Non valid domain	Non valid (all)	Non valid (%)
Win7	1488	20	742	158	920	61.83%
Win8.1	820	16	0	247	263	32.07%
Win10	1853	37	0	368	405	21.86%
macOS	4979	13	0	1 271	1 284	25.79%
Average	2429	16.33	247.33	558.67	822.33	39.90%

Table 3. Filtered out credit card numbers by FEA

	Credit card numbers	Non valid (LUHN)	Non valid (%)
Win7	35 477	4 271	12.04%
Win8.1	17 331	11 617	67.03%
Win10	59 855	45 866	76.63%
macOS	41 077	26 807	65.26%
Average	38 435	22 140.25	55.24%

5.1   Main Results

Email Addresses. Main results of FEA filtering of candidate email addresses are shown in Table 2. The columns show, respectively, the total of candidate email addresses (*Email Addresses*), addresses discarded due to bad syntax (*Non valid syntax*), addresses with non-existent TLD (*Non valid TLD*), addresses with non valid domains (*Non valid domain*). Finally, the last two columns hold the total of non valid addresses that were removed by FEA, as well as, the respective percentage.

Syntax validation has a residual contribution to filter out addresses, contributing with less than 1% of the eliminated email addresses. Surprisingly, filtering out addresses with non-existent top level domains (TLD) has extreme results across the forensic images: Win7 has nearly half of the candidate email addresses with bogus TLD, while all others have none. Domains not found on DNS have a major role in filtering out addresses, spotting an average of roughly 559 false addresses across the four images. Overall, FEA marked roughly 39.90% of the potential email addresses as invalid.

Credit Card Numbers. Table 3 shows the results of FEA validation of credit card numbers. The number of potential credit card number is quite significant, ranging from 17 331 (Win8.1) to 59 855 (Win10). By simply applying LUHN's checksum validation, FEA discards as false positives around 55% of the potential credit card numbers. Nonetheless, due to the huge absolute number of potential credit card numbers, the remaining 45% still represent a large volume of data.

6   Conclusion

FEA is a report module for the Autopsy digital forensic software. FEA eliminates some of the false positives email addresses and credit card numbers harvested by Autopsy. To detect false email addresses, FEA resorts to several validation techniques, such as syntax verification, assess the existence of top level domains, and DNS queries of email domains. For credit card numbers, FEA applies LUHN's checksum validation, removing credit card candidates that fail the verification. Additionally, FEA also harvests digital forensic images for potential public addresses and private keys of the Bitcoin cryptocurrency. It verifies that the found artifacts not only obey Bitcoin addresses and keys rules, but also checks for the existence of transactions involving the Bitcoin addresses in the blockchain. Being a report module, FEA presents its results through text, CSV and XLS files.

When tested with four digital forensic images, FEA marked as many as 40% of candidate email addresses as false positives and slightly more than 55% of potential credit card numbers as invalid. All of this is achieved through low demanding computational techniques, with the major used resource being queries to DNS and to the Internet Archive.

Although FEA's results are interesting, since they remove stale data from a digital forensic analysis, the large volume of remnant data, meant that other techniques need to be added to filter out more false positive artifacts. FEA's ability to filter out valid but non interesting addresses can be improved by supplying *remove lists*. Examples include *contoso.**, which is a fictitious company that appears in some Microsoft software, as well as, the non usable *example.** domains. We plan to address these and other issues in future work.

Acknowledgements. This work was partially supported by FCT, Instituto de Telecomunicações under project UID/EEA/50008/2013 and CIIC under project UID/CEC/04524/2016.

References

1. Paul, P.K., Bhuimali, A., Shivraj, K.S.: Internet corporation for assigned names and numbers: an overview. Asian J. Eng. Appl. Technol. **5**(2), 40–43 (2016)
2. Bahnsen, A.C., Aouada, D., Stojanovic, A., Ottersten, B.: Feature engineering strategies for credit card fraud detection. Expert Syst. Appl. **51**, 134–142 (2016)
3. Duchamp, D., et al.: Prefetching hyperlinks. In: USENIX Symposium on Internet Technologies and Systems, pp. 12–23 (1999)
4. Elz, R., Bush, R.: Clarifications to the DNS specification. Technical report (1997)
5. Eskandari, S., Leoutsarakos, A., Mursch, T., Clark, J.: A first look at browser-based Cryptojacking. arXiv preprint arXiv:1803.02887 (2018)
6. Garfinkel, S.: AFF and AFF4: where we are, where we are going, and why it matters to you. In: Sleuth Kit and Open Source Digital Forensics Conference (2010)
7. Garfinkel, S.L.: Digital media triage with bulk data analysis and bulk_extractor. Comput. Secur. **32**, 56–72 (2013)

8. Jung, J., Sit, E., Balakrishnan, H., Morris, R.: DNS performance and the effectiveness of caching. IEEE/ACM Trans. Netw. **10**(5), 589–603 (2002)
9. Klensin, J.: RFC 5321: simple mail transfer protocol (2008). https://tools.ietf.org/html/rfc5321
10. Liao, K., Zhao, Z., Doupé, A., Ahn, G.J.: Behind closed doors: measurement and analysis of CryptoLocker ransoms in Bitcoin. In: 2016 APWG Symposium on Electronic Crime Research (eCrime), pp. 1–13. IEEE (2016)
11. Nakamoto, S.: Bitcoin: a peer-to-peer electronic cash system (2008)
12. Panchal, E.P.: Extraction of persistence and volatile forensics evidences from computer system. Int. J. Comput. Trends Technol. (IJCTT) **4**(5), 964–968 (2013)
13. Postel, J.: Domain name system structure and delegation (1994)
14. Resnick, P.: RFC 5322: Internet message format (2008). https://tools.ietf.org/html/rfc5322
15. Rowe, N.C., Schwamm, R., McCarrin, M.R., Gera, R.: Making sense of email addresses on drives. J. Digit. Forensics Secur. Law: JDFSL **11**(2), 153 (2016)
16. Wachira, W., Waweru, K., Nyaga, L.: Transposition error detection in Luhn's algorithm. Int. J. Pure Appl. Sci. Technol. **30**(1), 24 (2015)

A Survey on the Use of Data Points in IDS Research

Heini Ahde, Sampsa Rauti$^{(\boxtimes)}$, and Ville Leppanen

University of Turku, Turku, Finland
sampsa.rauti@utu.fi

Abstract. In today's diverse cyber threat landscape, anomaly-based intrusion detection systems that learn the normal behavior of a system and have the ability to detect previously unknown attacks are needed. However, the data gathered by the intrusion detection system is useless if we do not form reasonable data points for machine learning methods to work, based on the collected data sets. In this paper, we present a survey on data points used in previous research in the context of anomaly-based IDS research. We also introduce a novel categorization of the features used to form these data points.

Keywords: Network security · Intrusion detection · Data points

1 Introduction

An *intrusion detection system* (IDS) is a system used to detect malicious activities in a specific network or system. A network intrusion detection system (NIDS) is placed in the network to monitor network traffic and to detect malicious activities by recognizing anomalous patterns in the incoming or outgoing packets. [1–3, 11] NIDS systems have existed already for quite a long time, and now host intrusion detection systems (HIDS) are becoming increasingly popular. Compared to NIDS, one can gather much more data in HIDS systems, because the system can detect all the internal events related to the inner workings of an operating system and its processes.

Signature-based IDSs detect attacks by observing previously recorded malicious patterns, such as malicious sequences of instructions. Anomaly-based IDSs, on the other hand, use machine learning to learn what normal activity in the network looks like and compare detected behavior to this profile. The strength of this approach is the ability to detect previously unknown attacks. [1, 11, 21] Today, anomaly-based intrusion detection is needed to deal with the increased amount of new variations of malicious programs and targeted attacks [7].

However, the data gathered by the system is useless if we cannot use it correctly to learn about the normal behavior [21]. We have to make a choice about which features in the data we are going to look at and how they are used in our analysis. The same is also true when we are configuring an IDS and building rules for it; we have to know what kind of features we are looking for.

© Springer Nature Switzerland AG 2020
A. M. Madureira et al. (Eds.): SoCPaR 2018, AISC 942, pp. 329–337, 2020.
https://doi.org/10.1007/978-3-030-17065-3_33

Specifically, it is very important to carefully choose the data points we use to draw conclusions from the data.

A *data point* is a single observed unit of information, a measured collection of features. For example, a simple data point in NIDS research could be a size of a specific network packet. On the other hand, a data point can be a more complex combination of several features, such as combining the IP addresses and port numbers of the sender and the receiver.

The contributions of this paper are as follows. In this survey, we review different types of data points used in literature in the context of anomaly-based IDSs. We look at how these data points have been aggregated from the gathered data and for what kind of purposes they have been utilized. We also provide a novel categorization of different features used to form the data points in IDS research. The categorization presented in this study can be used when choosing a meaningful set of features for data points in NIDS systems and research.

This research is a part of a project where we have had a change to utilize a new commercial HIDS system by F-Secure to collect a lot of data on internal events with the aim of creating new kind of machine learning solutions for (ab)normal behaviour profiling. Moreover, we are working towards an open-source implementation of system enabling one to collect data related to internal events and thus build such systems. As such systems produce a vast amount of data, efficient machine learning based methods are needed. Therefore, the purpose of this paper is to survey what kinds of features have been used in this context in the existing literature.

The rest of the paper is organized as follows. Section 2 presents the study setting and the research questions we aim to answer in this study. Section 3 presents a survey on the types of features used to form data points in NIDS research. Section 4 presents our classification for different types of features in NIDS research and their usual purposes. Section 5 concludes the paper.

2   The Study Setting and the Research Questions

The data collected by an IDS often consists of a large amount of events, each having lots of features or attributes. One approach would be to feed all of this data to a machine learning algorithm as such. However, the setting would easily become very complex as the data points would have a high number of dimensions.

Although one can use a multidimensional dataset to create profiles, this approach is often ineffective. The dataset often contains several irrelevant features that do not have a significant effect on the end result. Moreover, handling high dimensional datasets obviously increases the calculation time. Hotho et al. [10] note that a low-dimensional space is more favorable for finding clear clusters in the data.

It is often worth considering more elementary approaches and only looking at narrow slices of data. As a practical example, such a narrow a view could be a set of day-wise or machine-wise points formed from port numbers used by a selected application.

Because of the obvious need to choose good and meaningful data points for low-dimensional datasets, we have formulated the following two research questions for this study:

RQ1 What are the types of data points used in NIDS research?

The first research question studies data points in NIDS research considering statistical and machine learning based methods, and how these data points have been aggregated from the data (which features are included?).

RQ2 For what kind of purposes are the data points used?

The second question inquires what kind of observations the discovered data points aim to derive from the data.

In what follows, we present a review of the types of data points proposed and used in network intrusion detection systems. The purpose is to find out what kind of different network-related data points exist and for what purposes they have been utilized. To search for relevant publications, we used Google Scholar and keywords "profiling network traffic", "data mining network traffic" and "internet traffic behaviour profiling for network security monitoring" to collect literature references. Moreover, snowballing (looking at the references of found papers) was performed to make the search more complete. The publications that did not discuss aggregation of data points in NIDS research were excluded.

3   Types of Network Traffic Data Points

Shadi et al. [18] presented a hierarchical clustering algorithm for network flow data. They generated their data points in a 2D space of source and destination IP addresses. This 2D space represented all possible values for the source and destination IP addresses. In a related studies, Estan et al. [5] introduced a multidimensional traffic clustering, for analyzing IP-based traffic. Their method can cluster traffic along multiple different dimensions including source and destination IP addresses, protocol, source and destination port numbers. Mahmood et al. [15] introduced a new clustering method for generating conclusions of significant traffic flows. The key attributes were source and destination IP addresses, protocol, source and destination port numbers.

Xu et al. [25] studied significant behavior patterns for network security monitoring (detecting anomalies). They collected Internet backbone traffic flow and constructed four collections of clusters based on following extracted features:

- The source IP address
- The destination IP address
- The source port number
- The destination port number

Feroz et al. [6] demonstrated that cluster labels increase the classifier accuracy. They presented an approach that classifies URLs based on their lexical and host-based features. In a related study, Ma et al. [13] extracted lexical and host-based features (such as IP address of the URL, the mail exchanger and the name server) from URLs.

Packet header traces are widely used in data mining and machine learning analysis. McGregor et al. [16] used the following features and expectation-maximisation clustering algorithm to group the traffic flows into a small number of clusters:

- Byte counts
- Connection duration
- Interarrival statistics
- Packet size statistics (minimum, maximum, quartiles, minimum as fraction of max and the first five modes)
- The number of transitions between transaction mode and bulk transfer mode (the time when there was more than three successive packets in the same direction without any packets carrying data in the other direction)
- The time spent and the idle (all periods of 2 seconds or greater when no packet was seen in either direction)

Liu et al. [12] examined supervised and unsupervised machine-learning techniques to classify network traffic by TCP-based applications. Their K-means approach took following statistical information as an input vector to build classifiers (a = client, b = server):

- The number of total-packects-b-a
- The number of actual-data-bytes-b-a
- The number of pushed-data-pkts-b-a
- Size of the mean-IPpacket-a-b and size of the mean-IPpacket-b-a
- Size of the max-IPpacket-a-b and size of the max-IPpackect-b-a
- Variant of the IPpacketsize-a-b and variant of the IPpacketsize-b-a
- Duration

Magdalinos et al. [14] examined on how to automatically create user profiles and how to use these profiles to enhance the performance of network control functions. They applied well-known data mining and machine learning tools, such as k-means and naive Bayes, and following *context extraction and profiling engine* (CEPE) information:

- The status of the user (name, age, gender, education, operating systems, screen width and height, etc.)
- The device (time, transmit power, lost packets etc.)
- The combination of the service type (web, ftp, video, etc.) and network information (lost packets, transmit power, cell type, etc.)
- A log file (all monitored parameters)

Erman et al. [4] applied both supervised and unsupervised machine learning methods to classify network traffic. The authors generated data points using statistical features, such as total number of packets, mean data packet size, flow duration and the mean inter-arrival time of packets.

Hammerschmidt et al. [9] built communication profiles using connection-level IP flow records, such as transport protocol of the flow, time since previous flow started, duration of the flow, count of packet exchanged and amount of data received. One of their main purpose was to extract the key behavior from the records and also reduce irrelevant behavior. They learned communication profiles with the *DFASAT software package* using an IP flow dataset that contains real communication from hosts running botnet malware as well as legitimate traffic.

Wang et al. [23] applied two benchmark datasets, 20Newsgroups (20NG) and RCV1, to evaluate domain dependent document clustering. 20NG contains 20000 newsgroups documents and RCV1 contains manually labeled newswire stories from Reuters Ltd.

Gonzalez et al. [8] presented a platform *Net2Vec*, that is able to capture raw network flow data, transform it into meaningful data points and apply the predictions over the data points in real time. To showcase the applicability of the Net2Vec they constructed a user profiling scheme. They generated a two-element data point consisting of source IP address and hostname.

Singh et al. [20] examined anomaly based intrusion detection systems that learn to distinguish normal behaviour and abnormal behaviour. They used *NLS-KDD dataset* and *Kyoto University dataset*. The features of NLS-KDD dataset included for example duration, protocol type, byte information, service information, and log information.

Sarmadi et al. [17] examined the feasibility of profiling internet users based on volume and time of usage. As an experimental dataset they used real internet usage data collected via NetFlow logs (metadata). The data was collected in one month from 66 university students. The traffic flow contains information about usage time, octets, packets, port numbers and protocols. However, the focus of the paper was to examine only the following combination:

- Duration (the amount of milliseconds from the start of the flow to the end)
- Octets (the number of layer 3 bytes of the flow)

4   A Categorization for Data Point Features

To find a meaningful categorization for the data point features reviewed in the previous section, internet traffic classification methods presented in the previous literature can give us some clues. Wang et al. [24] and Singh [19] divide the internet traffic classification methods to following groups:

- *Port-based internet traffic classification.* In this classification method, port numbers in headers of the transport layer protocol, such as TCP, are examined and used when forming data points.
- *Payload-based internet traffic classification.* This approach, also called deep packet inspection, inspects packet payloads for characteristic signatures of known applications. It can be used to check that the content is supplied in the correct format and to make sure payload is not malicious.

Table 1. Distribution of NIDS data point related research within our feature categorization.

Category	Example features	Papers
Network addresses and ports	IP addresses, ports, hostnames, domains	$[5, 6, 8, 13, 15, 18, 25]$
Protocols and service types	HTTP, voice, voip, video	$[5, 9, 15, 20]$
Statistical features	connection duration, packet size, idle time	$[4, 9, 12, 14, 16, 17, 20]$
Network information	Lost packets, delay log information	$[14, 20]$
External features	Screen type, user's education	$[14, 23]$

- *Classification based on statistical traffic properties.* Statistical characteristics of internet traffic at the network layer, such as packet length, flow duration, inter arrival time of packet, standard deviation, are used to classify traffic.
- *Internet traffic classification using machine learning.* A dataset consisting of number several data points (that in turn consist of one or several features or attributes) is created. The output will be some specific pattern discovered in the data.

A similar taxonomy is used by Valenti et al. [22], as they also list port-based classification, payload-based classification, statistical classification as traffic classification techniques. In addition, the authors also include stochastic packet inspection and behavioural traffic classification in their taxonomy. Stochastic packet inspection uses the statistical fingerprints in the application layer headers and uses them to recognize formats of different application protocols. Behavioral classification generally monitors traffic of one host as a whole and examines traffic patterns (such as which transport level protocols are used and how many hosts are contacted), trying to profile applications that are executed on the target host.

The two categorizations do not directly fit for classifying data point features into groups. For example, they seem to exclude many potential types of features that may be important in profiling hosts or specific users, such as IP addresses or information about the user. Also, some categories, such as "machine learning" or "behavioral classification" are too coarse and make no sense in the context of categorizing features (after all, all features can be used for machine learning and behavioral analysis). However, some categories like data related to ports, packet payload or traffic statistics, seem to make more sense for feature types. Our categorization for NIDS research data points is presented in Table 1.

The categorization consists of the following categories:

Network Addresses and Ports. Network addresses and port defining the hosts and applications at the endpoints are important features when profiling hosts based on network traffic. Port numbers have traditionally been used to identify applications. Recently, however, an increasing number of programs use nonstandard ports, and malicious programs might use well-known protocol ports (such as 80 reserved for HTTP) in order to disguise their presence. Therefore, port numbers are often combined with other features, such as IP addresses [25] or used protocol [5] to form more informative data points. Domain names associated with IP addresses may also be useful as features, for instance when trying to identify malicious websites [13].

Used Protocol or Service Type. The used protocol (such as HTTP, FTP, SMTP...) can be used as a feature to profiling applications and clustering multidimensional network traffic [15]. A feature similar to the used protocol is service type, for example web, video or voip. Information about the protocol and service along with statistical features is useful for creating a model to profile hosts based on their traffic statistics [9] or giving predictions on the expected behavior of end users.

Statistical Features. Statistical data point features are related to traffic flow. They include for example statistics of interarrival times, connection durations, packet sizes and number of packets from host a to host b. Statistical features such as mean interarrival time and accumulated idle time are aggregated from other values. Several of these statistical features are often bundled together to form data points, for example in order to build a profiles for applications on a specific host [12].

Network Information. Network related features such as the transmit power of the network, network cell related information or delays in the network can also be used in profiling. However, when trying to profile hosts, applications or users, these features are usually combined with features from other categories to form data points.

External Features. An external feature refers to a feature not directly related to network traffic or properties of packets. Instead, this category includes data on the users and devices they use. Examples of such features include type of processor, operating system, coordinates of a mobile device, and completely non-technical features such as age and gender of the user. These features are usually used to build profiles for users, like Magladinos et al. [14] do by combining for instance features of users, devices and service types to form data points.

5   Conclusion

In this paper, we have reviewed different types of data points utilized in NIDS research. We have studied what sort of features have been previously used to aggregate data points in the existing literature. Based on these findings, we proposed a novel categorization for features used to form data points.

A data point is often multidimensional and consists of many features, often from several categories outlined previously. Since clustering high-dimensional datasets is difficult and computationally ineffective, we believe the new categorization we have provided in this study can help in choosing meaningful features for low-dimensional data points. While the features and data points vary in each individual system, we have seen that many feature categories that are used regularly.

In the future research, it might be interesting to experiment how well our classification works when building a NIDS system and classifying data gathered in the system. This would paint a clearer picture of how our classification can be used in NIDS research and how effective it is. The future work also includes reviewing features and data points used in the research related to HIDS systems, as this area grows more mature and including them in our categorization.

References

1. Al-Jarrah, O., Arafat, A.: Network intrusion detection system using attack behavior classification. In: 5th International Conference on Information and Communication Systems (ICICS), pp. 1–6. IEEE (2014)
2. Alanazi, H., Noor, R., Zaidan, B., Zaidan, A.: Intrusion detection system: overview. arXiv preprint arXiv:1002.4047 (2010)
3. Bhuyan, M.H., Bhattacharyya, D.K., Kalita, J.K.: Network anomaly detection: methods, systems and tools. IEEE Commun. Surv. Tutor. **16**(1), 303–336 (2014)
4. Erman, J., Mahanti, A., Arlitt, M.: Qrp05-4: Internet traffic identification using machine learning. In: 2006 Global Telecommunications Conference, GLOBECOM 2006, pp. 1–6. IEEE (2006)
5. Estan, C., Savage, S., Varghese, G.: Automatically inferring patterns of resource consumption in network traffic. In: Proceedings of the 2003 Conference on Applications, Technologies, Architectures, and Protocols for Computer Communications, pp. 137–148. ACM (2003)
6. Feroz, M.N., Mengel, S.: Phishing URL detection using URL ranking. In: 2015 IEEE International Congress on Big Data (BigData Congress), pp. 635–638. IEEE (2015)
7. Garcia-Teodoro, P., Diaz-Verdejo, J., Maciá-Fernández, G., Vázquez, E.: Anomaly-based network intrusion detection: techniques, systems and challenges. Comput. Secur. **28**(1–2), 18–28 (2009)
8. Gonzalez, R., Manco, F., Garcia-Duran, A., Mendes, J., Huici, F., Niccolini, S., Niepert, M.: Net2Vec: deep learning for the network. In: Proceedings of the Workshop on Big Data Analytics and Machine Learning for Data Communication Networks, pp. 13–18. ACM (2017)
9. Hammerschmidt, C., Marchal, S., State, R., Pellegrino, G., Verwer, S.: Efficient learning of communication profiles from IP flow records. In: 2016 41st Conference on Local Computer Networks (LCN), pp. 559–562. IEEE (2016)
10. Hotho, A., Maedche, A., Staab, S.: Ontology-based text document clustering. Künsliche Intelligenz (KI) **16**(4), 48–54 (2002)
11. Kemmerer, R.A., Vigna, G.: Intrusion detection: a brief history and overview. Computer **35**(4), supl27–supl30 (2002)
12. Liu, Y., Li, W., Li, Y.C.: Network traffic classification using k-means clustering. In: 2007 Second International Multi-Symposiums on Computer and Computational Sciences, IMSCCS 2007, pp. 360–365. IEEE (2007)
13. Ma, J., Saul, L.K., Savage, S., Voelker, G.M.: Beyond blacklists: learning to detect malicious web sites from suspicious URLs. In: Proceedings of the 15th ACM SIGKDD International Conference on Knowledge Discovery and Data Mining, pp. 1245–1254. ACM (2009)

14. Magdalinos, P., Barmpounakis, S., Spapis, P., Kaloxylos, A., Kyprianidis, G., Kousaridas, A., Alonistioti, N., Zhou, C.: A context extraction and profiling engine for 5G network resource mapping. Comput. Commun. **109**, 184–201 (2017)
15. Mahmood, A.N., Leckie, C., Udaya, P.: An efficient clustering scheme to exploit hierarchical data in network traffic analysis. IEEE Trans. Knowl. Data Eng. **20**(6), 752–767 (2008)
16. McGregor, A., Hall, M., Lorier, P., Brunskill, J.: Flow clustering using machine learning techniques. In: International Workshop on Passive and Active Network Measurement, pp. 205–214. Springer (2004)
17. Sarmadi, S., Li, M., Chellappan, S.: On the feasibility of profiling internet users based on volume and time of usage. In: 2017 9th Latin-American Conference on Communications (LATINCOM), pp. 1–6. IEEE (2017)
18. Shadi, K., Natarajan, P., Dovrolis, C.: Hierarchical IP flow clustering. In: Proceedings of the Workshop on Big Data Analytics and Machine Learning for Data Communication Networks, pp. 25–30. ACM (2017)
19. Singh, H.: Performance analysis of unsupervised machine learning techniques for network traffic classification. In: 2015 Fifth International Conference on Advanced Computing & Communication Technologies (ACCT), pp. 401–404. IEEE (2015)
20. Singh, R., Kumar, H., Singla, R.: An intrusion detection system using network traffic profiling and online sequential extreme learning machine. Expert Syst. Appl. **42**(22), 8609–8624 (2015)
21. Tsai, C.F., Hsu, Y.F., Lin, C.Y., Lin, W.Y.: Intrusion detection by machine learning: a review. Expert Syst. Appl. **36**(10), 11994–12000 (2009)
22. Valenti, S., Rossi, D., Dainotti, A., Pescapè, A., Finamore, A., Mellia, M.: Reviewing traffic classification. In: Data Traffic Monitoring and Analysis, pp. 123–147. Springer (2013)
23. Wang, C., Song, Y., El-Kishky, A., Roth, D., Zhang, M., Han, J.: Incorporating world knowledge to document clustering via heterogeneous information networks. In: Proceedings of the 21th ACM SIGKDD International Conference on Knowledge Discovery and Data Mining, pp. 1215–1224. ACM (2015)
24. Wang, Y., Xiang, Y., Zhang, J., Zhou, W., Wei, G., Yang, L.T.: Internet traffic classification using constrained clustering. IEEE Trans. Parallel Distrib. Syst. **25**(11), 2932–2943 (2014)
25. Xu, K., Zhang, Z.L., Bhattacharyya, S.: Internet traffic behavior profiling for network security monitoring. IEEE/ACM Trans. Netw. **16**(6), 1241–1252 (2008)

Cybersecurity and Digital Forensics – Course Development in a Higher Education Institution

Mário Antunes[1,2,3(✉)] and Carlos Rabadão[1,3]

[1] School of Technology and Management, Polytechnic of Leiria, Leiria, Portugal
{mario.antunes, carlos.rabadao}@ipleiria.pt
[2] INESC-TEC, CRACS, University of Porto, Porto, Portugal
[3] Computer Science and Communication Research Centre,
CIIC, Leiria, Portugal

Abstract. Individuals and companies have a feeling of insecurity in the Internet, as every day a reasonable amount of attacks take place against users' privacy and confidentiality. The use of digital equipment in illicit and unlawful activities has increasing. Attorneys, criminal polices, layers and courts staff have to deal with crimes committed with digital "weapons", whose evidences have to be examined and reported by applying digital forensics methods. Digital forensics is a recent and fast-growing area of study which needs more graduated professionals. This fact has leveraged higher education institutions to develop courses and curricula to accommodate digital forensics topics and skills in their curricular offers. This paper aims to present the development of a cybersecurity and digital forensics master course in Polytechnic of Leiria, a public higher education institution in Portugal. The authors depict the roadmap and the general milestones that lead to the development of the course. The strengths and opportunities are identified and the major students' outcomes are pointed out. The way taken and the decisions made are also approached, with a view to understanding the performance obtained so far.

Keywords: Higher education institutions · Digital forensics · Cybersecurity · Course development

1 Introduction

Information security can be seen as a branch of computer science scientific area and usually covers topics related with information and network security, cryptography, system administration and operating systems, just to mention a few. Recently, the nature and amount of crimes that involves information and computers security has changed its nature and the issues covered. This is mainly due to the growing number of existing assets (software and hardware) that are able to be involved in crimes and may end up in court.

The wide nature of assets bring the need to educate even more IT professionals in information and computers security into three distinct dimensions: prevention, detection and forensics. Regarding prevention and detection, the topics are roughly covered in 1st and 2nd cycle courses of higher education in cybersecurity fields [1]. With respect to the digital forensics topics, the overall scenario in Portugal is that they are barely and

© Springer Nature Switzerland AG 2020
A. M. Madureira et al. (Eds.): SoCPaR 2018, AISC 942, pp. 338–348, 2020.
https://doi.org/10.1007/978-3-030-17065-3_34

shallowly covered in higher education courses, giving the students the potential to apply self-study methodologies in these matters. That is, IT professionals are mostly challenged to have skills that enable them to deal with both prevention, and detection, being digital forensics topics usually relegated to a second stage.

Given this scenario and having in mind that digital forensics is an intrinsic part of cybersecurity, higher education institutions have to include those topics in the next coming course curriculum in computers security, that may match the expected requirements of IT security professionals. This motivation fueled Polytechnic of Leiria to start a set of educational and research initiatives with the aim of highlighting digital forensics topics in the well-established cybersecurity area of study.

The aims of this paper are two-fold: to frame the computer science and information security course curriculum development; to present the Master of Science (MSc) in Cybersecurity and Digital Forensics (MCDF) at Polytechnic of Leiria, highlighting its organization, functioning and performance evaluation.

The paper is organized as follows. In Sect. 2 the authors describe the background about courses development and accreditation process, focusing on the Portuguese case study. Section 3 describes the MCDF developed at Polytechnic of Leiria, namely the course organization, the history behind the MSc degree recently developed and the laboratory design to support students' activities. In Sect. 4 the authors present some preliminary results regarding the number of enrolled students and elaborate on the analysis from the experience obtained. Finally, in Sect. 5 some conclusions are delineated, as well as topics of future work.

2   Background

Higher education courses in Portugal can be generally classified in two distinct groups: as having an associated degree (e.g. BSc or MSc) or not (e.g. post-graduation or advanced courses). The former have to be submitted to an accreditation agency that have to approve their functioning and student's enrollment. The later are more informal and may be defined locally by each institution. The Portuguese accreditation agency is "Agência de Avaliação e Acreditação do Ensino Superior" (A3ES[1]) and is responsible by analyzing the course development process and make a proposal whether the course have conditions to proceed or not.

Typically, a course proposal dossier starts by characterizing the course, the institution and the teaching staff. Then, several institutional approvals have to be put in, as well as the detailed study plan and syllabus of all the curricular units and the intended learning outcomes expected to be achieved by the students. Finally, besides a SWOT characterization, it is mandatory to include a benchmark analysis with the intended learning outcomes of similar study programs offered by reference institutions of the European Higher Education Area. That is, the approval process is methodic and at the end aims to approve courses that may give the students the qualifications that are relevant to the society as a whole and in line the best practices in similar institutions all over the world.

---

[1] http://www.a3es.pt/en.

The relevant information gathered about the courses and institutions usually includes, when available, state-of-the-art organizations. Regarding digital forensics we may evidence the European Network of Forensic Science Institutes (ENFSI[2]) and the Forensic Science Education Programs Accreditation Commission (FEPAC[3]). Renowned institutions like ACM have also comprehensive and widely adopted guidelines to prepare new courses that respect the best practices [2, 3, 6–8]. They should thus to be considered on the designs and development of new courses.

The analysis of reports and reflections about the development of new courses in the digital forensics area can also give important insight on how to enhance study plans, methodologies and expected students' outcomes. Some examples can be found in [4, 5, 9].

The amount of courses including the word "security" in their designation is increasing, as the concerns about network and information security is becoming of interest to all the practitioners. By analyzing the formal designations of existing MSc courses in public higher institutions in Portugal, the courses credited are mainly in information security engineering, although they do not explicitly address digital forensic and data recovery. A simple search with the word "security" return eight courses, in which some of them probably include digital forensics topics. When we apply other terms like "informatics engineering" or "computer" we are able to retrieve more results. The returned courses belongs to computer science scientific area and are not expected the lecture of digital forensics topics[4]. These results reveal a low concern on developing cybersecurity and digital forensics courses in higher education institutions, at least by analyzing the formal designations of the courses.

Some countries started earlier and have invested a deep more on defining digital forensics curricula to prepare IT professionals, being UK a good example. It is possible to find various study programs dedicated specifically to digital forensics or together with information security, being the MSc in Computer Security and Forensics from University of Bedfordshire[5] and the MSc in Digital Investigation and Forensic Computing from University College Dublin[6] two good examples.

The course described in this paper innovates in certain way the panorama in Portugal, since integrates in the same study program two highly complementary areas of study: cybersecurity and digital forensics. The designation integrates both terms, "cybersecurity" and "digital forensics", which reflects the investment in both topics and the aim to form, with an advanced academic degree, IT professionals highly trained to operate in the three dimensions previously introduced: prevention, detection and digital forensics.

---

[2] http://enfsi.eu/.

[3] http://fepac-edu.org/accredited-universities.

[4] http://www.a3es.pt/en/accreditation-and-audit/accreditation-process-results/accreditation-study-programmes – values retrieved by October, 13, 2018.

[5] https://www.beds.ac.uk/howtoapply/courses/postgraduate/next-year/computer-security-and-forensics.

[6] http://www.ucd.ie/cci/education/prospective_students/msc_difc.html.

3   Case Study – MSc in Cybersecurity and Digital Forensics

This section describes the main issues related with the creation of an MSc degree in Cybersecurity and Digital Forensics (MCDF) in the Polytechnic of Leiria, in Portugal.

The course started at academic year 2017/2018 and culminated a tough but rewarding walk started in 2012, with the development of non-degree courses and the implementation of relevant partnerships. The details about the course organization, the expected students' outcomes and technical resources allocated are presented in this section, on trying to extract the benefits of the approach used to the development of similar courses.

3.1   The Institution - Polytechnic of Leiria

Polytechnic of Leiria (IPLeiria[7]) is a Portuguese public institution of higher education committed to the education of citizens, lifelong learning, research, dissemination and transfer of knowledge. It has around 12,500 students distributed by five schools that accommodate a wide set of courses in the following areas of study: engineering and management, arts, health, tourism and social sciences.

The School of Technology and Management (ESTG) is located in the city of Leiria and is the largest school, both in the number of students (around 6,000) and diversity of courses. Its mission is to train highly qualified people in fundamental engineering and management areas. The Department of Informatics Engineering (DEI), where MCDF belongs, is the largest department with around 900 students spread by three distinct levels: short technical courses (2 years), undergraduate courses (3 years) and MSc courses (2 years). The staff is highly qualified with a high rate of professors that hold a PhD and/or have relevant and proven experience in IT industry.

3.2   Background Activities

The MCDF is grounded in the sequence of a wide set of well succeeded initiatives, both of research and educational, in cybersecurity and digital forensics areas. These activities were carried on by ESTG-IPLeiria since 2012, from which the following items are highlighted:

- Bilateral agreements established between IPLeiria and relevant public institutions, namely Portuguese Judiciary Police, its school and laboratory, the cybercrime attorney's office of the Attorney's General Office and the forensics office of territorial polices (e.g. Guarda Nacional Republicana). These agreements enabled the sharing of experiences, the setup of short and intensive courses and the organization of technical events.
- Setup, organization and lecture of five editions, since 2015, of a post-graduation course on cybersecurity and digital forensics. The students came from different institutions, namely criminal polices, public and private institutions.

---

[7] http://www.ipleiria.pt.

- Setup of a forensics laboratory low-cost and fully equipped. This lab started its activity on May 2015 and enabled us to carry on forensics expertise related to public cases emanated from the Attorney's General Office.
- Setup and lecture of short courses for technicians and criminal detectives from criminal polices.
- Teachers' involvement on internships' supervision, projects and dissertations carried on mainly at Portuguese Judiciary Police.
- Teachers' involvement on lecturing at a multidisciplinary post-graduation course named "Multinational Smart Defense Project on Cyber Defense Education & Training (MN CD E&T)", promoted and organized by NATO.
- Organization of events that brought together practitioners from criminal polices, companies, academia and public-sector institutions. The workshops covered a wide range of topics in cybersecurity and digital forensics, namely legal issues, technical procedures and methodologies, case studies and sharing of experiences.

3.3   Students' Outcomes

Figure 1 depicts a list of the main outcomes that have been identified and that should be considered when developing a course curriculum in cybersecurity and digital forensics. These topics are spread in curricular units of knowledge that cover the three major general topics: prevention, detection and forensics. The outcomes are in line with the needs identified by the employers and the renowned recommendations [6, 8].

Prevention	Detection	Forensics
Policy and risk analysis	Penetration testing lab	Digital forensics
Computers network security	Management and analysis of security reports	Mobile forensics
Secure software development		Recovery systems lab
Cyberlaw and data protection regulations	Computer incident handling	Biometrics
Mobile devices and IoT security	Secure administration of computer systems	Ethical hacking
Fundamentals of information security Fundamentals of cryptography Fundamentals of networking		

Fig. 1.   Major outcomes expected to be achieved by the students.

Besides the need to invest in the fundamentals of information security, cryptography and networking, we may find concerns regarding new types of equipment (e.g. mobile and IoT), cyberlaw and software development.

Regarding digital forensics the efforts are on procedures, methodologies and tools. In the detection dimension the outcomes are mainly on setting up emergency research teams and applying techniques to easily manage incident reports.

The following are general learning outcomes, such as knowledge, skills and competences, intended to be developed by the students enrolled in MCDF[8]:

- To apply skills and comprehension ability to solve real world problems.
- To have strong skills on applications, methodologies and technologies related with planning, implementation and monitoring of cybersecurity.
- To have strong skills on applications, methodologies and technologies related with digital forensics.
- To integrate skills to deal with complex scenarios, to develop solutions or issue a judgment in situations of limited or incomplete information.
- To communicate the conclusions, the knowledge and reasoning to specialists and no specialists in the area of study.
- To possess ethical, professional and scientific research skills in cybersecurity and digital forensics.

3.4   Goals

The following generic goals are applied to MCDF and should guide the development of other similar courses:

- To enable the progress of the 1st cycle undergraduate students in computers engineering courses and similar areas of study.
- To produce highly qualified computers engineers and IT technicians in the areas of cybersecurity and digital forensics.
- To develop partnerships with companies and public institutions that may accommodate projects and dissertations to solve real world problems.
- To develop applied research in the areas of cybersecurity and digital forensics by the completion of projects and dissertations.
- To promote the advanced knowledge transfer to the society of the topics covered by the MSc course.
- To promote autonomous and long-life learning study.

3.5   General Organization

The MCDF is a 2nd cycle degree on cybersecurity that emphasizes one important and complementary topic: digital forensics. It is an advanced course with a practical nature and mostly oriented to solve real problems in the IT infrastructures and digital forensics of electronic and digital equipment.

Students are able to enroll both in a full-time and part-time (labor) schedules. The former is adjusted for students that want to continue their studies after completing the 1st cycle, while the later aims to give a chance for IT professionals already to attend the classes and complete their academic MSc degree.

---

[8] Official webpage of the MSc in cybersecurity and Digital Forensics at IPLeiria-ESTG: https://www.ipleiria.pt/cursos/course/mestrado-em-ciberseguranca-e-informatica-forense/?

The course is two years duration and its study plan is depicted in Fig. 2. It is organized in modular curricular units which encompass topics within prevention, detection and forensics in a balanced number of hours.

Regarding European Credits Transfer System (ECTS), the course has 120 ECTS, divided by 60 ECTS in each academic year. In the first year there are ten curricular units (five in each semester), each one having 6 ECTS. The second year is fully dedicated to prepare a dissertation, develop a project or apply for an internship in the company. Regardless the typology chosen by the students in the second year, there are always 60 ECTS involved and the major outcome to be achieved is to solve real world problems.

	Prevention	Detection	Forensics
	Policy and risk analysis for information security	Penetration testing lab	Digital Forensics – unit 1
	Computers Network Security	Management and analysis of security reports	Digital Forensics – unit 2
1st year	Secure administration of computer systems	Computers incidente handling	
	Project – unit 1		
	Project – unit 2		
	Dissertation		
2nd year	Project		
	Internship		

Fig. 2. Study plan of MCDF.

In both semesters of the 1st year there is a curricular unit for the development of projects in cybersecurity and digital forensics. The underpinning idea to develop integrated projects with the topics approached in the remaining units in each semester. Particularly in the curricular unit of "Project – unit 2", that happens in the 2nd semester, the students are invited to prepare the subject they want to develop in the second year of the course. During this curricular unit the students gather the most relevant bibliography, prepare the state-of-the-art and plan the major activities to be developed in the second year.

3.6   Main Entities

Figure 3 depicts the overall relationship between the main entities that interact between each other, namely the students, the courses offered and the employers.

The students came from distinct sources which gives a rich heterogeneity and transfer learning in the classroom. As an example, we may have in a classroom students with no professional experience (newly 1st year graduated students), private IT security technicians and digital forensics investigators.

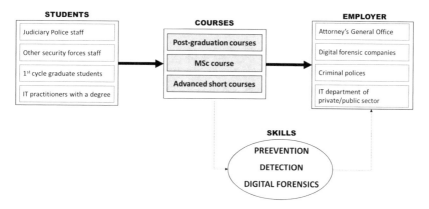

Fig. 3. Main entities involved and their relationship

The students have enrolled since 2012 in the existing courses. The skills and competences achieved by the students in the dimensions already described, have been successfully applied in the public/private institutions and companies to which they belong.

3.7 Digital Forensics Laboratory

Digital forensics classes need an adequate and well equipped laboratory. Prior to the development of MCDF course, ESTG-IPLeiria set up a laboratory to accommodate digital forensics activities in the scope of processes emanated from courts through Attorney's General Office. In that sense we have set up a room fully dedicated to this purpose, with the following equipment:

- Write blockers for several types of external device connectors (USB, SATA and IDE, PCIe).
- Forensics kit with a set of cables for mobile devices.
- Digital forensics software both proprietary and open-source (Autopsy[9]).
- Digital forensics open-source software "FTK Imager"[10] to produce an image of a digital device (e.g. disks).
- Digital forensics Linux stations with SIFT workstation software[11].
- A safe box to store equipment.

The overall idea was to have a low-cost and functional laboratory that could be used both to answer to attorneys' requests and also to support practical and hands-on lectures activities planned to MCDF.

---

[9] https://www.sleuthkit.org/autopsy/.
[10] https://accessdata.com/.
[11] https://digital-forensics.sans.org/.

4   Results Analysis

Table 1 presents the general overview with the amount of students enrolled in the distinct courses since 2014. The values are grouped by the three courses typologies, highlighting those related with the students enrolled in the MCDF.

The beginning was shy and only with the lecturing of editions of the post-graduation course. After 2017, with the first edition of MCDF, we have simultaneously two classes in others such distinct courses. MCDF in now in its second edition with an accumulated value of 60 students, being expected to reach around 100 students in 2019.

Table 1. Number of enrolled students.

Type of course	2014	2015	2016	2017	2018	2019*	Total
Post-graduation (6 months, no degree)	12	20	18	12	12		**74**
MCFD (2 years, MSc degree)				21	39	40	**100**
Advanced short courses (2 weeks, no degree)						45	**45**
Total	**12**	**20**	**18**	**33**	**51**	**85**	**219**

*Planned

In post-graduation courses, around 99% of the enrolled students have successfully completed all the curricular units and the final projects. With respect to MCDF, since we are in the second year of functioning, there is not yet students that have obtained the MSc degree. The second year students (enrolled in 2017) are developing the works that will lead to their dissertations or projects reports.

Since 2017 the post-graduation course was gradually replaced by short advanced courses. This decision reflects a shift in students' needs and requests, essentially those that came from private companies and police forces (Judiciary police and others). In these cases the courses have a short duration, the degree is not mandatory and the students invest in acquiring competences and skills.

5   Conclusions and Future Work

This paper described an MSc course curriculum development in cybersecurity and digital forensics, carried out at IPLeiria-ESTG. A broad description of the major steps taken is made, as well as a performance evaluation of the MSc course.

The course is now in its second edition and the students of the first edition are now starting their dissertation. The market accepted very well the course and the balance is very positive. The number of candidates has exceeded for the second consecutive year the number of available vacancies. The students' background is natively heterogeneous which enrich the sharing of ideas and the quality of projects made during the curricular units.

Looking back to the genesis of this course and to the decisions made, some conclusions arise. Firstly, digital forensics is an emergent and very relevant research topic, which is part of cybersecurity area and where it is worth investing. Secondly, an MSc degree in such a broad topic like "information security" or "computers security" doesn't necessarily covers "digital forensics". The need to have digital forensics experts should push the institutions to offer dedicates MSc courses. Finally, the path is made by walking and some actions are worth executing before setting up an MSc degree course, especially because we are dealing with a brand new area of studies.

The case study presented in this paper reveals the benefits of having carried out a bunch of initiatives related with digital forensics. Among those initiatives, there are two that stand out: to build a network of relevant partners with which we may establish solid partnerships; to setup a physical lab to experiment the digital forensics methodologies and procedures.

Being an emergent and worth learning area of study, a final word to the benefits of having a heterogeneous and knowledge transfer oriented teaching staff. This is notorious by incorporating highly specialized digital forensics experts in the teaching staff, namely coming from criminal polices, which fed both students and academics with relevant and up-to-date knowledge.

As future work, three directions are identified: to invest on international certification for the MSc degree and the correlated advanced courses; to create new partnerships and to strengthen the existing ones; to turn classroom experience on a deep hands-on and problem-solving experience, as much as possible.

Acknowledgments. This publication is funded by FCT - Fundação para a Ciência e Tecnologia, I.P., under the project UID/CEC/04524/2016.

References

1. Heitmann, G.: Challenges of engineering education and curriculum development in the context of the Bologna process. Eur. J. Eng. Educ. **30**(4), 447–458 (2005)
2. ACM/IEEE-CS Joint Task Force on Computing Curricula. Computer Science Curricula 2013 [Technical report]. ACM Press and IEEE Computer Society Press, December 2013
3. Cooper, P., Finley, G.T., Kaskenpalo, P.: Towards standards in digital forensics education. In: Proceedings of the 2010 ITiCSE WG Reports, pp. 87–95. ACM, June 2010
4. Lang, A., et al.: Developing a new digital forensics curriculum. Dig. Investig. **11**, S76–S84 (2014)
5. Srinivasan, S.: Computer forensics curriculum in security education. In: 2009 Information Security Curriculum Development Conference, pp. 32–36. ACM, September 2009
6. Dafoulas, G.A., Neilson, D., Hara, S.: State of the art in computer forensic education-a review of computer forensic programmes in the UK, Europe and US. In: 2017 International Conference on New Trends in Computing Sciences (ICTCS), pp. 144–154. IEEE, October 2017
7. Cabaj, K., Domingos, D., Kotulski, Z., Respício, A.: Cybersecurity education: evolution of the discipline and analysis of master programs. Comput. Secur. **75**, 24–35 (2018)

8. ACM/IEEE: Curricula guidelines for post-secondary degree programs in cybersecurity (2017). https://www.acm.org/binaries/content/assets/education/curricula-recommendations/csec2017.pdf
9. Santos, H., Pereira, T., Mendes, I.: Challenges and reflections in designing cyber security curriculum. In: World Engineering Education Conference (EDUNINE), IEEE, pp. 47–51. IEEE, March 2017

Model Driven Architectural Design of Information Security System

Ivan Gaidarski[1,3]($\boxtimes$), Zlatogor Minchev[1,2], and Rumen Andreev[3]

[1] Joint Training Simulation and Analysis Center, Institute of ICT,
Bulgarian Academy of Sciences, Sofia, Bulgaria
i.gaidarski@isdip.bas.bg, zlatogor@bas.bg
[2] Institute of Mathematics and Informatics,
Bulgarian Academy of Sciences, Sofia, Bulgaria
[3] Institute of Information and Communication Technologies, Sofia, Bulgaria
rumen@isdip.bas.bg

Abstract. The main objective of the paper is to present model-driven approach to development of Information Security System. We use data centric models, in which the main focus is on the data and we define a conceptual model of Information Security System architecture using the main information security concepts. Its construction is based on the domain analysis organized around the viewpoints "Information Security" and "Information Processing". The meta-models based on these viewpoints concern different aspects of the data and data protection. They are based on the summary of our practical experience in information security activities. Then the conceptual model is transformed to system design model with the help of UML – class, activity and deployment diagrams that transform the conceptual model of system architecture into actual solution or physical system.

Keywords: Conceptual modeling · Data protection · Architecture framework · UML

1 Introduction

Information security (IS) activities applied within an organization concern compliance with certain legislative regulations, including IS standards such as ISO 27000 [1, 2], ISACA's COBIT [3, 4], NIST "800 series" [5], sector-specific regulations – the Gramm-Leach-Bliley Act (GLBA) [6], Sarbanes-Oxley Act (SOX) [7, 8], Health Insurance Portability and Accountability Act (HIPAA) [9] and Payment Security Industry Data Security Standard (PCI DSS) [10]. These regulations can be considered as a set of good practices related with the most important aspects of IS. The cases where the IS are developed methodically and all requirements of the standards are satisfied but rare. The most common practice is to undertake single actions to solve certain IS problems as certain security incidents (attack on infrastructure, data leakage, loss of information, etc.) or to comply with new regulations, for example recently adopted regulation EU GDPR [11] for protection personal data of EU Citizens.

© Springer Nature Switzerland AG 2020
A. M. Madureira et al. (Eds.): SoCPaR 2018, AISC 942, pp. 349–359, 2020.
https://doi.org/10.1007/978-3-030-17065-3_35

The development of information security systems (ISS) can be done using two approaches – bottom-up and top-down. The first approach concerns a gradual protection of the information assets – day-to-day activities for improving the security of the organization's information, devices or applications. This is an engineering-oriented architecting of ISS, in which system analysis is component orientated. The big disadvantage of this analysis is that it provides a single view of the system, as the approach concentrates only on the "Do" ingredient of the domain analysis. It only covers the technology and deployment aspects of the system development. The realization of the information infrastructure is described with it platform model in terms of its available communication infrastructure and implementation technology.

These shortcomings can be avoided by the use of the top-down approach, which is based on goals set by the management, compliance with existing policies, procedures and processes, clearly defined responsibilities and roles of the involved participants and aiming specific results. As this approach has strong support of the senior management with allocated budget, approved plan and clear deadlines, it has better chances for success [2].

The main purpose of the ISS is to protect and secure the organization's information assets. The requirements to the development of ISS are to ensure the achievement of important properties such as reusability of the components, interoperability, scalability and easy deployment. The engineering-oriented architecting of systems is not able to guarantee their achievement. Since the top-down approach to systems development is very suitable for this objective, we suggest the usage of model-driven architecting of information security systems. This approach is suitable for creation of a reference methodology for ISS development which bases on determination of a domain analysis framework that serves for construction of domain models. The main objectives of this framework coincide with the goals of the framework for architectural description of systems that is presented in the standards IEEE 1471 and IEEE 42010 [22, 23]. They introduce concepts as View, Viewpoint, Stakeholders, Concerns and Environment which is related to the architecture description [12, 13, 15]. These concepts are workable in domain analysis and they guarantee a context for definition of a common conceptual framework, allowing the construction of conceptual models of ISS when the domain concerns information security [20, 24, 25].

At present, there is much interest in using Unified Modeling Language (UML) for architectural description, since it provides tools for sketching, analyzing, modeling and documenting knowledge and solutions about the architecture of a software-intensive system as ISS. The UML support techniques that enable system developers to record what they are doing, modify or manipulate suggested architectures, to reuse parts of them that exist and to communicate the architectural information collected during system development. The goal of UML [14] is to provide a standard notation that can be used by all object-oriented methods. It provides constructs for a broad range of systems and activities (analysis, system design and deployment). UML consists of several diagram types providing different views of the system model. The first type – Structural Diagrams represent the static structure of the system. They include Class Diagram, Component Diagram, Composite Structure Diagram, Deployment Diagram, Object Diagram and Package Diagram. The Behavior Diagrams represent general types of behavior. They include: Use Case Diagram, Activity Diagrams and State chart Diagram. The last type is Interaction Diagrams, which include: Sequence Diagram, Communication Diagram, Timing Diagram and Interaction Overview Diagram.

The main objective of the paper is to present model-driven approach to development of ISS that transform a conceptual model of system architecture into design model of the system, which is described with the help of UML. In Sect. 2 we define a conceptual model of ISS architecture using the main information security concepts. Its construction is based on the domain analysis organized around the viewpoint "Information Security". Then the conceptual model is transformed to system design model with the help UML – class, activity and deployment diagrams.

2   Meta Models of Information Security System

The traditional and still currently used security models are host or domain based. The main function of ISS is to mitigate the various threats to the data, including data theft, data loss and altering of the data. The threats can be result from different causes [18]: Intentional Actions (intentional deletion of a file or program, theft, hacking, sabotage, malicious act – virus, worm, malware, ransomware, etc.), Unintentional actions (accidental deletion of a file or program, misplacement of USB Flash drives or DVDs, administration errors), Failure (power failure, hardware failure, software crash/freeze, software bugs, data base corruption) or Disaster (fire, flood, earthquake).

Instead of Host/Domain based models, it is more suitable to use data centric models, in which the main focus is on the data. The meta-models discussed in the paper concern different aspects of the data and data protection. They are based on the summary of our practical experience in information security activities.

The most important questions to which an ISS has to answer from the "Information Security" viewpoint are What, How and Why. From that viewpoint we suggest a meta-model of ISS (see Fig. 1), which consists of 6 conceptual blocks: Endpoint protection, Communications protection (What), Monitoring and analysis, Management and configuration (How), Data protection and Security model and policy (Why).

The Endpoint Protection is responsible for defensive capabilities of devices. The main functionality, provided by this component includes authoritative identity, cyber and physical security. The Communications and Connectivity Protection component is responsible for protecting the communication between the endpoints via implementing

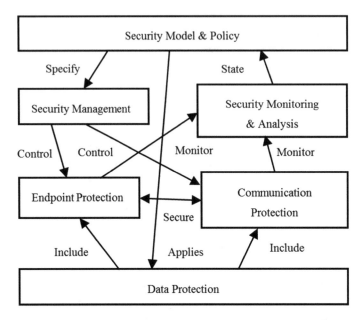

Fig. 1. Meta-model of "information security" viewpoint of ISS.

different methods as authentication and authorization of the traffic, cryptographic techniques for integrity and confidentiality and information flow control techniques. After endpoints and communications are secured, the state of the system must be preserved through the whole lifecycle by monitoring, analyzing and controlling all components of the system, which is done by the next two components – Security Monitoring and Analysis, and Security Configuration and Management.

The Security Model and Policy component defines how security policies are implemented to ensure confidentiality, integrity and availability of the system. It directs all other modules to work together and to deliver end-to-end security of the system. The Data Protection component directly and indirectly supports the first four modules. The scope of the module extends from data-at-rest in the endpoints to data-in-motion in the communications, the data gathered as part of monitoring and analysis function and all the system configuration and management data [16].

The data can be in one of three states at any given time: it can be at rest (on storage device), in motion (communications) or in use (processed in applications) [18].

The second meta-model is based on "Information processing" viewpoint in architecture description of ISS (see Fig. 2).

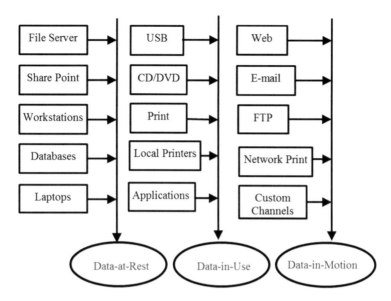

Fig. 2. Meta-model of "information processing" viewpoint of ISS.

To protect the different types of data it is necessary to be implemented specific Information Security Techniques (IST) in the basic components from the "information security" meta-model. The data have to be protected against loss, theft, unauthorized access and uncontrolled changes by applying IST such as confidentiality controls, integrity controls, access control, isolation and replication [17, 18]. On that base we suggest Multi-Layered conceptual model of ISS (see Fig. 3) which contains the meta-models representing "Information Security" and "Information Processing" viewpoints respectively and the relations between them:

- The Endpoint Protection component protects Data-at-Rest and Data-in-Use of the endpoints through relevant IST, as access control and passwords, antivirus, audit trails, physical security measures and etc.;
- Communications Protection component protects Data-in-Motion using cryptographic techniques, network segmentation, perimeter protection, gateways, firewalls, intrusion detection, network access control, deep packet inspection and network log analysis;
- Security Management component protects all configuration, monitoring, and operational data, using cryptography.
- Security Monitoring & Analysis component is responsible for the protection of the data for current system state, the monitoring of key system parameters and indicators. The typical ISTs in this component are cryptographic techniques.

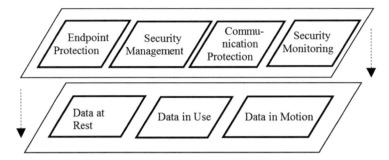

Fig. 3. Multi-layered conceptual model of ISS.

Depending from the requirements from different Stakeholders, we can add more meta-models for different viewpoints to satisfy their needs from the ISS. Later that multi-layered conceptual models will be transformed in real physical implementation of the ISS.

3   UML Design Model

System development lifecycle consists from the following processes:

- Requirements gathering processes – resulting in requirements model.
- Analysis processes to understand the requirements – resulting in an analysis model, describing implementation-independent solution which satisfies the requirements model.
- Design processes – resulting in design model, which describes implementation-specific solution, specified by the analysis model.
- Implementation processes to build a system – resulting in implementation model of physical system that satisfies the design model.
- Testing processes – to verify that a system satisfies its requirements.
- Deployment processes, to make the system available to the users.

Each process results in the corresponding model, which satisfy the previous requirements and is a base for the next level model [19, 24]. The whole process of system development is a model-to-model transformation. One of the most convenient ways is to use view- and object-oriented tool as UML. In model driven architectural modelling approach we will show the transformation from conceptual model into UML design model. For the description of the proposed Multi-layered meta model of ISS it is necessary to describe the system concepts, the different activities and system implementation. For description of the system concepts can be used UML Class Diagram, for description of the dynamic aspects of the system – UML Activity Diagram and for the implementation of the ISS's components – the UML Deployment Diagram [21].

UML Class Diagram

The purpose of Class Diagrams is to show the static structure of classifiers in system. The diagram provides basic notation, which can be used by other UML structure diagrams. The UML Class Diagrams consists of set of classes and relationships between the classes. We will define 3 classes – Management class, Protection class and Data class:

- The Management class has the following attributes: System state, Specification, Control, Monitor and class operations (methods): Security Model, Security Policy, Management, Configuration, Monitoring, Analysis.
- The Protection class has the following attributes: Data channels, Data in rest, Data in use, Data in motion, Analysis Data, Configuration data and class operations (methods): Endpoint protection and Communication protection.
- The Data class has the following attributes: Include, Exception, Data channel, Sensitive and class operations (methods): Pass, Block and Report.

The representation of the meta model of ISS in UML class diagram is shown on Fig. 4. We use the same concepts as in the meta-model from Fig. 1, which are separated as methods of the 3 main classes. For Security Model & Police concept we have methods Security Model and Security Policy, for Management & Configuration concept we have Management and Configuration methods, for Monitoring & Analysis concept we have the Monitoring and Analysis methods in in the Management class. The Endpoint Protection and Communication protection concepts are presented as Endpoint Protection and Communication methods in Protection class and finally Data Protection concept have the equivalent Pass, Block or Report methods in Data class. All these methods perform the relevant functionality as concepts from the meta-model. The Security Model and Security Policy define and manages the other classes methods with the help of Specification, Control, Monitor and System state attributes. The Endpoint Protection, Communication protection and Data protection methods secures the Data class and Protection class attributes and so on.

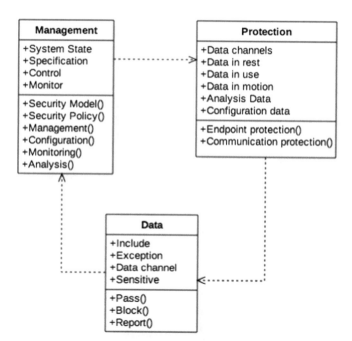

Fig. 4. UML Class Diagram.

UML Activity Diagram

The Activity diagram shows flowchart to represent the flow from one activity to another activity of the system objects. It captures the dynamic behavior of the system.

On Fig. 5 we represent the UML Activity diagram of one of the methods of Protection class – Endpoint Protection, which correspond to the concept Endpoint Protection from the meta-model. Similar Activity Diagrams must be created for all methods from the UML Class Diagram.

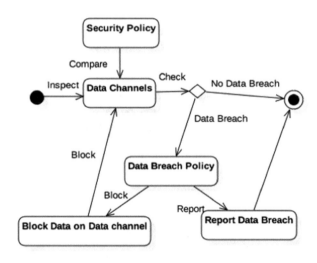

Fig. 5. UML Activity Diagram.

First the Data Channels are inspected for sensitive information by comparing the data with the Security Policy. If there is a match, then there is a Data breach, and the system must perform an action, defined in Data Breach Policy. That can be Report or/and Block. In the first case the Data Breach is being reported and the data can pass through the Data Channel. Otherwise the Data channel is blocked and the data cannot pass through the relevant data channel.

UML Deployment Diagram

Models the physical deployment of system components. As Component diagrams are used to describe the components, the deployment diagrams shows how they are deployed in the real environment. The hardware components (e.g. DLP (Data Leak Prevention) solution, web server, mail server, application server) are presented as nodes, with the software components that run inside the hardware components presented as artefacts.

The transformation of the meta model of ISS in UML deployment diagram is shown on Fig. 6. As the ISS usually is complicated system, composed by big number of security components we will show limited hardware components such DLP system on the server, which is connected with the software agents, running on each user workstations. Through the agents, the DLP server (Endpoint protection concept) monitor, control and manage the data on all data channels on the endpoint (workstations and laptops) – LAN, Wi-Fi, Bluetooth, USB ports, email, chat communications and etc. The DLP system is able to control the communication channels, to enable and disable the flow of the data through them. The data can be inspected as it shown on Fig. 5, compared with the security policy and if a data breach has occurred, it can be blocked and reported. This way the ISS can control the data breaches of sensitive information.

With the Deployment diagram can be illustrates the implementation and deployment of other security hardware/software components as Encryption System, Intrusion Detection Systems (IDS) and all of the rest physical components of the ISS.

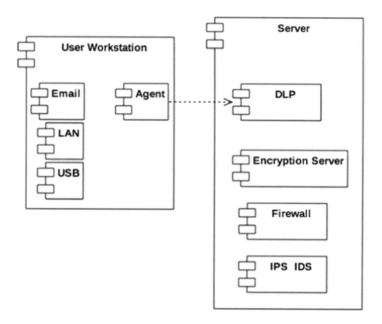

Fig. 6. UML Deployment Diagram.

4   Conclusion

The conceptual modeling is presented as a starting point in architecture description of an ISS that is used in definition of different viewpoints.

The goal of architecture model construction is to ensure usability, interoperability, easy deployment and scalability of ISS across organizations. One of the most important advantages of implementing this approach is to achieve model independence when changing external or internal conditions.

A more complete architectural description of the ISS could be achieved, if we assume more viewpoints. It is possible to define a two-layered conceptual model, organized around the viewpoints "Information Security" and "Information Processing".

Starting with a solution model of ISS, using conceptual modelling, our goal is to transform the solution model to design model with the help of UML. In this way the implementation of the designed ISS will be more easy and complete with transformation to UML diagrams, assuming one and the same language in construction of design models and implementation models of ISS.

The usage of UML for description of the implementation model as a prototype that can be presented with an available platform will simplify the transition from design to implementation model and the development process will be significantly simplified.

Acknowledgements. This study is partially supported by a research project grant "Modelling the Architecture of Information Security Systems in Organizations", Ref. No: 72-00-40-230/10.05.2017, ICT Sciences & Technologies Panel, Program for Young Scientists and PhD Students Support – 2017, Bulgarian Academy of Sciences. Additional gratitude is also given to Information and Communication Technologies for a Single Digital Market in Science, Education and Security (ICTinSES) program of the Ministry of Education and Science, Republic of Bulgaria.

References

1. Hintzbergen, J., Hintzbergen, K.: Foundations of Information Security Based on ISO27001 and ISO27002, p. 149. Zaltbommel, Van Haren (2010)
2. ISO 27001 Official Page. https://www.iso.org/isoiec-27001-information-security.html. Accessed 9 Nov 2018
3. IT Governance Institute: COBIT Security Baseline: An Information Survival Kit, 2nd edn, p. 14. IT Governance Institute (2007)
4. COBIT Resources: http://www.isaca.org/COBIT/Pages/default.aspx. Accessed 9 Nov 2018
5. NIST Special Publications (800 Series): http://www.csrc.nist.gov/publications/PubsSPs.html. Accessed 9 Nov 2018
6. Gramm-Leach-Bliley Act (GLBA) Resources. https://www.ftc.gov/tips-advice/business-center/privacy-and-security/gramm-leach-bliley-act. Accessed 9 Nov 2018
7. Anand, S.: Sarbanes-Oxley Guide for Finance and Information Technology Professionals, p. 93. Wiley, Hoboken (2006)
8. Sarbanes-Oxley Act SOX Resources. https://www.sec.gov/about/laws/soa2002.pdf. Accessed 9 Nov 2018
9. Beaver, K., Herold, R.: The Practical Guide to HIPAA Privacy and Security Compliance, 2nd edn, p. 4. Auerbach, Boca Raton (2011)
10. PCI Security Standards. https://www.pcisecuritystandards.org/pci_security/. Accessed 9 Nov 2018
11. EU General Data Protection Regulation Official Page. http://ec.europa.eu/justice/data-protection/reform/index_en.htm. Accessed 9 Nov 2018
12. IEEE Std 1471, IEEE Recommended Practice for Architectural Description of Software-Intensive Systems (2000)
13. ISO/IEC/IEEE 42010:2011 – Systems and Software Engineering – Architecture Description. https://www.iso.org/standard/50508.html. Accessed 9 Nov 2018
14. OMG. Unified Modeling Language (UML), V. 1.5. https://www.omg.org/spec/UML/1.5/About-UML/. Accessed 9 Nov 2018

15. Hilliard, R.: Aspects, concerns, subjects, views. In: First Workshop on Multi- dimensional Separation of Concerns in Object-Oriented Systems (OOPSLA 1999), pp. 1–3 (1999)
16. Industrial Internet of Things Volume G4: Security Framework, pp. 46–61, May 2017. http://www.iiconsortium.org/pdf/IIC_PUB_G4_V1.00_PB.pdf. Accessed 2018/11/9
17. Killmeyer, J.: Information Security Architecture: An Integrated Approach to Security in the Organization, pp. 203–240. CRC Press, Taylor & Francis Group, LLC, Boca Raton (2006)
18. Rhodes-Ousley, M.: Information Security the Complete Reference, 2nd edn, pp. 303, 234–238. The McGraw-Hill, New York City (2013)
19. Alhir, S.: Understanding the model driven architecture (MDA). Methods Tools **11**(3), 17–24 (2003)
20. Fernandez, E.: Security Patterns in Practice, pp. 25–50. Wiley, Hoboken (2013)
21. Dennis, A., Wixom, B., Tegarden, D.: System Analysis & Design – An Object-Oriented Approach with UML, 5th edn, pp. 19–52. Wiley, Hoboken (2015)
22. Perroud, T., Inversini, R.: Enterprise Architecture Patterns, pp. 18–22. Springer, Heidelberg (2013)
23. Hilliard, R.: Using the UML for architectural description. In: Proceedings of UML 1999. Lecture Notes in Computer Science, vol. 1723, pp. 1–15. Springer (1999)
24. Breu, R., Grosu, R., Huber, F., Rumpe, B., Schwerin, W.: Systems, views and models of UML. In: Schader, M., Korthaus, A. (eds.) The Unified Modeling Language, Technical Aspects and Applications, pp. 3–8. Physica Verlag, Heidelberg (1998)
25. Kong, J., Xu, D., Zeng, X.: UML-based modeling and analysis of security threats. Int. J. Softw. Eng. Knowl. Eng. **20**(6), 875–897 (2010)

An Automated System for Criminal Police Reports Analysis

Gonçalo Carnaz[1,2]([⊠]), Vitor Beires Nogueira[1,2], Mário Antunes[3,7],
and Nuno Ferreira[4,5,6]

[1] Informatics Department, University of Évora, Évora, Portugal
d34707@alunos.uevora.pt, vbn@di.uevora.pt
[2] LISP, Évora, Portugal
[3] School of Technology and Management, Polytechnic Institute of Leiria,
Leiria, Portugal
mario.antunes@ipleiria.pt
[4] Institute of Engineering of Coimbra, Polytechnic Institute of Coimbra,
Coimbra, Portugal
nunomig@isec.pt
[5] INESC-TEC, Porto, Portugal
[6] GECAD, Institute of Engineering, Polytechnic Institute of Porto, Porto, Portugal
[7] INESC-TEC, CRACS, University of Porto, Porto, Portugal

Abstract. Information Extraction (IE) and fusion are complex fields
and have been useful in several domains to deal with heterogeneous data
sources. Criminal police are challenged in forensics activities with the
extraction, processing and interpretation of numerous documents from
different types and with distinct formats (templates), such as narrative
criminal reports, police databases and the result of OSINT activities,
just to mention a few. Such challenges suggest, among others, to cope
with and manually connect some hard to interpret meanings, such as
license plates, addresses, names, slang and figures of speech. This paper
aims to deal with forensic IE and fusion, thus a system was proposed to
automatically extract, transform, clean, load and connect police reports
that arrived from different sources. The same system aims to help police
investigators to identify and correlate interesting extracted entities.

Keywords: Information extraction · Fusion · Criminal forensics ·
ETL · NER systems · Criminal police reports

1 Introduction and Motivation

Criminal police departments deal with forensic information from heterogeneous
data sources during a criminal investigation that generates a deluge of data, such
as narrative police reports, crime scene reports, spreadsheets, police databases or
Open Source Intelligence (OSINT[1]). In everyday activities, police investigators

---

[1] Data collected from publicly available sources to be used in an intelligence context.

© Springer Nature Switzerland AG 2020
A. M. Madureira et al. (Eds.): SoCPaR 2018, AISC 942, pp. 360–369, 2020.
https://doi.org/10.1007/978-3-030-17065-3_36

manually create criminal police reports that describe the crimes investigated with relevant forensics assets to be analyzed. Therefore, structured, semi-structured and unstructured data produced could benefit from an information fusion approach to deal with heterogeneous police data sources that could be used for a forensic decision support system. All issues regarding information fusion must be analyzed and solved to be applied appropriated computational methods to extract relevant information.

The Portuguese Internal Security Report[2] compiles all crimes investigated and reported by the institutions that compose the Internal Security System. According to the 2016 edition of such report, in that year there were 330,872 investigated crimes. Considering that for each crime the police investigators must not only produce several reports and forensic evidence but also analyze and process such elements, this leads to an extremely time-consuming task for the police forces. Therefore, our motivation is to resort to a computational approach to help in analyzing and processing the vast amount of police reports.

We focus our study in one Police Institution and their criminal police reports, with different file types and formats (templates). Therefore, our contribution in this paper is to present a system that automatically extracts, transforms, cleans, loads and connects police reports that arrived from different sources and additionally, aims to help police investigators to identify and correlate interesting extracted entities.

The rest of the paper is organized as follows: in Sect. 2 we describe the related works useful to support our work; the Sect. 3 details the system by which we proposed to extract, transform, clean and load the narrative police reports into a common format and finally in Sect. 4, we expose our conclusions and future work.

This work was partially funded by the Agatha Project SI&IDT n° 18022.

2   Literature Review

The motivation for our work is two-fold: *"how to extract, transform and load relevant information?"* and *"how to identify forensic information from police reports?"*. Therefore, we have investigated literature with related works for Extract, Transform and Load (ETL) approaches and the named-entity recognition (NER) systems for the Portuguese language, that could help to create an information fusion environment.

ETL concept is defined as *"The ETL process extracts the data from source systems, transforms the data according to business rules, and loads the results into the target data warehouse."* [1]. There are several proposed approaches, e.g. from a real-time data ETL framework [2] that processes historical data and real-time data separately; or a technique for data streams handling that synchronizes data streams from incoming data sources [3]; or the proposal of a Integration Based Scheduling Approach (IBSA) that deals with data synchronization

---

[2] www.ansr.pt.

issue [4]; using an ontology-based methodology approach to resolve homogeneity regarding data sources and the integration of data by its meaning [5]; [6] using a domain ontology (stored inside a data warehouse as metadata) and all findings in data sources are semantically analyzed, and finally [7] proposed an ontology-based approach to support the conceptual design of the ETL processes, based on a graph-based representation, to support structured and semi-structured data extraction. There are several ETL tools, from commercial tools, e.g. IBM Infosphere, Oracle Warehouse Builder or Informatica PowerCenter and Open Source tools, e.g. Talend OpenStudio, Pentaho Kettle or CloverETL that enables the development of ETL tasks.

Regarding NER systems, Nadeau et al. [8] defined NER as *"...is a subproblem of information extraction and involves processing structured and unstructured documents and identifying expressions that refer to peoples, places, organizations, and companies..."*.

In 2014, Dozier et al. [9] proposed a NER system to extract entities from legal documents, e.g. judges, companies, courts or others; Arulanandam et al. [10] proposed a system to extract crime information from online newspapers is proposed, focused in the "hidden" information related to the theft crime.

Shabat et al. [11] proposed in 2015 a system for crime information extraction from the Web, with an NER task, using classification algorithms, like Naive Bayes, Support Vector Machine and K-Nearest Neighbor. An indexing module was added to crime type identification, using the same classification algorithms. Yang et al. [12] proposed an approach based on raw text to extract semi-structured information using text mining techniques.

Bsoul et al. [13] proposed in 2016, a system to extract verbs and their use, using two datasets: a real dataset from crime and industrial datasets with benchmarks. Additional, the Porter stemming algorithm for word stems was used, for verbs identification.

In 2017, Schraagen et al. [14], proposed an NER system, named as *Frog*, using a manually annotated corpus, created from 250 criminal complaints reports, where domain experts identified: entities, like location, person, organization, event, product, and others. Al-Zaidy et al. [15] proposed several methods to be applied to discovering criminal communities, analyzing their relations, and extract useful information from criminal text data.

3   Proposed Approach

Figure 1 illustrates a high-level design of our system proposal for information fusion related with criminal police reports analysis to achieve a path for reducing police investigators time-consuming and accuracy on forensic activities.

Our approach is divided into four main blocks: (1) *Police Departments*, represents the police investigations heterogeneous data sources, which produces all criminal police reports that will be processed by the following blocks; in (2) *Data Repository Block*, we centralize all criminal police reports in their different formats and types, and originated from each police department; the (3) *Preprocessing Block*, responsible for data extract, parse, clean, filter and integrate into a common format; (4) *Information Retrieval Block*, aim to retrieve relevant information related to criminal investigations from criminal police reports, e.g. as license plates, addresses, names, slang, and emotions. Finally, the (5) represents the file which the relevant information for Police decision making.

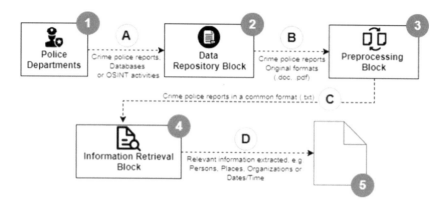

Fig. 1. Criminal police reports analysis - a system for information fusion.

3.1   Police Departments and Data Repository Block

The (A) and (B) represents the data flow between blocks. In this case, the crime police reports in their original format, e.g. Microsoft™ Word file type. The Table 1 shows the number of pages, words, characters, paragraphs and lines for each four criminal police reports analyzed.

Table 1. Criminal police reports summary.

Police reports	Pages	Words	Characters	Paragraphs	Lines
Police01	14	4866	24883	174	458
Police02	3	516	2763	66	120
Police03	32	11811	71864	310	864
Police04	47	16678	85099	484	1307

The (2) *Data Repository Block* works as a staging area for non-processed documents. We need this block because our sources are from different police departments (narcotics, homicides or economic crime) and this is a confluent point to archive the police crime reports.

3.2   Preprocessing Block

Figure 2 depicts the *Preprocessing* block that aims to propose a system that automatically extract, transform, clean, load and connect criminal police reports that arrived from different sources into a common format. To that purpose, we have defined a group of ETL tasks that could decrease the "dirty" data and facilitate the retrieval of police information assets, such as crime types, criminals name, locations, vehicles, and so on. Therefore, the ETL proposed is divided: (1) *Extract/Parse* task aims to extract and parse the criminal police reports from its original format into a common format (.txt); in (2) *Data Preparation* task we proposed to normalize the extracted text, following (A) Regex (Regular Expressions) rules based on domain lexicon for text normalization, and cleaning rules; in (3) *Data Loading* task aims to load the normalized file into a target (4) Data Staging Area.

To developed the proposed system, we used Java[3] programming language and other toolkits, namely Apache™ Tika[4] used to extract and parse documents from different formats, e.g. word and pdf files formats.

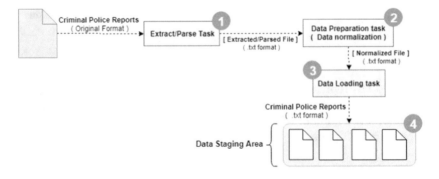

Fig. 2. Criminal police reports analysis - our ETL proposal.

Extract and Parse. We extract all data from criminal police reports, not any subset of interest, assuming that the relevant data identification will be done at later in time. To execute this task the following features have been added: (a) File formats identification; (b) Extract data, fetching data with appropriated connector to the external sources; (c) Parse data, the parsing task was performed by reading the data stream and building an in-memory model to facilitate the data transformation. In our case, and during the parse task, we add two different tags: <PLAIN TEXT> *Unstructured data*</PLAIN TEXT>, that identifies unstructured data, and <TABLES> *Semi-structured data*</TABLES>, that identifies semi-structured data.

---

[3] https://docs.oracle.com/javase/7/docs/technotes/guides/language/.
[4] https://tika.apache.org/.

Table 2. Criminal police reports after extraction/parsing task summary.

Police reports	Words	Characters	Lines
Police01	6187	36314	254
Police02	1110	6113	293
Police03	11901	71906	326
Police04	18083	108719	572

Table 2 shows the results obtained after the extract and parse tasks, each criminal police report were transformed into a common file format, like a text document. We choose this format, because reduced formatting, easy to read in every text editor and manipulate in any programming language, and also the file size is reduced.

Data Preparation. We build a *Data Preparation* task supported by different sub-tasks: cleaning, remove duplicates, filters, and transformation. We developed our tasks with a set of regular expressions, that could be updated along the process for a more accurate data preparation and normalization. Therefore, text analysis needs some tasks to increase the regularity, e.g. each sentence contains end-mark, commas are followed by a space, cleaning double spaces, replace nulls or adding structure, e.g. splitting extracted text into different sections that have a special meaning, such as witnesses or suspects. The filter and transformation tasks, reads the extracted text seeking for two tags: <PLAIN TEXT> *Unstructured data*</PLAIN TEXT> and <TABLES> *Semi- structured data*</TABLES>. The desired outcome is to detect and separate the unstructured data (plain text) from the semi-structured data (tables). Figure 3 shows how the narrative police report will be after *Data Preparation* task.

Fig. 3. Portuguese crime police reports after data preparation.

Data Loading and Data Staging Area. We proposed in this phase to populate a target location, named by *Data Staging Area*, with a common file format and information from police investigations. from a set of information sources and after sequential grouped tasks, e.g. Extract/Parse and Data Preparation, controlled by a set of rules. Therefore, we have the *Data Staging Area* were all documents are stored to be processed by the natural language processing module, after all, preprocessing tasks.

3.3   Information Retrieval Block

The *Information Retrieval* block looks for relevant information from crime police reports. Then the information obtained integrating this forensic information by identifying in each criminal police report relevant information. Figure 4 represents our approach based on a named-entity recognition (NER) system that recognizes relevant named-entities in our criminal police reports, e.g. Persons, Locations, Organization and Time/Date. To retrieve named-entities from our dataset, we have implemented the following modules: (1) a module that detects sentences and tokens (words), called Sentence Detector, that aim to detect sentences in each text, using a trained model for the Portuguese language. The chosen model use a machine learning algorithm, called Maxent [16] already trained and downloaded[5]. The Tokenization module performs tokens (words) detection, and also uses a trained model retrieved from the same website, also using the Maxent algorithm (the pre-defined algorithm in both cases).

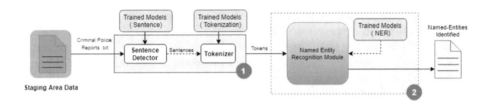

Fig. 4. Our named-entity recognition (NER) system proposal.

The (2) Named-Entity Recognition module detects named-entities, such as persons, places, organizations, and dates. To detect named-entities. We have trained a corpus, called *Amazonia* Corpus[6], that have 4.6 millions of words (about thousand sentences) retrieved from Overmundo[7] website, written in Portuguese-Brazilian language, and automatically annotated by PALAVRAS [17]. To train the *Amazonia* Corpus, we realize the following steps:

- Download the Apache OpenNLP[8] toolkit;
- Use the *TokenNameFinderConverter* tool to convert the Amazonia corpus into OpenNLP format;
- To train the model, we used the *TokenNameFinderTrainer* tool, with the pre-defined Maxent algorithm to generate the model, named by *ner-amazonia.bin*, that will be used as the trained model in our NER system.

---

[5] http://opennlp.sourceforge.net/models-1.5/.

[6] http://www.linguateca.pt/floresta/ficheiros/gz/amazonia.ad.gz.

[7] http://www.overmundo.com.br/.

[8] https://opennlp.apache.org/.

3.4 Obtained Results

To perform the experiments, we obtained a dataset with four documents that are police reports from a real process, described in Table 1. We have select two other systems for dataset evaluation based on the following criteria: (1) multilingual support; (2) focused on the Portuguese language; (3) Open source tools. The systems selected were:

- Linguakit: a multilingual toolkit for natural language processing with a NER system incorporated, created by the ProLNat@GE Group[9] (CITIUS, University of Santiago de Compostela);
- RAPPort - A Portuguese Question-Answering System [18]: that uses a NLP pipeline with a NER system;

Table 3 shows the results obtained. We have used information retrieval metrics [19] to measure NER systems performance, namely: (1) P: Precision, (2) R: Recall and (3) $F1$ or $F - Measure$. *Precision* is defined by the ratio of correct answers (True Positives) among the total answers produced (Positives),

$$P = \frac{TP}{TP + FP} \tag{1}$$

$$R = \frac{TP}{TP + FN} \tag{2}$$

$$F1 = \frac{2 * P * R}{P + R} \tag{3}$$

where *TP - True Positive*, a predicted value was positive and the actual value was positive and *FP - False Positive*, predicted value was positive and the actual value was negative [20]. *Recall* is defined as a ratio of correct answers (True Positives) among the total possible correct answers (True Positives and False Negatives), where *FN - False Negative*, a predicted value was negative and the actual value was positive [20]. Finally, $F1$ or $F - Measure$ - is a harmonic mean of precision and recall.

Globally, our system has reached the highest F-measure result for each entity for the detected entities, having the best trade-off regarding both measures

Table 3. Test evaluation results.

	Person			Organization			Place			Date/Time		
	P	R	F1	P	R	F1	P	R	F1	P	R	F1
Our proposal	0.98	0.93	0.95	0.73	0.27	0.39	0.98	0.97	0.96	0.95	0.91	0.93
RAPPort	0.91	0.96	0.94	0.33	0.33	0.33	0.92	0.94	0.93	0.86	0.98	0.92
Linguakit	0.67	0.28	0.39	0.12	0.82	0.21	0.50	0.78	0.60	0.90	0.90	0.90

---

[9] https://gramatica.usc.es/pln/.

(precision and recall). The results obtained were quite better, because we have cleaned and transformed the used dataset, before been processed by the NLP pipeline, which removes some existent entropy from the dataset that is passed to the NER systems.

4   Conclusion and Future Work

The work developed in this paper tries to achieve a forensic information fusion for criminal police reports retrieved from heterogeneous data sources. We have focused on the obtained results, mainly in the *Information Retrieval* block as it allow us to prove the necessity of a process to integrate relevant information from police heterogeneous sources and associated information security. Therefore, we achieve relevant results about information extraction regarding named-entities, e.g. persons, places, organizations, and dates, for criminal analysis and to support a forensic decision support system for police investigations. For future work, we define the following goals:

- To integrate information from other Police Institutions, where criminal police reports follow different document templates;
- To apply a machine learning algorithm to recognize "dirty" data without human intervention and correct data errors or anomalies;
- To improve our natural language processing system in Portuguese by adding new named-entities, e.g. vehicles, license plates or drugs;
- To identify relations or events among relevant information extracted from criminal police reports.

References

1. Kakish, K., Kraft, T.A.: ETL evolution for real-time data warehousing. In: Proceedings of the Conference on Information Systems Applied Research ISSN, vol. 2167, pp. 1508 (2012)
2. Li, X.: Real-time data ETL framework for big real-time data analysis. In: IEEE International Conference on Information and Automation, no. August, pp. 1289–1294 (2015)
3. Majeed, F., Mahmood, M.S., Iqbal, M.: Efficient data streams processing in the real time data warehouse. In: 2010 3rd International Conference on Computer Science and Information Technology, vol. 5, pp. 57–61, July 2010
4. Song, J., Bao, Y., Shi, J.: A triggering and scheduling approach for ETL in a real-time data warehouse. In: 2010 10th IEEE International Conference on Computer and Information Technology, pp. 91–98, June 2010
5. Chávez, J.V., Li, X.: Ontology based ETL process for creation of ontological data warehouse. In: 2011 8th International Conference on Electrical Engineering, Computing Science and Automatic Control, pp. 1–6, October 2011
6. Jiang, L., Cai, H., Xu, B.: A domain ontology approach in the ETL process of data warehousing. In: 2010 IEEE 7th International Conference on E-Business Engineering, pp. 30–35, November 2010

7. Skoutas, D., Simitsis, A.: Ontology-based conceptual design of ETL processes for both structured and semi-structured data. Int. J. Semant. Web Inf. Syst. (IJSWIS) **3**(4), 1–24 (2007)
8. Nadeau, D., Sekine, S.: A survey of named entity recognition and classification. Lingvisticae Investigationes **30**(1), 3–26 (2007)
9. Dozier, C., Kondadadi, R., Light, M., Vachher, A., Veeramachaneni, S., Wudali, R.: Named entity recognition and resolution in legal text. In: Francesconi, E., Montemagni, S., Peters, W., Tiscornia, D. (eds.) Semantic Processing of Legal Texts, pp. 27–43. Springer, Heidelberg (2010)
10. Arulanandam, R., Savarimuthu, B.T.R., Purvis, M.A.: Extracting crime information from online newspaper articles. In: Proceedings of the Second Australasian Web Conference - Volume 155, AWC 2014, Darlinghurst, Australia, pp. 31–38. Australian Computer Society, Inc. (2014)
11. Shabat, H.A., Omar, N.: Named entity recognition in crime news documents using classifiers combination. Middle-East J. Sci. Res. **23**(6), 1215–1221 (2015)
12. Yang, Y., Manoharan, M., Barber, K.S.: Modelling and analysis of identity threat behaviors through text mining of identity theft stories. In: 2014 IEEE Joint Intelligence and Security Informatics Conference, pp. 184–191, September 2014
13. Bsoul, Q., Salim, J., Zakaria, L.Q.: Effect verb extraction on crime traditional cluster. World Appl. Sci. J. **34**(9), 1183–1189 (2016)
14. Schraagen, M.: Evaluation of Named Entity Recognition in Dutch online criminal complaints. Comput. Linguist. Neth. J. **7**, 3–15 (2017)
15. Al-Zaidy, R., Fung, B.C.M., Youssef, A.M.: Towards discovering criminal communities from textual data. In: Proceedings of the 2011 ACM Symposium on Applied Computing, SAC 2011, pp. 172–177. ACM, New York (2011)
16. Sharnagat, R.: Named entity recognition: a literature survey. Center For Indian Language Technology (2014)
17. Bick, E.: The parsing system "PALAVRAS": automatic grammatical analysis of Portuguese in a constraint grammar framework. 2000. 412 f. Ph.D. thesis, Aarhus University, Denmark University Press (2000)
18. Rodrigues R., Gomes, P.: Rapport–a Portuguese question-answering system. In: Portuguese Conference on Artificial Intelligence, pp. 771–782. Springer (2015)
19. Mansouri, A., Affendey, L.S., Mamat, A.: Named entity recognition approaches. J. Comput. Sci. **8**(2), 339–344 (2008)
20. Konkol, I.M.: Named entity recognition. Ph.D. thesis, University of West Bohemia (2015)

Detecting Internet-Scale Traffic Redirection Attacks Using Latent Class Models

Ana Subtil[1]([✉]), M. Rosário Oliveira[2], Rui Valadas[1], Antonio Pacheco[1], and Paulo Salvador[3]

[1] Instituto de Telecomunicações and Instituto Superior Técnico, Universidade de Lisboa, Lisbon, Portugal
anasubtil@tecnico.ulisboa.pt
[2] CEMAT and Instituto Superior Técnico, Universidade de Lisboa, Lisbon, Portugal
[3] Instituto de Telecomunicações and DETI, Universidade de Aveiro, Aveiro, Portugal

Abstract. Traffic redirection attacks based on BGP route hijacking has been an increasing concern in Internet security worldwide. This paper addresses the statistical detection of traffic redirection attacks based on the RTT data collected by a network of probes spread all around the world. Specifically, we use a Latent Class Model to combine the decisions of individual probes on whether an Internet site is being attacked, and use supervised learning methods to perform the probe decisions. We evaluate the methods in a large number of scenarios, and compare them with an empirically adjusted heuristic. Our method achieves very good performance, superior to the heuristic one. Moreover, we provide a comprehensive analysis of the merits of the Latent Class Model approach.

Keywords: Traffic redirection attack · BGP security · Statistical learning · Latent Class Model

1 Introduction

Internet-wide connectivity is currently provided by the Border Gateway Protocol (BGP) [12]. However, BGP is inherently insecure and allows the deployment of route hijacking attacks [1,2,9]. Since, in 2008, Pilosov and Kapela [10] proposed a BGP exploit to implement Internet-scale man-in-the-middle traffic redirection attacks, multiple security reports have acknowledged their occurrence [3,4,8]. There are several proposals to enhance the BGP security [2,6], but there is no perspective of deploying them in the near-future. Currently the only defense against redirection attacks is the constant monitoring of the BGP updates and the filtering of suspicious network prefixes, but this is no easy task. The difficulty in detecting redirection attacks is further increased by occasional changes in the routes of regular traffic (e.g due to router failures or BGP reconfigurations).

© Springer Nature Switzerland AG 2020
A. M. Madureira et al. (Eds.): SoCPaR 2018, AISC 942, pp. 370–380, 2020.
https://doi.org/10.1007/978-3-030-17065-3_37

Customer networks are the most affected by this type of attacks: it is difficult for such networks to detect the attacks and, even after detection, their countermeasures are limited to requesting prompt actions from their ISPs or implementing temporary extreme security policies, such as terminating or limiting sensible communications and increasing the required encryption level for communications services.

Zhang et al. [15] and Liu et al. [7] proposed solutions to identify and characterize BGP route changes. However, these solutions require the periodical analysis of the BGP RIBs (Routing Information Bases), which is computationally demanding and difficult to implement in real time. Salvador and Nogueira [13] proposed a low-cost monitoring infrastructure for customers, with their limited access to network information, to detect traffic redirection attacks towards specific Internet sites, consisting of probes spread all around the world, which measure the RTT to the Internet site being monitored. The authors also proposed an heuristic method based on the collected RTT measurements to decide when a monitored Internet site is being attacked.

Our work builds upon the monitoring infrastructure proposed in [13]. We introduce the use of statistical methods to detect the attacks. Specifically, we use a Latent Class Model (LCM) to combine the decisions of individual probes on whether an Internet site is being attacked, and use supervised learning methods to perform the probe decisions. Our method achieves very good results, superior to the heuristic ones.

The paper is structured as follows. Section 2 describes the measurement methodology. Section 3 introduces the various detection methods. Section 4 compares the results of the various methods. Finally, Sect. 5 presents the main conclusions of the work.

2   Measurement Methodology

The monitoring infrastructure proposed in [13] consists of a set of virtual private servers with modest computational requirements, and installed in datacenters spread all around the world. The role of these servers is to measure the RTT to specific targets, to detect if these targets are suffering traffic redirection attacks. The *targets* are hosts/routers located at some ISP premises. We will refer to the servers performing the RTT measurements as *probes*. Moreover, the locations redirecting the traffic (i.e. the man-in-the-middle agents performing the attacks) will be referred to as *relays*.

Under regular circumstances, a given probe will measure the RTT of the regular path from the probe to the target. When an attack is being perpetrated, this traffic is diverted through the relay. The RTT is then the sum of three delay components—probe to relay, relay to target, and target to probe—which may deviate (significantly or not) from the RTT of the regular path.

Detecting traffic redirection attacks is based on the deviations between the RTTs of the regular and relayed paths. The effectiveness of the detection methods depends on the relative location of probe, target, and relay. Specifically, the detection becomes more difficult when the relay is either close to the probe or to the target, in which case the RTTs of the regular and relayed paths will have similar values. Thus, the performance of the detection methods may be improved if the target is evaluated using several probes dispersed geographical, and if the results of the various RTT measurements are properly combined. That is where statistical methods may play an important role.

3   Detection Methods

This section presents the detection methods addressed in this work. Besides the heuristic proposed in [13], we consider three well-known supervised classifiers combined using LCMs.

Heuristic Approach. The heuristic [13] uses a single feature—the average RTT ($avgRTT$) computed over 10 RTT measurements—and uses a moving average of past $avgRTT$ to calculate a decision threshold. When an observation is evaluated, its $avgRTT$ is compared with the decision threshold, which is set at $\epsilon = 1.2$ times the average of the past $h = 480$ $avgRTTs$ that were not classified as attacks (also not including the observation being evaluated). However, an attack is only declared by a probe when $K = 10$ consecutive observations exceed the decision threshold. Finally, a target is considered under attack at a specific instant using a voting rule: an attack is declared when $\rho = 50\%$ or more probes detect the attack.

LCM-Combined Supervised Classifiers. In this approach, the observations of each probe are first classified (either as regular or an attack) using a supervised classifier. Subsequently, LCM is employed to combine the classification outcomes of multiple probes, and determine if the corresponding target is under attack. In this case, each observation is characterized by three features: the average, minimum, and maximum RTT statistics, computed over 10 RTT measurements.

We tested a large set of supervised classifiers, and selected the ones that led to the best results: k-nearest neighbors (kNN), C5.0 decision trees, and random forest. The first is a lazy classifier, very popular by its simplicity and its overall good results. In our case, the number of neighbors is the smallest value that maximizes the overall accuracy estimated using the labeled training set. C5.0 and random forest are decision trees; the latest considers several decision trees in order to make the method more accurate [14].

We adopted a LCM with $m = 4$ binary input variables, X_i ($i = 1, ..., m$), expressing if classifier i, associated with probe i, declared an attack ($X_i = 1$) or not ($X_i = 0$). Let Y be the binary variable, called latent variable, whose categories, called latent classes, are $\{0, 1\}$ and indicate if the target is under regular traffic conditions or under a redirection attack, respectively. The LCM estimates the probability of the latent class, $p_y = P(Y = y)$, $y = 0, 1$, and the probability of an attack declared by classifier i given that the latent class is y,

Table 1. Targets, probes, and relays.

Probes	Targets	Relays
Chicago, USA	Amsterdam, Netherlands	Los Angeles 1 (LA1), USA
Frankfurt, Germany	Viña del Mar, Chile	Madrid, Spain
Hong Kong, China	Los Angeles 2 (LA2), USA	Moscow, Russia
London, UK	Johannesburg, South Africa	São Paulo, Brazil

$\pi_{iy} = P(X_i = 1 | Y = y)$, using

$$
\begin{aligned}
P(\mathbf{X} = \mathbf{x}) &= \sum_{y=0}^{1} P(Y = y) \prod_{i=1}^{m} P(X_i = x_i | Y = y) \\
&= \sum_{y=0}^{1} p_y \prod_{i=1}^{m} \pi_{iy}^{x_i} (1 - \pi_{iy})^{(1-x_i)},
\end{aligned}
\tag{1}
$$

where $x_i = 0, 1$, $\mathbf{X} = (X_1, \ldots, X_m)$, and $\mathbf{x} = (x_1, \ldots, x_m)$. This model assumes the so-called hypothesis of conditional independence, which states that the observed variables X_i are conditionally independent given the level of the latent variable Y.

The LCM parameters are estimated through the Expectation-Maximization (E-M) algorithm, using an iterative procedure specific to LCM [5]. Since the solution reached may be a local maximum, it is important to repeat the iterative process with different starting values to obtain the global maximum.

Once the parameters are estimated, an observation can be assigned to a class, based on the estimated posterior probabilities of the latent classes conditional on the patterns of probe classifications, $\mathbf{x}$,

$$
P(Y = y | \mathbf{X} = \mathbf{x}) = \frac{p_y \prod_{i=1}^{m} \pi_{iy}^{x_i} (1 - \pi_{iy})^{(1-x_i)}}{P(\mathbf{X} = \mathbf{x})}.
\tag{2}
$$

Specifically, a pattern $\mathbf{x_0}$ is assigned to the class y with higher posterior probability, i.e.

$$
y_0 = \arg \max_{y \in \{0,1\}} P(Y = y | \mathbf{X} = \mathbf{x_0}).
\tag{3}
$$

For example, if the observed pattern of probe classifications is $(1, 1, 0, 1)$, and the LCM estimates that $P(Y = 0 | \mathbf{X} = (1, 1, 0, 1))$ is 0.2 and $P(Y = 1 | \mathbf{X} = (1, 1, 0, 1))$ is 0.8, the LCM model decides that the target is under attack ($y = 1$).

The LCM combination of classifiers can be seen as the counterpart of the voting rule of the heuristic. However, unlike the heuristic, where an attack to a target is declared when 50% or more probes declare the attack, there is no deterministic rule in LCM, and the decision depends on the estimated performance of each classifier being combined, and on the estimated probability of an attack.

The heuristic, the three supervised classifiers, and the LCM were all implemented in R [11].

4 Results

4.1 Data Collection

In our study, we have used a monitoring infrastructure with 12 servers rented in different locations across the Internet. To emulate a real network with traffic redirection attacks, each server was given a specific role of probe, target or relay. Table 1 indicates which locations were used as probe, target, or relay. Note that, under operational conditions, all servers would be used as probes.

The RTT measurements were obtained by sending UDP packets to a predefined closed port on the target and measuring the time until the reception of an ICMP port unreachable packet. Each redirection attack was emulated using two GRE IPv4 tunnels, one between probe and relay, and another between relay and target.

RTT measurements were gathered between May 22nd and June 17th, 2017. During this period of around 25 days, we measured the RTT from each probe to each target, simultaneously through the regular path and through its four relayed paths. Thus, altogether we have characterized 80 distinct paths during this 25 day period.

To evaluate the RTT, each probe made 10 RTT measurements every 120 s (by sending 10 UDP packets to the target). Then, we computed average, minimum, and maximum over the 10 RTT measurements, and these are the features that we work with. Thus, in our datasets there is a multivariate RTT observation per probe every 120 s, consisting of average, minimum, and maximum RTT computed over 10 individual measurements.

4.2 Dataset Composition

Using the collected data, we composed a total of 16 datasets, one for each probe-target pair. The datasets were composed such that periods of regular traffic alternate with periods of relayed traffic (attacks), while maintaining the chronological sequence of the observations. Thus, in relation to the original RTT measurements, periods defined as regular include only the measurements of the regular path, and periods defined as redirection attacks through some relay include only measurements of the corresponding relayed path. The duration of the attack periods was set to 6 h (180 observations), and was the same for all probe-target pairs.

Moreover, each dataset was split into two parts with the same number of observations, where the first part was used for training the detection methods and the second one for testing them. The training phase includes only three attack periods, leaving one relay aside, to make the detection problem more challenging. The testing phase includes four attack periods, one for each available relay.

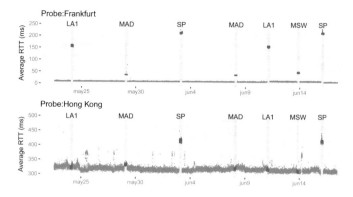

Fig. 1. Traffic datasets from Frankfurt and Hong Kong probes to target Amsterdam. Regular traffic depicted in blue. Relayed traffic (attacks) in red, with a shaded background band and the corresponding relay identification.

Figure 1 gives an example of two datasets having Amsterdam as target where, in the first dataset, the probe is Frankfurt and, in the second one, the probe is Hong Kong. The figure shows the average RTTs of the datasets. Blue data represents regular traffic and red data represents relayed traffic (attacks). The first three attack periods, having LA1, Madrid (MAD), and São Paulo (SP) as relays, are the ones of the training phase, and the remaining attacks are the ones of the testing phase. As mentioned above, training with only 3 relays—leaving the Moscow (MSW) relay aside—poses an additional challenge to the supervised classifiers, because the data used to build the classifiers in the training phase does not comprise data associated with this relay.

The two datasets of Fig. 1 illustrate two extreme scenarios. In the first scenario (Frankfurt probe), there is a significant difference between the average RTT values of the regular and relayed traffic, for all relays. This is when the attack detection is relatively easy. The second scenario (Hong Kong probe) is more challenging, since only the average RTT values associated with the relay São Paulo stand out from the remaining values. This is where statistical methods may be particularly helpful.

For the detection methods considered in this paper, we assume the use of $m = 4$ probes per evaluated target. Thus, the methods combine measured data gathered by four probes to decide whether a specific target is being attacked by a specific relay. In the case of the heuristic, only the data collected during the testing phase is considered, since no training is required.

4.3   Performance Evaluation Metrics

We evaluate the performance of the classifiers using metrics associated with the attack class, namely the recall (Re), the False Positive Rate (FPR), and the precision (Pr). The recall is the probability of an attack being correctly classified as an attack, the FPR is the probability that a regular observation is wrongly assigned as an attack, and the precision is the probability that an observation

Table 2. Classifier performance measures.

Methods	Targets	Recall	FPR	Precision	Global
Heuristic	Amsterdam	1	0	1	0
	Viña del Mar	0.750	0	1	0.250
	LA2	0.750	0	1	0.250
	Johannesburg	0.750	0	1	0.250
LCM-combined supervised classifiers					
kNN	Amsterdam	1	0	1	0
	Viña del Mar	0.921	0.002	0.978	0.082
	LA2	0.750	0	1	0.250
	Johannesburg	1	0	1	0
C5.0	Amsterdam	1	0	1	0
	Viña del Mar	0.910	0.001	0.985	0.092
	LA2	0.750	0	1	0.250
	Johannesburg	1	0	1	0
Random forest	Amsterdam	1	0	1	0
	Viña del Mar	0.926	0.002	0.978	0.077
	LA2	0.750	0	1	0.250
	Johannesburg	0.996	0	1	0.004

classified as an attack is, in fact, an attack. In addition, to summarize these three performance measures, we determine the Euclidean distance between their estimated values and the ones we would obtain using a perfect classifier, i.e., Re = 1, FPR = 0, and Pr = 1. We refer to this performance metric as the *global* metric.

4.4 Discussion

We start by discussing the global results obtained with the performance metrics described in Sect. 4.3, and then analyze in detail some special cases.

Global Results. Table 2 shows the recall, FPR, precision, and global metric, obtained by each classifier for each target under analysis.

The heuristic performs optimally in the Amsterdam target (global metric = 0), but has lower performance in the remaining targets. The recall is 0.750 for all these targets, because for each target the heuristic completely fails to detect the traffic redirected through one relay. Specifically, the heuristic fails relay São Paulo in the Viña del Mar target, relay LA1 in the LA2 target, and relay Madrid in the Johannesburg target, and that is why the global metric is always 0.250 in these cases.

Regarding the LCM-combined supervised classifiers, all of them perform better than the heuristic for targets Viña del Mar and Johannesburg, exhibiting

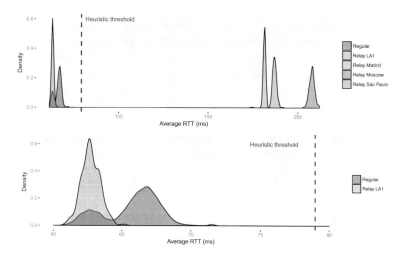

Fig. 2. Average RTT densities for target LA2 and probe London. Regular traffic and traffic redirected through all the relays (above); Zoom of the upper plot, showing regular traffic and traffic relayed only through LA1 (below).

higher recall and lower global metric values. When the target is Amsterdam, the performance of the supervised classifiers is optimal, as in the heuristic case. The worst performance is for the LA2 target, and equals the heuristic one. In fact, all methods fail to detect the traffic redirected through relay LA1, due to their close proximity.

Proximity Between Targets and Relays. Figure 2 illustrates the difficulties associated with cases where the targets and relays are closely located. The figure shows the densities of the average RTT (encompassing only the observations of the testing phase) for LA2 target and London probe. It can be seen that densities of the regular traffic and of the traffic relayed through LA1 are markedly superposed; this also happens in the case of the other two features (minimum and maximum RTT). Thus, detecting attacks in this case is extremely difficult, at least with this set of features.

Figure 2 also reveals the bimodal distribution of the average RTT corresponding to the regular traffic. Route changes resulting from causes other than redirection attacks may explain the heterogeneity of the regular observations, which may pose additional difficulties for the detection of attacks. Since our datasets are real measurements obtained in the Internet, all these difficulties are present in the analysis.

The LCM Advantage. In a distributed infrastructure of monitoring probes, some may provide accurate classification results, while others may not, depending on the relative location of probes, relays, and targets; as discussed above, the exception is when the relay is close to the target. Therefore, when combining the classification results of each probe, it makes sense to take into account its accuracy, giving more weight to the more accurate decisions. This is precisely the strength of the LCM approach. Note that the voting rule used in the heuristic method makes no distinction between individual probe decisions.

We illustrate the LCM advantage using an example. We concentrate on an attack period performed by the São Paulo relay over the Viña del Mar target. Note that an attack period has 180 observations. In Table 3 we show the number of classification errors of the heuristic and the random forest methods. In the heuristic method, the Chicago probe detects all 180 attacks, but the other probes detect none. Thus, due to its voting rule, the heuristic never detects this attack. In the random forest method, the number of errors is also high in all probes except London, which makes only 6 errors. However, the LCM model recognizes that the London probe is the most accurate and makes only 4 errors in deciding whether Viña del Mar is under attack.

The LCM estimates posterior probabilities for each pattern of probe classification, and decides based on these probabilities. In this example, we know for sure that all patterns should be classified as attack. In the estimated LCM this is true for all patterns, except $(0,0,0,0)$—all four probes declare regular traffic—and $(0,0,1,0)$—all probes except Hong Kong declare regular traffic; these patterns occur only 4 times.

It is interesting to analyze the cases where the LCM correctly decided on an attack, when only one probe declared an attack. These are cases where the heuristic would decide wrongly, due to its voting rule. In our case, this occurred only with the London probe, i.e. on pattern $(0,0,0,1)$. For this pattern, the posterior probabilities estimated by the LCM, according to Eq. (2), are $P(Y = 0|\mathbf{X} = (0,0,0,1)) \simeq 10^{-16}$ and $P(Y = 1|\mathbf{X} = (0,0,0,1)) \simeq 1$ and, therefore, the LCM classifies the pattern as an attack. The extremely low value of $P(Y = 0|\mathbf{X} = (0,0,0,1))$ is mostly determined by the term $P(X_4 = 1|Y = 0) \simeq 10^{-18}$, which indicates that the probability that the London probe declares an attack when there is no attack is very low. Moreover, $P(X_4 = 1|Y = 1) = 0.94$, which indicates that the London probe declares an attack when in fact an attack occurs with high probability. Clearly, the LCM considers the decisions made by the London probe to be more trustful, and this is what allows making correct decisions in most cases.

Table 3. Absolute number of classification errors of the heuristic and random forest methods, for relay São Paulo and target Viña del Mar.

Probes	Heuristic	Random forest
Chicago	0	99
Frankfurt	180	135
Hong Kong	180	161
London	180	6
Final classification	180	4

5 Conclusions

Traffic redirection attacks caused by BGP route hijacking can be detected by a monitoring infrastructure of probes, spread all around the world, measuring the RTT to Internet sites under analysis. However, detection methods for this type of attacks must be carefully designed, due to the possibility of having the attackers located near the probes or the targets. In this work, we propose a statistical method where a LCM is used to combine the decisions of individual probes on whether a site is being attacked, and where these decisions are performed using supervised learning classifiers; specifically, we have considered kNN, C5.0, and random forest as supervised classifiers. We also compared our method with an empirically adjusted heuristic. Our method achieves very good performance, superior to the heuristic one. Moreover, we provide a comprehensive analysis of the merits of the LCM approach.

The monitoring infrastructure and statistical detection method have simple technical requirements and low cost, providing the hitherto powerless customer networks, totally dependent on their ISPs surveillance, with a tool for detecting redirection attacks.

Acknowledgments. This research was supported by Instituto de Telecomunicações, Centro de Matemática Computacional e Estocástica, and Fundação Nacional para a Ciência e Tecnologia, through projects PTDC/EEI-TEL/5708/2014, UID/EEA/50008/2013, and UID/Multi/04621/2013. A. Subtil was funded by the FCT grant SFRH/BD/69793/2010.

References

1. Ballani, H., Francis, P., Zhang, X.: A study of prefix hijacking and interception in the internet. ACM SIGCOMM CCR **37**(4), 265–276 (2007). https://doi.org/10.1145/1282427.1282411
2. Butler, K., Farley, T., McDaniel, P., Rexford, J.: A survey of BGP security issues and solutions. Proc. IEEE **98**(1), 100–122 (2010)
3. Cimpanu, C.: DNS Poisoning or BGP Hijacking Suspected Behind Trezor Wallet Phishing Incident. Bleeping Computer News (2018). https://www.bleepingcomputer.com/news/security/dns-poisoning-or-bgp-hijacking-behind-trezor-wallet-phishing-incident/
4. Cowie, J.: The New Threat: Targeted Internet Traffic Misdirection. Blog - Renesys-The Internet Intelligence Authority (2013). http://www.renesys.com/2013/11/mitm-internet-hijacking/
5. Goodman, L.A.: Analyzing Qualitative/Categorical Data: Log-linear Models and Latent Structure Analysis. Abt Books, Cambridge (1978)
6. Huston, G., Rossi, M., Armitage, G.: Securing BGP - a literature survey. IEEE Commun. Surv. Tutor. **13**(2), 199–222 (2011)
7. Liu, Y., Luo, X., Chang, R., Su, J.: Characterizing inter-domain rerouting by betweenness centrality after disruptive events. IEEE J. Sel. Areas Commun. **31**, 1147–1157 (2013)
8. Madory, D.: BGP/DNS Hijacks Target Payment Systems. Oracle+Dyn Blog (2018). https://dyn.com/blog/bgp-dns-hijacks-target-payment-systems/

9. Murphy, S.: BGP Security Vulnerabilities Analysis, RFC 4272 (Informational). Internet Engineering Task Force (2006)

10. Pilosov, A., Kapela, T.: Stealing the internet - an internet-scale man in the middle attack. In: DEFCON 16 (2008)

11. R Core Team: R: A Language and Environment for Statistical Computing. R Foundation for Statistical Computing, Vienna, Austria (2018). https://www.R-project.org/

12. Rekhter, Y., Li, T., Hares, S.: A Border Gateway Protocol 4 (BGP-4), RFC 4271(Draft Standard). Internet Engineering Task Force (2006). http://www.ietf.org/rfc/rfc4271.txt

13. Salvador, P., Nogueira, A.: Customer-side detection of internet-scale traffic redirection. In: 2014 16th International Telecommunications Network Strategy and Planning Symposium, pp. 1–5 (2014). https://doi.org/10.1109/NETWKS.2014.6958532

14. Trevor, H., Robert, T., Friedman, J.H.: The Elements of Statistical Learning: Data Mining, Inference, and Prediction. Springer, Heidelberg (2009)

15. Zhang, Z., Zhang, Y., Hu, Y.C., Mao, Z.M., Bush, R.: iSPY: detecting IP prefix hijacking on my own. IEEE/ACM Trans. Netw. **18**(6), 1815–1828 (2010)

Passive Video Forgery Detection Considering Spatio-Temporal Consistency

Kazuhiro Kono[1(✉)], Takaaki Yoshida[2], Shoken Ohshiro[2],
and Noboru Babaguchi[2]

[1] Faculty of Societal Safety Sciences, Kansai University,
7-1 Hakubai, Takatsuki, Osaka, Japan
k-kono@kansai-u.ac.jp
[2] Graduate School of Engineering, Osaka University,
2-1 Yamadaoka, Suita, Osaka, Japan
{yoshida,ohshiro}@nanase.comm.eng.osaka-u.ac.jp,
babaguchi@comm.eng.osaka-u.ac.jp

Abstract. This paper proposes a method for detecting forged objects in videos that include dynamic scenes such as dynamic background or non-stationary scenes. In order to adapt to dynamic scenes, we combine Convolutional Neural Network and Recurrent Neural Network. This enables us to consider spatio-temporal consistency of videos. We also construct new video forgery databases for object modification as well as object removal. Our proposed method using Convolutional Long Short-Term Memory achieved Area-Under-Curve (AUC) 0.977 and Equal-Error-Rate (EER) 0.061 on the object addition database. We also achieved AUC 0.872 and EER 0.219 on the object modification database.

Keywords: Video forgery detection · Dynamic scene ·
Convolutional LSTM · Modification database

1 Introduction

Recently, many video contents are increasingly delivered over the internet. One of the reasons is the development of video editing technologies. When we want to perform advanced editing tasks, we can edit videos automatically/semi-automatically at high speed with high accuracy by using commercially available software tools and open libraries. It does not require professional skills for video editing.

On the other hand, the spread of video editing technologies has established the situation where everyone can produce forged videos efficiently. For example, we consider that someone eliminates a specific person from a video maliciously. In this case, he can create fake information that the person was not here then.

In order to guarantee the integrity of videos, technologies for detecting forged videos are strongly required [1]. Note here that there are various kinds of tampering and two types of videos, i.e., dynamic or static videos. We often impose

© Springer Nature Switzerland AG 2020
A. M. Madureira et al. (Eds.): SoCPaR 2018, AISC 942, pp. 381–391, 2020.
https://doi.org/10.1007/978-3-030-17065-3_38

some restrictions on the kinds of video/image tampering and the scenes of videos because it is difficult to detect all the tampering videos at once.

The purpose of our research is to detect and identify whether objects in the frames of a video are forged or not. Such a video tampering attack is known as spatial tampering attack. We treat dynamic videos that mean videos with dynamic scenes such as non-stationary scenes or dynamic background regions. We also take a passive based approach where we do not require preprocessing.

Many passive video/image forgery detection frameworks consist of two parts: feature extraction and classification. Up to the present, there exist many issues to be solved in both parts. In feature extraction, there is an issue that numerous previous methods use features that depend on the kind of tampering techniques and the type of videos. For example, Su et al. [2] extract features by using background subtraction and address only videos taken by fixed cameras, i.e., static videos whose viewpoint cannot move. Another method is on the assumption that forged videos are created by copy-move attack [3]. In classification, many previous works use support vector machine [4] or k-means [2] as discriminators, and they do not consider the temporal nature of videos. Recently, several methods using Convolutional Neural Network (CNN) [5] and Recurrent Neural Network (RNN) [6,7] have been proposed. The former method by Rota et al. is a method for detecting image tampering, and their method cannot treat the temporal structure of videos in the case of applying to video forgery detection. The latter methods do not consider the spatial consistency of videos entirely instead of finding the temporal consistency of videos.

Another major issue in the field of video forgery detection is that the type of tampering to be researched is limited. Sowmya et al. [1] divide spatial tampering attacks into three types of attacks: object addition attack, object removal attack, and object modification attack. Most related works address either object addition attack or object removal attack, and there is no work for object modification attack. Besides, several studies target stationary scenes, which are not realistic.

In this paper, we detect several types of tampering for realistic videos like the videos uploaded to video sharing sites. We assume videos with the dynamic background such as tree fluctuation and snowfall and with swinging perspectives such as camera shake. We also address two types of spatial tampering attacks: object removal attack and object modification attack. We remove objects in videos by using inpainting techniques [8]. We also assume modification attack where we change the kinds and tones of the color of objects in videos.

The novelty of this paper is to detect video tampering by using a model combining CNN and RNN to consider the spatial and temporal aspects of videos. CNN provides spatial consistency, and RNN offers temporal consistency. Hence, we expect that we can detect and identify forged objects in dynamic videos with high accuracy. We also prepare not only an object removal database created from 53 videos in the database CDnet2014 for object detection tasks but also a new object modification database created from 34 videos that we acquire from a video hosting site. Using the databases, we verify the performance of our method.

2   Related Work on Video/Image Forgery Detection

In this section, we describe passive video/image forgery detection methods and several databases for video tampering.

2.1   Passive Methods for Detecting Forged Images

In the case of treating each frame in a video as each image, we can use image forgery detection methods to detect forged regions in the video. Farid et al. [9] introduce a method using the characteristics of cameras. They also describe another technique using specific patterns appeared in the forged regions when compressing the forged images. Recently, several methods based on the neural network have been proposed [5,10]. For example, Rota et al. [5] suggest a way of learning an input image for each divided region by CNN. This method needs no feature extraction and applies to several types of scenes. These methods, however, cannot examine the temporal structures of videos. If we use such techniques in video forgery detection, there is still room for improvement.

2.2   Passive Methods for Detecting Forged Videos

In the case of targeting static videos consisting of stationary scenes such as the videos taken by fixed cameras, we extract features by using background subtraction, and it is known that we can detect forged regions with high accuracy [2,11]. However, detection accuracy decreases in non-stationary scenes where it is difficult to apply background subtraction.

In order to achieve forgery detection for dynamic videos with non-stationary scenes, we need to detect inconsistency between the frames of a video [12]. For example, Saxena et al. [4] use optical flow as a feature, and their method does not require background subtraction. They do not take into account the temporal structure of videos in the step of classification although they examine the temporal consistency between each frame in the part of feature extraction.

Recently, several methods using neural network have been proposed in the field of video forgery detection. In the case of detecting forgery objects by using CNN, we only consider the features between the adjacent frames of videos [13]. Thus, most researchers often use RNN that is superior in time series analysis. For example, D'Avino et al. [7] present an approach that learns the model of real videos (i.e., source videos) through the architecture based on AutoEncoder and RNN. This model works as an anomaly detector. Karita et al. [6] propose a video forgery detection method in dynamic scenes by using DeepFlow feature and RNN. There is, however, an issue that these methods do not consider the spatial aspect of videos sufficiently whereas they focus on the time series analysis.

2.3   Databases for Video Tampering

The number of existing databases for video tampering is limited in contrast to databases for image tampering. The number of tampering videos in the databases is tiny. At present, databases used and published in the existing work are SULFA [14] or a database created by D'Avino et al. [7]. The number of videos in these databases is 10 individually. Besides, the type of spatial tampering in videos is limited because these databases address copy-paste tampering or splicing. In this paper, we not only reconstruct an inpainting database for object removal created by Karita et al. [6] but also construct a new database for object modification. The content of object modification is the color change of objects.

3   Video Forgery Detection Method Considering Spatio-Temporal Consistency

In this paper, we use CNN considering the spatial structure of the video and RNN considering the temporal structure. We attempt to detect forgery regions by adopting a model using Convolutional LSTM (ConvLSTM) network [15] where contains the convolution operation inside LSTM cells. This is called a video forgery detection method with ConvLSTM (VFD with ConvLSTM). An overview of VFD with ConvLSTM is shown in Fig. 1.

ConvLSTM is used for precipitation nowcasting [15]. ConvLSTM is a variant of Long Short-Term Memory (LSTM) containing the convolution operation inside the LSTM. ConvLSTM enables us to learn complex patterns with contrast to LSTM. The reason is that spatial information is lost because we handle images converted to one-dimensional vectors as input data when addressing the images on LSTM.

Let the input on time t be $\mathbf{X}_t$, the output be $\mathbf{H}_t$, the previous hidden state be $\mathbf{H}_{t-1}$, and the cell state be $\mathbf{c}_t$, respectively. The terms $\mathbf{a}_{\mathrm{in},t}$, $\mathbf{a}_{\mathrm{cell},t}$, $\mathbf{a}_{\mathrm{out},t}$ are the input gate, the forget gate, the output gate, respectively. These are all 3D tensors. Then, the key equations of ConvLSTM are given as follows:

$$\mathbf{a}_{\mathrm{in},t} = \sigma\left(\mathbf{W}_{x_{\mathrm{in}}} * \mathbf{X}_t + \mathbf{W}_{h_{\mathrm{in}}} * \mathbf{H}_{t-1} + \mathbf{W}_{c_{\mathrm{in}}} \circ \mathbf{c}_{t-1} + \mathbf{b}_{\mathrm{in}}\right), \tag{1}$$

$$\mathbf{a}_{\mathrm{cell},t} = \sigma\left(\mathbf{W}_{x_{\mathrm{cell}}} * \mathbf{X}_t + \mathbf{W}_{h_{\mathrm{cell}}} * \mathbf{H}_{t-1} + \mathbf{W}_{c_{\mathrm{cell}}} \circ \mathbf{c}_{t-1} + \mathbf{b}_{\mathrm{cell}}\right), \tag{2}$$

$$\mathbf{c}_t = \mathbf{a}_{\mathrm{cell},t} \circ \mathbf{C}_{t-1} + \mathbf{a}_{\mathrm{in},t} \circ \tanh\left(\mathbf{W}_{x_c} * \mathbf{X}_t + \mathbf{W}_{h_c} * \mathbf{H}_{t-1} + \mathbf{b}_c\right), \tag{3}$$

$$\mathbf{a}_{\mathrm{out},t} = \sigma\left(\mathbf{W}_{x_{\mathrm{out}}} * \mathbf{X}_t + \mathbf{W}_{h_{\mathrm{out}}} * \mathbf{H}_{t-1} + \mathbf{W}_{c_{\mathrm{out}}} \circ \mathbf{c}_t + \mathbf{b}_{\mathrm{out}}\right), \tag{4}$$

$$\mathbf{H}_t = \mathbf{a}_{\mathrm{out},t} \circ \tanh\left(\mathbf{c}_t\right), \tag{5}$$

where $*$ denotes the convolution operator, $\circ$ denotes the Hadamard product, and σ is the sigmoid function. The terms $\mathbf{W}_{x_{\sim}}$ and $\mathbf{W}_{c_{\sim}}$ are weight matrices, and $\mathbf{b}_{\sim}$ is a bias vector.

ConvLSTM provides a learning framework that produces a sequence from another sequence by using RNN. As shown in Fig. 1, we copy the last state of the encoder network to the decoder outputting the frames predicted to be tampered, and we predict several forged region frames. The input data is the normalized RGB images of 64 frames whose size is resized to 160×160. Based on [15], we construct each component of ConvLSTM, as shown in Table 1.

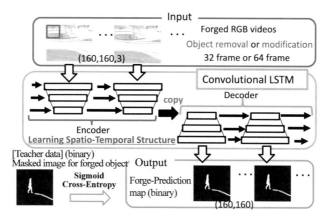

Fig. 1. An overview of video forgery detection with Convolutional LSTM.

Table 1. Each component of the detection model with ConvLSTM.

Layer	Layer type	Filter size	Stride	The number of filters	Activation function
Layer 1	ConvLSTM	3×3	1×1	16	tanh
Layer 2	ConvLSTM	3×3	1×1	12	tanh
Layer 3	ConvLSTM	3×3	1×1	12	sigmoid

Because of the 2-class classification problem, the loss function is given by

$$E(\mathbf{x}, \hat{\mathbf{y}}, W) = -\sum_t \sum_i \sum_j \{\hat{y}_{t,i} \log p(y_{t,i,j} = 1 | \mathbf{x}_1, \ldots, \mathbf{x}_t, W)$$

$$+ (1 - \hat{y}_{t,i,j}) \log p(y_{t,i,j} = 0 | \mathbf{x}_1, \ldots, \mathbf{x}_t, W)\}. \tag{6}$$

Here, $y_{t,i,j} \in \{0,1\}$ is a variable representing whether the (i, j) region is forged or not on time t when input $\mathbf{x}_t$ is given. A variable $\hat{y}_{t,i,j} \in \{0,1\}$ is teacher data representing whether the (i, j) region is forged or not on time t. As shown in Fig. 1, we use masked images for forged objects as teacher data. We train our network so that the prediction results approach the masked images by using the loss function (6).

For training, we use the Adam learning algorithm which is an extension to stochastic gradient descent to update $w \in W$. We also use backpropagation through time (BPTT) algorithm to calculate a gradient $\frac{\partial E}{\partial w}$. We set a learning rate $\alpha = 0.001$, exponential decay rates for moment estimation $\beta_1 = 0.9$ and $\beta_2 = 0.999$, and numerical stability $\epsilon = 10^{-8}$. The training is terminated when the loss of the Eq. (6) does not decrease over ten rounds.

4   Detection Experiments for Forged Videos

We describe two forged video databases and compared methods in details. We then present experimental results and verify the detection accuracy of our method.

4.1   Creation of Forged Video Databases and Evaluation Metric

When we construct a database for object removal, we use the database CDnet2014 for object detection tasks. CDnet2014 includes 53 original videos and masked videos. We then reconstruct the original videos to 89 videos that consist of 750 frames on average because there are many frames with no masks in the videos. All the reconstructed videos include more than 100 frames with masks.

By performing inpainting for masked objects, we remove the objects from the videos. Such a tampering database is called Inpainting-CDnet2014. In the phase of training, we use an inpainting method by Bertalmio et al. [16]. In the phase of verifying, we use another inpainting method by Telea et al. [17].

We also develop an object modification database that consists of 34 forged videos[1]. We change the colors of objects in the original videos to other colors by using Adobe After Effects CC®. Tampered objects are moving objects such as birds and cars. This database is called the Modification Database.

We use binary images whose size is resized to 160×160 as teacher data. In the 89 videos of Inpainting-CDnet2014, we use 71 videos for learning, other seven videos for verifying, and the other 11 videos for evaluating. In the 34 videos of Modification Database, we use 27 videos for learning and the other seven videos for verification and evaluation. The changed colors of the objects in the seven videos for verifying are different from those of the seven videos for evaluating.

As an evaluation metric, we use the Area Under Curve (AUC) which is the area under the Receiver-Operating-Characteristic curve (ROC curve). ROC curve plots True Positive Rate (TPR) against False Positive Rates (FPR) for every possible threshold. We also use Equal-Error-Rate (EER) that indicates a value when FPR is equal to 1-TPR.

4.2   Compared Methods

We implement two methods by Karita et al. [6] and Rota et al. [5]. Karita's method is a video forgery detection method with LSTM (that is, RNN) that uses raw images (the frames of videos) and dense optical flows as input.

Rota et al. detect forged regions in an image by dividing the image into several fixed patches and learning these patches with CNN. We change the output of the final layer to obtain an output of patch size $40 \times 40 = 1600$ because we get the probability of being tampered with each pixel. We also use a sigmoid function because of the 2-class classification problem whereas the output of the

---

[1] We obtain original videos from Pixabay (https://pixabay.com/, 2018/08/31 access).

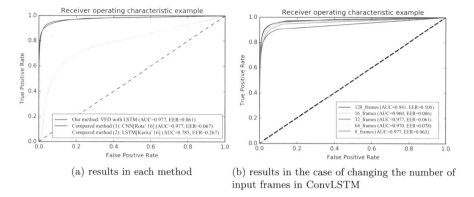

(a) results in each method

(b) results in the case of changing the number of input frames in ConvLSTM

Fig. 2. ROC curve, AUC, and EER in the case of experimenting for Inpainting-CDnet2014.

final layer is passed to a softmax function in [5]. Note that, in the phase of learning, Rota's method uses the forged frames of the videos only because their method is overfitting in the case of using all the frames of the videos.

4.3   Experimental Results for Inpainting-CDnet2014

Figure 2(a) shows the ROC curve, AUC, and EER when applying our method and two compared methods to Inpainting-CDnet2014. From Fig. 2(a), VFD with ConvLSTM has the best performance for both AUC and EER. However, Rota's CNN method achieves almost the same performance. This implies that, in the case of inpainting, the detection methods considering spatial information are superior to those of considering temporal information. The reason is that unnatural regions appear firmly in the spatial domain because the object-removed areas are interpolated by propagating information from the surrounding areas.

Examples of the detection results by each method in Inpainting-CDnet2014 are shown in Fig. 3. This video includes a dynamic background like the fountain. VFD with ConvLSTM can detect the forged car in the most frames whereas the detection system by RNN cannot detect this car as a result of responding the fountain. The system by CNN does not detect the forged car in several frames.

In order to examine the influences of parameters of ConvLSTM in Inpainting-CDnet2014, ROC curve, AUC, and EER in the case of changing the number of input frames in VFD with ConvLSTM are shown in Fig. 2(b). From Fig. 2(b), we observe that the detection accuracy of VFD with ConvLSTM does not depend on the number of the input frames except for the case of inputting 128 frames. This is one evidence that spatial information is more important than temporal information in the case of inpainting tampering. VFD with ConvLSTM where the number of the input frames is 128 is overfitting.

4.4　Experimental Results for Modification Database

Figure 4(a) shows the ROC curve, AUC, and EER when applying each method to Modification Database. From Fig. 4(a), VFD with ConvLSTM has the best performance for both AUC and EER. We observe that the detection accuracy of Karita's RNN approach is the same as that of Rota's CNN approach. We develop the forged videos in Modification Database semi-automatically and politely. In the case of such elaborate tampering, we infer that we cannot detect the forged regions fully unless we use both temporal information and spatial information.

Examples of the detection results by each method in Modification Database are shown in Fig. 5. VFD with ConvLSTM can detect the forged car in almost the frames by considering the temporal consistency whereas CNN method does not identify the car in several frames and does not track the car. RNN system cannot detect this car clearly although detecting the surroundings of the car.

To examine the influences of parameters of ConvLSTM in Modification Database, ROC curve, AUC, and EER in the case of changing the number of input frames in VFD with ConvLSTM are shown in Fig. 4(b). From Fig. 4(b), we observe that the detection accuracy of VFD with ConvLSTM is the best

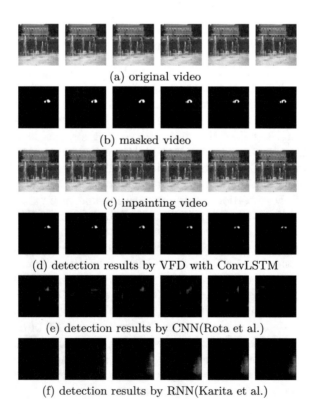

(a) original video

(b) masked video

(c) inpainting video

(d) detection results by VFD with ConvLSTM

(e) detection results by CNN(Rota et al.)

(f) detection results by RNN(Karita et al.)

Fig. 3. Example results of forgery detection in the frame 695–700 of the video fountain01 in Inpainting-CDnet2014.

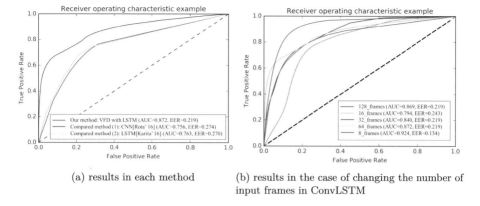

(a) results in each method

(b) results in the case of changing the number of input frames in ConvLSTM

Fig. 4. ROC curve, AUC, and EER in the case of experimenting for Modification Database.

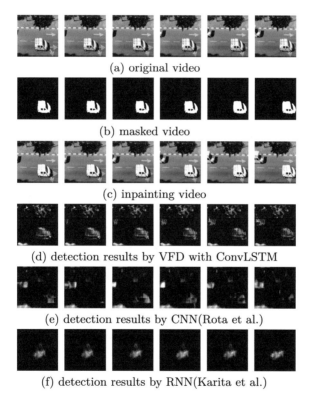

(a) original video

(b) masked video

(c) inpainting video

(d) detection results by VFD with ConvLSTM

(e) detection results by CNN(Rota et al.)

(f) detection results by RNN(Karita et al.)

Fig. 5. Example results of forgery detection in the frame 25–30 of the video car06 in Modification Database.

when inputting eight frames. There is, however, a possibility that we can create a good model by chance. The detection accuracies are high when 32 frames and 64 frames except for eight frames are input. This is because the length of frames where forged objects appear is between 32 frames and 64 frames at the longest.

5 Conclusion

In this paper, we propose a passive forgery detection method against spatial video tampering with the neural network. The characteristics of our method is to use Convolutional LSTM in order to consider the spatio-temporal consistency of videos. We also challenge the detection of object modification attack as well as that of object removal attack in the videos including dynamic scenes. As a result, we can improve the detection accuracy of forged objects by our method.

In our experiments, the performance of the video forgery detection against object modification is not enough. One of the causes is that the number of forged videos are not enough. We plan to experiment on more massive databases.

Acknowledgments. This work was supported by JSPS KAKENHI Grant Number JP16H06302.

References

1. Sowmya, K.N., Chennamma, H.R.: A survey on video forgery detection. Int. J. Comput. Eng. Appl. **9**(2), 17–27 (2015)
2. Su, L., Huang, T., Yang, J.: A video forgery detection algorithm based on compressive sensing. Multimedia Tools Appl. **74**(17), 6641–6656 (2015)
3. Bestagini, P., Milani, S., Tagliasacchi, M., Tubaro, S.: Local tampering detection in video sequences. In: 2013 IEEE International Workshop on Multimedia Signal Processing, pp. 488–493 (2013)
4. Saxena, S., Subramanyam, A., Ravi, H.: Video inpainting detection and localization using inconsistencies in optical flow. In: 2016 IEEE Region 10 Conference, pp. 1361–1365 (2016)
5. Rota, P., Sangineto, E., Conotter, V., Pramerdorfer, C.: Bad teacher or unruly student: can deep learning say something in image forensics analysis? In: 23rd International Conference on Pattern Recognition, pp. 2503–2508 (2016)
6. Karita, S., Kono, K., Babaguchi, N.: Video forgery detection using a time series model in dynamic scenes. IEICE Technical Report, EMM2015-80, vol. 115, no. 479, pp. 25–30 (2016)
7. D'Avino, D., Cozzolino, D., Poggi, G., Verdoliva, L.: Autoencoder with recurrent neural networks for video forgery detection. Electron. Imaging **2017**(7), 92–99 (2017)
8. Newson, A., Almansa, A., Fradet, M., Gousseau, Y., P'erez, P.: Video inpainting of complex scenes. SIAM J. Imaging Sci. **7**(4), 1993–2019 (2014)
9. Farid, H.: Image forgery detection. IEEE Signal Process. Mag. **26**(2), 16–25 (2009)
10. Cozzolino, D., Poggi, G., Verdoliva, L.: Recasting residual-based local descriptors as convolutional neural networks: an application to image forgery detection. In: ACM Workshop on Information Hiding and Multimedia Security, pp. 159–164 (2017)

11. Richao, C., Gaobo, Y., Ningbo, Z.: Detection of object-based manipulation by the statistical features of object contour. Forensic Sci. Int. **236**, 164–169 (2014)
12. Lin, C.-S., Tsay, J.-J.: A passive approach for effective detection and localization of region-level video forgery with spatio-temporal coherence analysis. Digit. Investig. **11**(2), 120–140 (2014)
13. Yao, Y., Shi, Y., Weng, S., Guan, B.: Deep learning for detection of object based forgery in advanced video. Symmetry **10**(3), 1–10 (2018)
14. Qadir, G., Yahaya, S., Ho, A.: Surrey university library for forensic analysis (SULFA) of video content. In: IET Conference on Image Processing, pp. 1–6 (2012)
15. Xingjian, S., Chen, Z., Wang, H., Yeung, D.-Y., Wong, W.-K., Woo, W.-C.: Convolutional LSTM network: a machine learning approach for precipitation nowcasting. In: Advances in Neural Information Processing Systems, pp. 802–810 (2015)
16. Bertalmio, M., Bertozzi, A.L., Sapiro, G.: Navier-stokes, fluid dynamics, and image and video inpainting. In: 2001 IEEE Computer Society Conference on Computer Vision and Pattern Recognition, vol. 1, no. C, pp. I.355–I.362 (2001)
17. Telea, A.: An image inpainting technique based on the fast marching method. J. Graph. Tools **9**(1), 23–34 (2004)

Author Index

States